Fungal Genomics

Advances in Genetics, Volume 57

Fungal Genomics

Edited by

Jay C. Dunlap

Department of Genetics
Dartmouth Medical School
Hanover, New Hampshire

ELSEVIER

AMSTERDAM • BOSTON • HEIDELBERG • LONDON
NEW YORK • OXFORD • PARIS • SAN DIEGO
SAN FRANCISCO • SINGAPORE • SYDNEY • TOKYO
Academic Press is an imprint of Elsevier

Academic Press is an imprint of Elsevier
525 B Street, Suite 1900, San Diego, California 92101-4495, USA
84 Theobald's Road, London WC1X 8RR, UK

This book is printed on acid-free paper.

For information on all Elsevier Academic Press publications visit our Web site at www.books.elsevier.com

ISBN-13: 978-0-12-017657-1
ISBN-10: 0-12-017657-2

PRINTED IN THE UNITED STATES OF AMERICA
07 08 09 10 9 8 7 6 5 4 3 2 1

Contents

3 Genomics of the Plant Pathogenic Oomycete *Phytophthora*: Insights into Biology and Evolution   97
Howard S. Judelson

4 Sex and Virulence of Human Pathogenic Fungi   143
Kirsten Nielsen and Joseph Heitman

Contributors

Numbers in parentheses indicate the pages on which the authors' contributions begin.

Lorena Altamirano (49) Department of Plant Pathology, University of California, Riverside, California 92521

Scott E. Baker (219) Fungal Biotechnology Team, Chemical and Biological Process Development, Environmental Technology Directorate, Pacific Northwest National Laboratory, Richland, Washington 99352

Meray Baştürkmen (49) Oregon Health & Science University, Beaverton, Oregon 97006

Bruce W. Birren (49) Broad Institute of MIT & Harvard, Cambridge, Massachusetts 02142

Katherine A. Borkovich (49) Department of Plant Pathology, University of California, Riverside, California 92521

Patrick Collopy (49) Department of Genetics, Dartmouth Medical School, Hanover, New Hampshire 03755

Hildur V. Colot (49) Department of Genetics, Dartmouth Medical School, Hanover, New Hampshire 03755

Sarah F. Covert (1) Daniel B. Warnell School of Forest Resources, University of Georgia, Athens, Georgia 30602

Matthew Crawford (49) Broad Institute of MIT & Harvard, Cambridge, Massachusetts 02142

Christopher Crew (49) Department of Plant Pathology, University of California, Riverside, California 92521

Susan Curilla (49) Department of Genetics, Dartmouth Medical School, Hanover, New Hampshire 03755

Ralph A. Dean (175) Department of Plant Pathology, North Carolina State University, Raleigh, North Carolina 27606

Dave DeCaprio (49) Broad Institute of MIT & Harvard, Cambridge, Massachusetts 02142

Jay C. Dunlap (49) Department of Genetics, Dartmouth Medical School, Hanover, New Hampshire 03755

James E. Galagan (49) Broad Institute of MIT & Harvard, Cambridge, Massachusetts 02142

Maria Garcia-Pedrajas (1) Departamento de Microbiología del Suelo y Sistemas Simbióticos, Estación Experimental del Zaidín, CSIC, 18080 Granada, Spain

N. Louise Glass (49) Department of Plant and Microbial Biology, University of California, Berkeley, California 94720

Scott E. Gold (1) Department of Plant Pathology, University of Georgia, Athens, Georgia 30602

Joseph Heitman (143) Department of Molecular Genetics and Microbiology, Department of Pharmacology and Cancer Biology, and Department of Medicine, Duke University Medical Center, Durham, North Carolina 27710

Matthew R. Henn (49) Broad Institute of MIT & Harvard, Cambridge, Massachusetts 02142

Heather M. Hood (49) Oregon Health & Science University, Beaverton, Oregon 97006

Howard S. Judelson (97) Department of Plant Pathology, Center for Plant Cell Biology, University of California, Riverside, California 92521

Takao Kasuga (49) Department of Plant and Microbial Biology, University of California, Berkeley, California 94720

Steven J. Klosterman (1) Department of Plant Pathology, University of Georgia, Athens, Georgia 30602

Michael Koerhsen (49) Broad Institute of MIT & Harvard, Cambridge, Massachusetts 02142

Randy Lambreghts (49) Department of Genetics, Dartmouth Medical School, Hanover, New Hampshire 03755

Lisa Larson (49) Broad Institute of MIT & Harvard, Cambridge, Massachusetts 02142

Liubov Litvinkova (49) Department of Plant Pathology, University of California, Riverside, California 92521

Jennifer J. Loros (49) Department of Genetics, Dartmouth Medical School, Hanover, New Hampshire 03755

Gregory S. May (263) Division of Pathology and Laboratory Medicine, The University of Texas M. D. Anderson Cancer Center, Houston, Texas 77030

Kevin McCluskey (49) Fungal Genetics Stock Center, School of Biological Sciences, University of Missouri, Kansas City, Missouri 64110

Phil Montgomery (49) Broad Institute of MIT & Harvard, Cambridge, Massachusetts 02142

Mary Anne Nelson (49) Department of Biology, University of New Mexico, Albuquerque, New Mexico 87131

Kirsten Nielsen (143) Department of Molecular Genetics and Microbiology, Duke University Medical Center, Durham, North Carolina 27710

Gyungsoon Park (49) Department of Plant Pathology, University of California, Riverside, California 92521

Matthew Pearson (49) Broad Institute of MIT & Harvard, Cambridge, Massachusetts 02142

Michael H. Perlin (1) Department of Biology, University of Louisville, Louisville, Kentucky 40292

Michael Plamann (49) Fungal Genetics Stock Center, School of Biological Sciences, University of Missouri, Kansas City, Missouri 64110

Carol Ringelberg (49) Department of Genetics, Dartmouth Medical School, Hanover, New Hampshire 03755

Matthew S. Sachs (49) Oregon Health & Science University, Beaverton, Oregon 97006

Taylor Schoberle (263) Division of Pathology and Laboratory Medicine, The University of Texas M. D. Anderson Cancer Center, Houston, Texas 77030

Mi Shi (49) Department of Genetics, Dartmouth Medical School, Hanover, New Hampshire 03755

Chaoguang Tian (49) Department of Plant and Microbial Biology, University of California, Berkeley, California 94720

Jeffrey P. Townsend (49) Department of Ecology and Evolutionary Biology, Yale University, New Haven, Connecticut 06520

B. Gillian Turgeon (219) Department of Plant Pathology, Cornell University, Ithaca, New York 14853

Gloria E. Turner (49) Department of Chemistry and Biochemistry, University of California, Los Angeles, California 90095

Richard L. Weiss (49) Department of Chemistry and Biochemistry, University of California, Los Angeles, California 90095

Jin-Rong Xu (175) Department of Botany and Plant Pathology, Purdue University, West Lafayette, Indiana 47907

Junhuan Xu (49) Department of Biology, University of New Mexico, Albuquerque, New Mexico 87131

Xinhua Zhao (175) Department of Botany and Plant Pathology, Purdue University, West Lafayette, Indiana 47907

Preface

The enormous diversity of life styles and metabolic capacity represented by the filamentous fungi have been appreciated for decades. While research on these organisms may have begun with the emergence of the important filamentous fungal model systems *Neurospora* and *Aspergillus* in the early to mid-twentieth century, it clearly took rapid flight with fundamental work showing the roles that fungi play in animal and plant pathogenesis. Despite the clear importance of fungi to life on Earth, however, many of these systems were difficult to work with and research lagged; even a decade ago it was often the case that biological systems with less biology but more traction were preferred for many studies. However, in this field, as in many, genomics is proving to be the great leveller, and as genomic sequences of important systems are emerging and genomic tools are appearing, both the amount and quality of research on fungi is increasing exponentially. Fungal genomics is having a pivotal impact on applied research in agriculture, food sciences, natural resource management, pharmaceuticals, and biotechnology, as well as in basic studies in the life sciences.

This volume highlights many of the premier fungal research systems chosen both for their utility as research models as well as for direct impact on understanding pathogenesis. The systems, including plant and animal pathogens along with the well-established filamentous fungal model system *Neurospora*, were chosen because they are at various stages in the development of genomic technologies. The case studies provide a primer for where the fields of fungal biology and pathogenesis have been and will go in the future.

Lastly, I join all of the authors in sadly noting the passing, on January 2, 2007, of David Perkins. Perkins was one of the premier fungal geneticists of this or any age. Active both intellectually and scientifically until the final weeks of his life at 87 years, Perkins gave generously of his time and effort in promoting the growth of fungal genetics and supporting fungal geneticists regardless of their research system. We remember him with fondness, acknowledge his many selfless and beneficial acts with gratitude, and dedicate this volume to him with due humility.

Thank you David.

Jay C. Dunlap

1

Genetics of Morphogenesis and Pathogenic Development of *Ustilago maydis*

Steven J. Klosterman,[*,1] **Michael H. Perlin,**[†]
Maria Garcia-Pedrajas,[‡] **Sarah F. Covert,**[§] **and**
Scott E. Gold[*]

[*]Department of Plant Pathology, University of Georgia
Athens, Georgia 30602
[†]Department of Biology, University of Louisville, Louisville, Kentucky 40292
[‡]Departamento de Microbiología del Suelo y Sistemas Simbióticos
Estación Experimental del Zaidín, CSIC, 18080 Granada, Spain
[§]Daniel B. Warnell School of Forest Resources, University of Georgia
Athens, Georgia 30602

[1]Current address: USDA-ARS, Salinas, California 93905.

Advances in Genetics, Vol. 57   0065-2660/07 $35.00
DOI: 10.1016/S0065-2660(06)57001-4

ABSTRACT

Ustilago maydis has emerged as an important model system for the study of fungi. Like many fungi, U. *maydis* undergoes remarkable morphological transitions throughout its life cycle. Fusion of compatible, budding, haploid cells leads to the production of a filamentous dikaryon that penetrates and colonizes the plant, culminating in the production of diploid teliospores within fungal-induced plant galls or tumors. These dramatic morphological transitions are controlled by components of various signaling pathways, including the pheromone-responsive MAP kinase and cAMP/PKA (cyclic AMP/protein kinase A) pathways, which coregulate the dimorphic switch and sexual development of U. *maydis*. These signaling pathways must somehow cooperate with the regulation of the cytoskeletal and cell cycle machinery. In this chapter, we provide an overview of these processes from pheromone perception and mating to gall production and sporulation *in planta*. Emphasis is placed on the genetic determinants of morphogenesis and pathogenic development of U. *maydis* and on the fungus–host interaction. Additionally, we review advances in the development of tools to study U. *maydis*, including the recently available genome sequence. We conclude with a brief assessment of current challenges and future directions for the genetic study of U. *maydis*.   © 2007, Elsevier Inc.

I. INTRODUCTION

Ustilago maydis belongs to the Ustilaginales, an order of basidiomycetes that includes semiobligate biotrophic plant pathogenic fungi that are commonly known as smuts. Although basidiomycete fungi of the Urediniomycetes (rusts) and Hymenomycetes (mushrooms) have a greater economic impact as plant pathogens and agricultural commodities, respectively; smut diseases remain a persistent problem (Alexopolous *et al.*, 1996). The characteristic disease cycle of the smuts alternates between growth in a haploid budding form and growth in a dikaryotic filamentous form; this pattern is exemplified by *U. maydis* (Fig. 1.1). In the plant, the filamentous dikaryon, resulting from fusion of mating-compatible haploid cells, ramifies locally in the host tissue producing spectacular symptoms. The most dramatic and defining symptom in the *U. maydis*–maize interaction is the induction of gall (tumor) formation in the infected host tissues. When mature, these galls produce copious, black, dusty masses of teliospores that *en masse* resemble soot or smut, from which the common name of the group is derived.

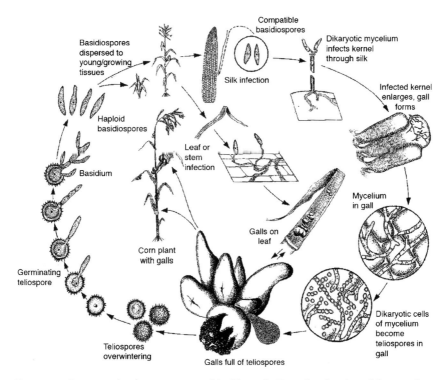

Figure 1.1. Disease cycle of corn smut caused by *U. maydis*. Reproduced with modifications from Agrios (1997) with the permission of Elsevier.

U. maydis is completely dependent on its host (maize, *Zea mays*) to complete its life cycle (Fig. 1.1) because it is incapable of *ex planta* sporulation.

U. *maydis* is among the most important fungal models for the study of morphogenesis, mating, and signaling. It remains the only genetically well-developed model plant pathogen among the basidiomycetes, representing a valuable comparative tool for the study of rust and other smut diseases. In this chapter, we explore the genetics of morphogenesis and pathogenic development of *U. maydis*. Two of the well-studied signal transduction pathways, the cAMP/PKA pathway and the pheromone-responsive MAP kinase cascades, are discussed in the context of their roles in mating and dimorphic growth. The regulation of the MAP kinase and cAMP/PKA pathways is somehow entwined with the cell cycle machinery and cytoskeletal components and hence we review these components involved in morphogenesis and pathogenic development. Where pertinent, we incorporate cytological and biochemical evidence to provide a contextual framework of understanding. Since the development of *U. maydis* is intricately intertwined with its host, review is given to studies assessing signaling in the fungus–host interaction from both perspectives. We also explore genome-wide approaches and advances in the tools available for the manipulation of *U. maydis*. Finally, we place the review in a broader context by discussing future challenges and directions in the study of *U. maydis*.

II. MATING

A. The mating loci

How does one find a mate? This is a perennial problem for all organisms. Fungi that are heterothallic, like *U. maydis*, must be able to distinguish between self- and nonself-mating partners. Among the heterothallic basidiomycetes, there are two compatibility systems: bipolar and tetrapolar. Bipolar systems depend on a single MAT (mating-type) locus that has at least two alleles. Tetrapolar systems have two MAT loci and at least two alleles at each locus. U. *maydis* is a tetrapolar species, with a distinct *a* locus that governs cell fusion between haploid partners and a *b* locus that controls sexual development after fusion and pathogenicity (Banuett, 1995; Bolker, 2001; Casselton and Olesnicky, 1998; Kronstad and Staben, 1997; Schirawski *et al.*, 2005).

B. The *a* locus: Self/nonself-recognition

The *a* locus in U. *maydis* has two different alleles, *a1* and *a2*. In reality, the two versions of this locus share very little homology and should, therefore, properly be referred to as "idiomorphs" (Bakkeren *et al.*, 1992). This locus allows different haploid cell types of the fungus to communicate. The give and take of the conversation that leads to mating is dependent on pheromones and their respective receptors. Each cell produces pheromones to communicate with potential-mating partners, and synthetically produced pheromone can also induce the formation of mating tubes in liquid (Basse and Steinberg, 2004). The *a1* locus is a 4.5-kb region that encodes a 40-amino acid (aa) pheromone precursor (from *mfa1*) and the corresponding receptor for the a2 pheromone (from *pra1*). The precursor produced from *mfa1* is processed to a mature tridecapeptide (13 aa: GRDNGSPIGYSSX). The third gene at the *a1* locus is *rba1*, which may be targeted to the mitochondrial membrane. The *a2* locus ($\sim$8 kb) contains at least four functional genes and one pseudogene. In addition to the 38-aa Mfa2 lipoprotein pheromone precursor, the locus encodes Pra2, the receptor for the *a1*-encoded pheromone, Lga2 and Rga2; a pseudogene, *rba2*, is also present. The mature *a2*-encoded pheromone produced from the Mfa2 precursor is a nonapeptide (9 aa:NRGQPGYYC). Both *a1*- and *a2*-encoded pheromones are carboxylated at the C-terminal Cys, as well as farnesylated (Szabo *et al.*, 2002). They both contain Cys-A-A-X, which is characteristic of prenylated proteins (like the yeast pheromones). In addition, some residues conserved between the *a1*- and *a2*-encoded pheromones (indicated in bold, above) are important for activity, possibly playing a role in folding into the binding pocket of the receptor (Szabo *et al.*, 2002). The seven-transmembrane receptors (Pra1, 357 aa and Pra2, 346 aa) contain an STE3 motif and share 24% aa similarity overall. The secreted pheromone from one mating type is recognized by the corresponding receptor of the opposite mating type. This "recognition" begins a cascade of signaling events that alters gene expression in the *a* and *b* loci. Lga2, although not directly related to mating, may be involved in the control of mitochondrial fusion (since after mating, mitochondria of both parents are present). Lga2 and Rga2, neither of which appears required for pathogenicity, localize to mitochondria. Interestingly, Lga2 and Rga2 function as suppressors of a pathogenicity defect encountered in mutants in the *mrb1* gene (Bortfeld *et al.*, 2004). The *mrb1* pathogenicity defect is dependent on the *a2* idiomorph. mrb1 was first detected as a constitutively expressed gene flanking a set of maize-induced genes (Basse *et al.*, 2002). Mrb1 is a mitochondrial matrix protein of the p32 family, whose human homologue is associated with the SR family splice factor ASF/SF2. The p32 proteins have been implicated in diverse regulatory processes, including pre-mRNA splicing and mitochondrial RNA editing (Bortfeld *et al.*, 2004; Hayman *et al.*, 2001; Krainer *et al.*, 1991). Compatible *mrb1* mutants have reduced proliferation in plant tissue and tumor development is severely reduced. Deletion of *rga2* partially relieves the pathogenicity defect of the *mrb1* mutant. Yeast two-hybrid analysis showed that Rga2 and Mrb1 interact. Conditional expression of Lga2 in haploids led to mitochondrial fragmentation, mtDNA degradation, reduced respiratory function, and reduced vegetative growth. If both Lga2 and Mrb1 are missing, pathogenesis is unaffected, showing that the original *mrb1* defect is dependent on a functional Lga2 protein (Bortfeld *et al.*, 2004).

C. The *b* locus: Self/nonself-recognition

The *b* locus was identified from a *b1* library transformed into a *b2* diploid to produce filamentous growth (Kronstad and Leong, 1989; Schulz *et al.*, 1990). Two divergently transcribed genes are found at the *b* locus, *bE* and *bW*, encoding 410- and 626-aa proteins, respectively. Both contain a central homeodomain (~60 aa), separating a variable N-terminal region from a conserved (>90%) C-terminal constant region. Any combination of differing *bE* and *bW* alleles (e.g., *bE1bW2* or *bE2bW1*) generates a successful mating reaction (Gillissen *et al.*, 1992) (Fig. 1.2). Each partner must have functional homeodomains. For the HD1 type in *bE*, this means a homeodomain of the HD1 type, which is also found in the MATα2 protein of *Saccharomyces cerevisiae*; *bW* contains an HD2-type domain as is found in the MATa1 protein of *Saccharomyces cerevisiae* (Hiscock and Kues, 1999). The *bE* and *bW* products must physically interact to form a functional heterodimer, but they can only do so if from alternate alleles. Specificity in their dimerization involves the variable regions in the N-terminus and this provides self/nonself-recognition (Kamper *et al.*, 1995). Yee and Kronstad (1993) carried out elegant structure/function analyses of this specificity through the use of chimeric *b* alleles and showed that this is due to polar–hydrophobic interactions. The *b* locus may be considered a central

Figure 1.2. 2b or not 2b that is the question. The upper portion of the photo shows an *a1b1* strain overlaid with an *a2b2* strain yielding the white filamentous dikaryotic growth of the compatible-mating reaction. The bottom portion shows an *a1b1* strain overlaid with an *a2b1* strain yielding yeast growth. These results are indicative of the feature that two different *b* alleles are essential for the formation of the filamentous (and pathogenic) dikaryon.

regulator of pathogenicity, since an active *b* complex is sufficient to render haploid strains pathogenic (Basse and Steinberg, 2004), though the level of virulence is likely determined by additional factors (see below).

Various reports in the literature indicate that there are >25–35 *b* alleles. More careful examination in field populations of the Americas shows about 20 alleles (G. May, personal communication). The biggest surprise is that all 20 were not recovered in Mexico, the likely birthplace of modern maize. But all 20 *b* alleles have been found in the rest of North and also in South America. It is possible that the Mexican populations of U. *maydis* have undergone a bottleneck recently, perhaps well after domestication (G. May, personal communication). What is the purpose of having so many alleles? Is one combination more or less virulent than another? These will be interesting questions for further study. Furthermore, does intratetrad mating occur? This phenomenon has been observed in the anther smut, *Microbotryum violaceum*, where it likely reduces the effect of haplolethal alleles linked to mating type (Hood and Antonovics, 2000). However, Tillet was unable to demonstrate that teliospores of U. *maydis* were infectious when dusted on the plant surface suggesting that the inoculum may be derived from already postmeiotic cells (Christensen, 1963; Walter, 1935; and J. K. Pataky, Department of Crop Sci., University of Illinois, http://www. sweetcorn-uiuc.edu/Common-smut/Huitlacoche-WSMBMP.doc).

D. Targets of *bE/bW* regulation

A large number of genes respond to the action of the bE/bW heterodimer. Most of the genes that are affected appear to be indirectly controlled, instead being activated or blocked by the direct targets for binding by the heterodimer (Kahmann and Kamper, 2004). The bE/bW heterodimer binds to *b*-binding sites (bbs) upstream of *b*-responsive genes (Brachmann *et al.*, 2001; Romeis *et al.*, 2000) (Fig. 1.4). The bbs1 sequence (TTCATGATGAGAAGTGTGACAGACTG-TGC) was identified upstream of the *lga2* gene. There is a repeated motif in bbs1 ($A^C/_GTGTG$) that is also found in the *mfa2* promoter. This binding site is also upstream of *mfa2* and is similar to a *Saccharomyces cerevisiae*-mating repeat, a motif (GATGN$_9$ACA) that defines the haploid specific gene consensus for repression by the MATa1-MATα2 heterodimer in *Saccharomyces cerevisiae* (Li *et al.*, 1995). The *b*-responsive genes include members of the *a* mating-type locus where *mfa* and *pra* are downregulated and *lga2* is induced (Romeis *et al.*, 2000). So far only three direct targets of the bE/bW heterodimer have been reported: *dik6*, *polX*, and *lga2* (Brachmann *et al.*, 2001; Romeis *et al.*, 2000). Presumably, the genes directly activated control the large number of genes (at least 246) indirectly regulated by or dependent on the b heterodimer (Kahmann and Kamper, 2004). Genes indirectly regulated include *kpp6*, encoding a MAP kinase required for plant penetration (Brachmann *et al.*, 2003) and *rep1* (Wosten *et al.*, 1996). The *rep1* gene encodes a protein that, on processing, yields a group of twelve 35- to 53-aa peptides called repellents which are functionally but not structurally related to hydrophobins. Disruption of *rep1* results in a wettable phenotype and a failure to produce aerial hyphae (Wosten *et al.*, 1996).

E. The response to mating: Activation of cAMP and MAPK pathways

Exposure of cells to pheromone of the opposite mating type leads to G2 arrest, upregulation of pheromone genes, and the elaboration of conjugation or mating hyphae (Garcia-Muse *et al.*, 2003). The upregulation of the *a* and *b* genes in this response through the transcriptional regulator Prf1 is dependent on activation of the MAPK and cAMP pathways (Kaffarnik *et al.*, 2003). But the connection between the pheromone receptors (Pra1/2) and the MAPK or cAMP pathway is presently unknown (Fig. 1.3). In *Cryptococcus neoformans* and *Saccharomyces cerevisiae*, heterotrimeric G-proteins activate cAMP and MAP kinase pathways via G-protein–coupled receptors (Wang *et al.*, 2000; Whiteway *et al.*, 1989). There are four Gα-subunits in U. *maydis* but only Gpa3 has been reported to have a function during mating and cAMP signaling (Kruger *et al.*, 1998; Regenfelder *et al.*, 1997). The single Gβ-subunit in U. *maydis*, Bpp1, has been studied for a potential role in mating (Muller *et al.*, 2004). Gpa3 and Bpp1, the Gα- and Gβ-subunits of a heterotrimeric G-protein, respectively, are apparently both involved in adenylate cyclase (Uac1) activation (Kruger *et al.*, 1998; Muller *et al.*, 2004). Similar to *gpa3* mutants and other mutants of the cAMP pathway, *bpp1* deletion mutants grow filamentously and exhibit reduced pheromone gene expression. These defects can be rescued by the addition of exogenous cAMP (Kruger *et al.*, 1998; Muller *et al.*, 2004). A constitutively active allele of *gpa3* can suppress the filamentous phenotype of *bpp1* deletion strains, indicating that Bpp1 and Gpa3 are components of the same heterotrimeric G-protein acting on Uac1 (Muller *et al.*, 2004). However, unlike *gpa3* mutants, *bpp1* mutants can induce gall production in plants, suggesting that Gpa3 acts independently of Bpp1 during pathogenic development (Kruger *et al.*, 1998; Muller *et al.*, 2004).

Activation of the pheromone-responsive MAP kinase cascade likely occurs through a Ras-mediated signaling mechanism but the connection to upstream Pra1/2-mediated signaling remains unclear (Fig. 1.3). It is known that *ras2* is epistatic to the MAPK pathway; constitutive expression increases pheromone gene expression and promotes filamentation dependent on the pheromone-responsive MAP kinase cascade (Lee and Kronstad, 2002). Overexpression of Sql2, a Cdc25-like Ras guanine nucleotide exchange factor (RasGEF) for Ras2, results in filamentous growth (Muller *et al.*, 2003a) similar to the expression of a dominant active Ras2 allele (Lee and Kronstad, 2002). Interestingly, Sql2 activates Ras2 in pathogenic development but not during mating (Muller

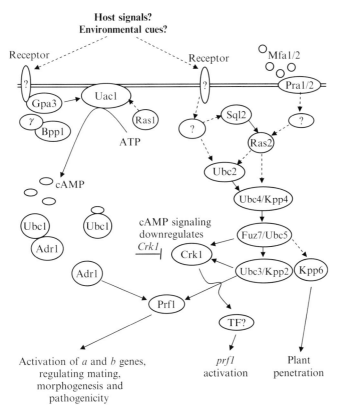

Figure 1.3. Interplay of the cAMP and MAP kinase pathways in U. *maydis*. Both pathways impinge on the regulation of the transcription factor Prf1 to modulate mating, morphogenesis, and pathogenicity. Solid arrows between shapes indicate that a genetic and/or physical interaction has been established while solid arrows elsewhere indicate input stimuli and/or output responses. Dashed lines between shapes and question marks indicate hypothetical interactions/proteins. The sizes of the ovals/circles do not reflect the relative sizes of the proteins. See the text for the references and additional information on these signaling pathways.

et al., 2003a), suggesting that another exchange factor may play a role in Ras2 activation during mating. The Ubc4/Kpp4 MAPKK kinase (Muller *et al.*, 2003b) and the Ubc2 adaptor protein each contain a Ras-association (RA) domain (Mayorga and Gold, 2001), a potential site for interaction with Ras proteins. Constitutively activated Ras1 increases pheromone (*mfa1*) production with no observed role in morphogenesis (Muller *et al.*, 2003a) and may interact with an RA domain present in Uac1 (Feldbrugge *et al.*, 2004). Additional signaling components, not shown in Fig. 1.3, include the PAK-like protein kinases, Smu1 and Cla4, which show possible differential effects on mating depending on the mating-type background. There is a greater effect of a *smu1* mutation (i.e., reduced *mfa2* expression) in the *a2* background (Smith *et al.*, 2004), whereas Cla4 deletion causes reduction in mating efficiency in the *a1* background (Leveleki *et al.*, 2004).

F. Interplay between the cAMP and MAPK pathways

Cross talk between the cAMP and MAP kinase cascade-signaling pathways regulating mating and morphogenesis has been well documented in fungi (Lee *et al.*, 2003; Mosch *et al.*, 1999). In *U. maydis*, haploid cells typically divide as budding yeast (Fig. 1.1) but deletion of a single gene, *uac1*, encoding adenylate cyclase, generated a constitutive filamentous haploid. Thus, knowledge of the interplay of the cAMP and MAPK pathways in *U. maydis* was advanced by analyzing a number of suppressors of the filamentous phenotype of the *uac1* mutant (Barrett *et al.*, 1993; Gold *et al.*, 1994a). These suppressor mutants were termed *ubc* for *Ustilago* bypass of cyclase (Gold *et al.*, 1994a; Mayorga and Gold, 1998). Complementation of these *ubc* suppressor mutations led to the restoration of the filamentous *uac1⁻* phenotype and enabled the isolation of five *ubc* genes discussed here, encoding components of both the cAMP and MAP kinase pathways (Gold *et al.*, 1994a, 1997; Mayorga and Gold, 1998). Three of these genes (*ubc3–5*) were also identified via their homology to *Saccharomyces cerevisiae* proteins involved in mating, and named *kpp2*, *kpp4*, and *fuz7*, respectively (Banuett and Herskowitz, 1994; Muller *et al.*, 1999, 2003b). The *ubc3/kpp2*, *ubc4/kpp4*, and *fuz7/ubc5* genes encode a MAP kinase, MAPKK kinase, and a MAPK kinase, respectively, and all are members of the pheromone-responsive MAP kinase cascade depicted in Fig. 1.3 involved in mating, morphogenesis, and pathogenic development (Andrews *et al.*, 2000; Banuett and Herskowitz, 1994; Mayorga and Gold, 1999; Muller *et al.*, 2003b). Mutants of *ubc1*, which encodes the regulatory subunit of PKA, exhibit defects in bud site selection (Gold *et al.*, 1997). Ubc2 (Mayorga and Gold, 1998, 2001), discussed in more depth below, shares homology with Ste50-type adaptor proteins that regulate pheromone-responsive MAP kinase pathways in yeast (Ramezani-Rad, 2003).

Integration of signaling between the cAMP and MAP kinase pathways hinges in part on the activation of the HMG-domain transcription factor Prf1, which is differentially phosporylated by PKA and the MAP kinase Ubc3/Kpp2 to activate *a* and *b* gene expression (Kaffarnik *et al.*, 2003) (Fig. 1.4). Induction of the *a* mating-type genes requires the PKA phosphorylation sites while induction of the *b* genes requires both the PKA and MAPK phosphorylation sites. Prf1 recognizes pheromone response elements (PREs: ACAAAGGGA) present upstream of both *a* and *b* locus genes. Prf1 itself may be autoregulated through PRE elements (Hartmann *et al.*, 1996, 1999). Another HMG-domain transcription factor, Rop1, binds to Rop1 recognition site (RRS) elements in the *prf1* promoter

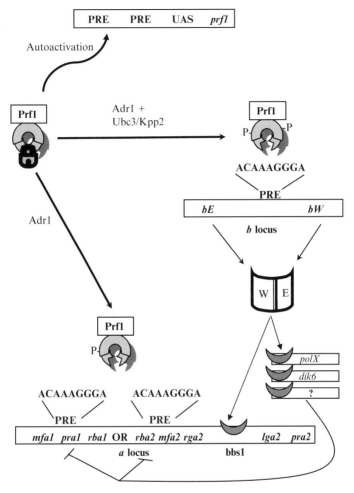

Figure 1.4. Regulation of mating-type specific genes initiated by the transcription factor Prf1. Phosphorylation by Ubc3/Kpp2 and the cAMP-dependent PKA, Adr1, kinase is required for activation of the *b* locus, while phosphorylation by Adr1 alone is crucial for activation of the *a* locus genes. Activation occurs by binding of Prf1 to PRE, upstream of the target genes, including the *prf1* gene itself. Additional controls of Prf1 expression occur at the UAS. The downstream events include activation of genes directly by the bE/bW heterodimer at *b*-binding sites (bbs), including at the *lga2*, *dik6*, and *polX* genes. Some of the genes activated directly then go on to activate or repress genes affected indirectly by bE/bW.

and is required to activate *prf1* as a response to pheromone in axenic culture (Brefort *et al.*, 2005). Methylation of proteins affects mating and may thus reflect an additional posttranscriptional control of Prf1 (Fischer *et al.*, 2001). Inhibitors of *b*-regulated genes also have been identified. Both Rum1 (Quadbeck-Seeger *et al.*, 2000) and Hda1 (histone deacetylase) inhibit some *b*-regulated genes in sporidia (Reichmann *et al.*, 2002). Some aspects of development, such as the induction of mating hyphae formation in response to pheromone, display independence from Prf1 (Muller *et al.*, 2003b), suggesting that other pathways are operational in this signaling.

Integration of the cAMP and MAP kinase signaling also has been demonstrated through the study of *crk1*, which encodes an Ime2-like kinase that is necessary for appropriate mating and development (Garrido and Perez-Martin, 2003). Both cAMP and Crk1 regulate *prf1* expression via the upstream-activating sequence (UAS) (Garrido *et al.*, 2004), possibly through an unknown transcription factor as suggested by Brefort *et al.* (2005) (Fig. 1.3). Inactivation of *crk1* suppresses the filamentous growth of a mutant defective in the cAMP pathway and blocks gall formation *in planta*. Additionally, strains deficient in *gpa3* produce highly elevated levels of the *crk1* message, suggesting that the cAMP pathway negatively regulates *crk1* expression (Garrido and Perez-Martin, 2003; Garrido *et al.*, 2004). The MAPK pathway counteracts this negative regulation of *crk1*. Immunoprecipitation experiments revealed that both Ubc3/Kpp2 and Fuz7/Ubc5 interact physically with Crk1 (Garrido *et al.*, 2004). Taken together, these findings indicate that Crk1 is a key integration point of the cAMP and MAP kinase pathways that modulate sexual development.

III. DIMORPHISM

Control of the dimorphic switch has been an area of intensive study in *U. maydis* and other pathogenic fungi (Sanchez-Martinez and Perez-Martin, 2001). The dimorphism that occurs in response to mating or in response to varying conditions has led to speculation on whether all filaments are created equal. Elevation of bE/bW levels in haploids leads to filamentous growth (Brachmann *et al.*, 2001; Hartmann *et al.*, 1996) but this does not necessarily mean that these strains will be pathogenic. For example, *gpa3* and *adr1* deletion mutants generated in solopathogenic haploid strains do not cause disease symptoms (Durrenberger *et al.*, 1998; Loubradou *et al.*, 2001). Moreover, differences in the levels of pathogenicity have been reported between filamentous haploid solopathogens, diploid solopathogens, and the dikaryons. Likely as a reflection of these differences, global variations in gene expression have been identified in these different cell forms (Babu *et al.*, 2005). In addition, other filamentous forms of *U. maydis* are observed in the laboratory, and although most of these forms are not, in and of themselves, capable of causing disease, it is instructive to explore the environmental signals that elicit such responses, as well as the signaling pathways involved. Several of these environmental signals are discussed in turn below.

A. Low nitrogen

Ammonium is a preferred source of nitrogen for many organisms and a family of well-defined and highly conserved transporters, known variously as AMT or MEPs, has been identified in organisms from bacteria to humans (Marini *et al.*, 1997). A bacterial member of this AMT/Mep family, AmtB, was crystallized (Khademi *et al.*, 2004; Zheng *et al.*, 2004) and shown to form a homotrimer in the membrane. The interface of the trimer appears to be a gas channel for transport of NH_3. The best characterized of the fungal NH_4 permeases (MEPs) are those from *Saccharomyces cerevisiae* (Lorenz and Heitman, 1998; Marini *et al.*, 1997). Mep2p is the high-affinity permease that also "senses" conditions of low ammonium (<5 mM), to trigger the switch to pseudohyphal growth (Marini and Andre, 2000). *U. maydis* sporidia likewise produce hyphae when grown on SLAD (50-mM ammonium). Two *U. maydis* Mep homologues were identified and their genes were cloned and characterized (Smith *et al.*, 2003). Disruption of either *ump1* or *ump2* alone had no apparent effect on pathogenicity nor did deletion of *ump1* yield any obvious phenotypes save a slight reduction in ammonium uptake. The high-affinity permease, Ump2, was identified in a screen for genes upregulated during budding growth (Garcia-Pedrajas and Gold, 2004). Both *ump1* and *ump2* could complement the growth defect of yeast triple *mep1mep2mep3* mutants on low ammonium (Smith *et al.*, 2003). However, only *ump2* complemented the ability to produce pseudohyphae under these conditions. Its disruption in *U. maydis* yielded cells that similarly were unable to produce filaments on SLAD. The *ump1ump2* double mutant clumped when grown in potato dextrose broth. As expected, the double mutant had no observable uptake of ammonium in *in vitro* assays but, surprisingly, grew on SLAD, at either pH 7 or pH 3, producing tangled filaments. This poses the question, how do such double mutants grow on low ammonium? A search of the *U. maydis* genome sequence for homologues to the yeast Mep2p revealed only sequences corresponding to Ump1 and Ump2. However, other possible transporters of nitrogen or nitrogenous compounds are predicted.

Possible clues to the role(s) of Meps/Umps in regulating filamentous growth in different fungi come from growth of wild type and mutants of the respective fungi on alternative nitrogen sources (Kuruvilla *et al.*, 2001; Martinez-Espinoza *et al.*, 2004). When nitrogen quality is low (e.g., Arg, Pro, urea, γ-aminobutyrate), intracellular Glu and Gln levels are low and this is important since 85% of total cellular nitrogen comes from glutamate and the remaining 15% is derived from the amide of glutamine (Magasanik and Kaiser, 2002). High-quality nitrogen sources thus include Gln and NH_4. The GATA transcription factors help regulate gene expression to discriminate between differences in nitrogen quality (Cooper, 2002; Magasanik and Kaiser, 2002). GATA-binding sites are located upstream of the MEP/UMP genes; GATA control has been demonstrated for *Saccharomyces cerevisiae* and *Candida albicans* Mep2

(Biswas and Morschhauser, 2005), and is likely at work for U. *maydis* Ump2 (Smith *et al.*, 2003). However, there are differences among these fungi with respect to their filamentation on different limiting nitrogen sources and the degrees to which filamentation on alternative sources is dependent on the respective Mep2 homologue. In *Saccharomyces cerevisiae*, Mep2 is not required for filamentation when other N sources are limiting. In contrast, in *Candida albicans* and U. *maydis*, Mep2/Ump2 may play a more central role. Wild-type or *ump1* mutant haploid U. *maydis* cells filament on plates with nitrogen provided only as 100 μM of the following compounds: NH_4, Glu, urea, Gln, His, and Pro. Limiting arginine fails to stimulate filamentation (R. Schmelz and M. H. Perlin, unpublished data). The *ump2* mutant has reduced filamentation or has no filamentation on all limiting nitrogen sources except His. The *ump1ump2* double mutant still filaments on limiting Gln, Pro, Urea, His, but fails to do so on Glu and forms tangled filaments on NH_4. *Candida albicans* wild type and mutants respond similarly, except the filamentous response on low Arg may reflect a stress response (Biswas and Morschhauser, 2005). Moreover, the switch to filamentous growth in response to pH was not dependent on Mep2 (Biswas and Morschhauser, 2005), as appears also to be the case for U. *maydis* Ump2 (Gupta, 2006).

Both the MAPK and cAMP/PKA pathways are implicated in filamentous growth of *Saccharomyces cerevisiae*, *Candida albicans*, and U. *maydis*. But how is Mep2/Ump2 involved? In *Candida albicans*, a C440 truncation of Mep2 yields to a hyperfilamentous phenotype that can activate filamentation in mutants defective in either the MAPK or cAMP pathway, but not both. Thus, Mep2 must signal through both; on the other hand, Ras1 is epistatic to both pathways for this response (Biswas and Morschhauser, 2005). In U. *maydis*, exogenously supplied cAMP reverses the filamentous phenotype on low nitrogen. Another observation that connects Ump2 to the cAMP/PKA pathway is the fact that transcription of the *ump2* gene is eightfold higher in budding cells compared to the *uac1* constitutively filamentous mutant (Garcia-Pedrajas and Gold, 2004). Moreover, fungal Meps contain a putative site for PKA phosphorylation that is lacking in nonfungal species. The ability to phosphorylate the target Thr or Ser is required to allow signaling in the filamentous response to low ammonium, but not for transport (Smith *et al.*, 2003). Thus, in U. *maydis*, the connection to the PKA pathway may occur at multiple levels.

B. pH

Another signal that can affect the dimorphic switch is pH. *Candida albicans* is inhibited from growing as filaments at acid pH (Soll, 1985). In contrast, the transition to filamentous growth is stimulated for U. *maydis* haploid cells by a shift to acid media (Martinez-Espinoza *et al.*, 2004; Ruiz-Herrera *et al.*, 1995): at pH 7.0, U. *maydis* grows as budding yeast cells, while at pH 3.0, it produces filaments. This ability to respond to acid pH appears dependent on several additional environmental factors, as well as functional components of the cAMP and MAPK signal transduction pathways. First, transfer of cells from ice to 37°C was important for induction of filaments, as was the type of nitrogen source provided (NH₄NO₃ for pH 3, KNO₃ for pH 7) (Martinez-Espinoza et al., 2004). When mutants in the MAPK- and cAMP-dependent PKA pathways were examined, it was determined that *ubc* mutants in the MAPK pathway (i.e., *ubc2*, *ubc5*, *ubc3*) were unable to respond to acid pH and formed yeast-like colonies on solid media at either pH 3 or pH 7. In contrast, *ubc1*, encoding the regulatory subunit of the cAMP-dependent PKA, was not required for the low-pH response. The Prf1 transcription factor was also not required for response to acid pH. As expected, mutants unable to activate the PKA catalytic subunit (*uac1* and *adr1*) were filamentous, regardless of pH and, unlike the other mutants, filamentation of the *adr1* mutant could not be blocked or reversed by addition of exogenously supplied cAMP. cAMP levels are an important determinant for the observed phenotypes at pH 7 and pH 3, as supported by measurements indicating high levels at pH 7 and consistently low (four- to sevenfold lower) levels at pH 3. Moreover, *uac1* filamentous mutants had very low-cAMP levels at either pH, while a strain with a constitutively active *gpa3* allele displayed levels comparable with wild type. Further, addition of cAMP to the medium suppressed the low-pH effect, both in wild-type and in the *ubc1* mutant (Martinez-Espinoza et al., 2004), suggesting that there are receptors of cAMP in the cell in addition to the regulatory subunit of PKA.

C. Additional stimuli: Air and polyamines

In *U. maydis*, other pathways may operate cooperatively in signaling the dimorphic switch. For example, it is known that exposure to air results in the production of aerial hyphae in *U. maydis* and in some fungi hydrophobins are thought to play a role in the formation of these hyphae (Elliot and Talbot, 2004; Gold et al., 1994a; Wosten and Willey, 2000). Guevara-Olvera et al. (1997) revealed a role of polyamine biosynthesis as a determinant of the dimorphic switch. Mutants of the ornithine decarboxylase (*odc*) gene, which encodes a product that catalyzes the first step in polyamine biosysnthesis, behave as polyamine auxotrophs. The dimorphic switch was inhibited in medium containing the minimum concentration of polyamines to support growth. Supplementation of the medium with additional polyamines led to the dimorphic transition (Guevara-Olvera et al., 1997). The signaling pathways required for this response have yet to be elucidated.

D. Lipids as signals

The haploid filaments induced by the environmental cues described so far are clearly not functionally equivalent to the filaments found *in planta* during productive infection of maize. Aside from ploidy considerations, this may reflect differences between the conditions shown to elicit filamentous growth in the laboratory compared with those present in the internal setting of the plant host. Klose *et al.* (2004) demonstrated that a variety of lipids were capable of inducing filamentous growth of both haploid and diploid strains. In general, the transition could be triggered by plant-derived triacylglycerides, including those from corn, sunflower, canola, and olive oils. Moreover, the filaments produced by such induction resembled strongly those found *in planta* and the hyphae were highly invasive in agar. The ability to respond was dependent on functional cAMP and MAPK pathways, as mutants (*ubc1*, *hgl1*, *ras2*, *ubc3*, *fuz7/ubc5*) in these pathways failed to respond. Interestingly, the response to lipids, like that to low pH, did not require functional Prf1 and thus differed from Ubc1, which was not necessary for acid-induced filamentation, suggesting the complexity of the function of these signaling components in regulating morphogenesis. However, the response to lipids and fatty acids could be suppressed by glucose, suggesting that lipids serve not only as signals, but also as a carbon source, such as might be found late in infection, after carbohydrates have been exhausted in the tumor and the fungus prepares for sporulation. Growth on lipids induced an extracellular triacylglycerol lipase activity. An exception was the *ubc1* mutant, which grew poorly on lipids and failed to exhibit the lipase activity. As mentioned earlier, filamentous growth induced by air, low pH, or low nitrogen is independent of a functional bE/bW heterodimer.

E. Chitin synthases

Chitin synthase (CHS) enzymes have often been proposed as important determinants of morphogenesis in fungi and potentially in dimorphism of U. *maydis*. As such, CHSs may function as effectors of environmental stimuli for morphogenesis. U. *maydis* has a total of eight CHS encoding genes falling into five classes (Weber *et al.*, 2005). CHSs are essential for production of the major cell wall constituent chitin, essential for maintenance of morphology and internal osmotic pressure. The cell wall composition of yeast and mycelia of U. *maydis* was reported to be quite similar, with some increase in detected chitin in hyphal cell walls (Xoconostle-Cazares *et al.*, 1996). Early work analyzed subsets of these genes. Gold and Kronstad (1994) studied two genes, *chs1* and *chs2*. These genes were expressed similarly in single haploid and mating cells, and when each gene was deleted there was a detectable reduction in CHS activity in cell extracts. The single and double deletion mutants exhibited no detectable mating, morphological or pathogenicity defect. Xoconostle-Cazares *et al.* (1996, 1997) reported identification of four genes, the disruption of which affected chitin content. For one gene, *umchs5*, disruption was reported to cause significant growth defects and reduced virulence. A thorough analysis of the entire set of

U. maydis CHS genes and their protein localization pointed out significant differences in the functions of the various members in affecting different portions of the life cycle even though all eight are expressed in both yeast and hyphal cells (Weber *et al.*, 2006). The level of induction in hyphal cells did not correlate to mutant phenotypes for the various genes. In fact, *chs1* and *chs4*, the most highly induced CHS genes in filaments, have no mutant phenotype while *chs7*, which is not induced, has a dramatic mutant phenotype in hyphae. Chs5 and Chs7, both CHS class IV members, play major roles in morphogenesis of sporidia, conjugation tubes, and dikaryotic filaments. These genes respectively play minor and major roles during infection. Most critical for infection however are Mcs1 and Chs6, both members of class V CHS enzymes. These data suggest that at least in some cases, rather than possessing redundant function of members within a class, often both members of a class are critical for a given overall function. The authors discuss that extrapolating function from one fungus to another for the various CHSs or even classes is impossible. Overall, for *U. maydis*, the class IV and V enzymes appear to play important roles in both morphogenesis and pathogenicity while the other classes are of relatively minor importance.

IV. CELL CYCLE AND CYTOSKELETAL REGULATION

A. Cell cycle

Recent advances in the knowledge of the regulation of cell cycle in *U. maydis* reveal similarities in these processes with other model fungi. In *Schizosaccharomyces pombe* and *Saccharomyces cerevisiae*, one cyclin-dependent kinase (CDK) in each of these fungi and a set of cyclins control events of the cell cycle such as the onset of mitosis and S phase (O'Farrell, 2001). The catalytic activity of CDK is regulated by inhibitory phosphorylation of specific residues by Wee1-like kinases. In *U. maydis*, the CDK, Cdk1 has been identified along with Wee1 and two B-type cyclins, Clb1 and Clb2 (Garcia-Muse *et al.*, 2004; Sgarlata and Perez-Martin, 2005a). *U. maydis* strains overexpressing a mutagenized form of Cdk1 that remains unphosphorylated produce aggregates of cells that divide by septation but remain attached. Inhibitory phosphorylation of Cdk1, likely by the essential Wee1 kinase, is also required for mating and pathogenicity (Sgarlata and Perez-Martin, 2005a). A Cdc25-related phosphatase likely counteracts Wee1-mediated phoshorylation of the Cdk1–Clb2 complex and thus regulates entry into mitosis (Sgarlata and Perez-Martin, 2005b). Levels of the *U. maydis* Clb1 and Clb2 cyclin proteins fluctuate during the cell cycle, falling in G1 and rising at entry to S/G2/M (Garcia-Muse *et al.*, 2004). Clb1 depletion in a conditional mutant strain resulted in cell cycle arrest at two points, one at which cells exhibited a prereplicative 1C DNA content and one in which the cells exhibited a postreplicative 2C DNA content. Clb2 depletion arrested the cells when they exhibited a 2C DNA content. These analyses suggested that Clb1 is required for the G1 to S and G2 to M transitions while Clb2 is required for entry into mitosis. This latter result was substantiated by the observation that overexpression of *clb2* caused early entry into mitosis. Additionally, cells overexpressing *clb2* divided by septation and grew filamentously but did not produce tumors in the plant, suggesting a tight linkage between cell cycling and pathogenic development (Garcia-Muse *et al.*, 2004).

Two additional genes that play a role in the control of cell separation have been identified. Weinzierl *et al.* (2002) identified *don1* and *don3* (donuts) by complementation of a mutant colony morphology that resembled a donut. *don1* encodes a protein containing a nucleotide exchange factor (GEF) domain, a pleckstrin homology (PH) domain, and an FYVE zinc finger domain while *don3* encodes a conserved Ser/Thr-protein kinase domain characteristic of Ste20-like kinases, targets of small GTP-binding proteins of the Rho/Rac family, including Cdc42. Two-hybrid assays revealed Don1 and Don3 both interact with the *U. maydis* Cdc42 protein, although, interestingly, Don3 lacks the CRIB domain that commonly mediates Cdc42–Ste20 interactions. Both the wild-type and *don1/don3* mutant budding cells form a primary septum at the bud neck between mother and daughter cells following bud formation and mitosis. However, *don1* and *don3* mutants are unable to form a secondary septum and consequently do not form a fragmentation zone which leads to a multiple bud phenotype. Thus, *don1* and *don3* appear necessary for secondary septum formation and cell separation (Weinzierl *et al.*, 2002).

B. Microtubules and actin

Transitions in the cell cycle and the maintenance of cell polarity in *U. maydis* and other fungi involve dynamic rearrangement of the microtubule (MT)- and F-actin-based cytoskeleton (Banuett and Herskowitz, 2002; Fuchs *et al.*, 2005; Steinberg and Fuchs, 2004; Steinberg *et al.*, 2001; Straube *et al*, 2003; Xiang and Plamann, 2003). Many of the studies to analyze MT and actin function take advantage of the inhibitory drugs such as benomyl and lantrunculin A, inhibitors of MT and F-actin, respectively. In addition, components of the actin and MT-based cytoskeletal network are now routinely "tagged" by GFP-variant fusion proteins and can be followed by time-lapse methodology. A number of the studies indicate linkage of MTs and F-actin to polar growth and secretion (Banuett and Herskowitz, 2002; Steinberg and Fuchs, 2004). Fuchs *et al.* (2005) detailed the essential roles of both actin and MTs in the early stages of sexual development of *U. maydis*. F-actin is required for budding growth, the formation of conjugation hyphae, cell fusion, and in the maintenance of *b*-dependent hyphae whereas MTs exhibit a less pronounced role in the formation of conjugation hyphae and are dispensable for cell fusion during mating. However, as previously established, MTs play a crucial role in mitosis and the extended growth of the *b*-dependent hyphae (Fuchs *et al.*, 2005). Although the amount of work on MT- and actin-based function in the past 8 years has been extensive, many questions remain such as the level of cooperativity between these components and the level of interplay with other signaling pathways in *U. maydis*.

C. Molecular motors and transport

Molecular motors such as dynein, kinesin, and myosin drive cytoskeletal re-arrangements coupled to cell cycle machinery and move cargo to the growing tip to support polar growth of the cell (Xiang and Plamann, 2003). Searches of *U. maydis* genomic sequence have revealed the presence of 14 such motors (Basse and Steinberg, 2004). The *dyn1* and *dyn2* genes encode the two components of the dynein heavy chain that play a role in nuclear migration (Straube *et al.*, 2001), evident in the fact that two or more nuclei begin to appear in conditional *dyn* mutants when their expression is repressed. These mutants also exhibit more and longer MTs, suggesting a function of dynein in MT dynamics (Straube *et al.*, 2001). Lehmler *et al.* (1997) identified the *kin2* gene encoding the heavy chain of kinesin, a plus-end-directed MT motor important for pathogenicity (Lehmler *et al.*, 1997; Steinberg *et al.*, 1998). *kin2* deletion mutants lack the large basal vacuole present in wild-type dikaryons (Steinberg *et al.*, 1998). Steinberg *et al.* (1998) postulated that Kin2 is involved in vacuole formation and that the accumulation of these vacuoles at the basal end of the tip plays a critical role in supporting cytoplasmic migration. A *U. maydis* class V myosin (Myo5) motor is apparently involved in actin-based transport of vesicles and organelles to actively growing regions of the buds and filaments (Weber *et al.*, 2003). Temperature-sensitive *myo5* mutants exhibit reduced dikaryon formation likely attributable to impaired pheromone perception and conjugation tube formation. As a result, *myo5* mutants are also reduced in mating and pathogenicity (Weber *et al.*, 2003). *kin3* encodes an N-terminal kinesin motor domain, a forkhead-associated domain, and a C-terminal PH domain (Wedlich-Soldner *et al.*, 2002). The *kin3* mutant cells form tree-like aggregates due to a cell separation defect and a defect in the bipolar-budding pattern. These morphological defects most likely arise from the disruption of bidirectional transport of early endosomes, evident from the finding that unlike wild-type cells, *kin3* mutants do not exhibit endosomal clustering at the septa and distal cell pole. Kin3-GFP was localized to endosomes, organelles likely involved in the delivery of components for cell growth, and its movement was associated with MTs in a plus-end-directed manner. Further analysis of a *dyn2* temperature-sensitive mutant in a *kin3* mutant background revealed a role of dynein in minus-end-directed movement of endosomes. Thus, a balance between Kin3 and dynein activity organizes polar endosomes in opposing directions during budding (Wedlich-Soldner *et al.*, 2002). Overall, these studies indicate a recycling of membanes via endo- and exocytosis and that this action, driven by dynein and Kin3, contributes to polar growth (Wedlich-Soldner *et al.*, 2000, 2002).

D. Differences with other fungi

Before leaving the topic of cell cycle regulation, it is of interest to point out some of the notable differences in the cell cycle and cytoskeletal rearrangements of *U. maydis* and compared with those of other fungi. First, when haploid *U. maydis* cells are exposed to mating pheromone, they undergo G2 arrest and produce conjugation tubes (Garcia-Muse *et al.*, 2003). This cycling is apparently different from the situation described in *Saccharomyces cerevisiae*, which generally lacks a G2 phase (O'Farrell, 2001). Second, MT nucleation in *U. maydis* occurs away from the nucleus at an MT-organizing center near the bud neck (Straube *et al.*, 2003). In many fungi, including *Saccharomyces cerevisiae*, MTs are nucleated at the spindle pole body (Steinberg and Fuchs, 2004; Xiang and Plamann, 2003). Additionally, unlike most fungi where the nuclear envelope (NE) persists or remains essentially intact during mitosis, the NE of *U. maydis* disassembles in mitosis. In *U. maydis*, the mechanism of NE breakdown has been correlated with dynein-mediated nuclear migration that strips off the envelope as chromosomes move from the mother cell to the bud cell (Straube *et al.*, 2001, 2005). Ras3, a Ras-like GTPase, has been identified as a participant in the removal of the NE (Straube *et al.*, 2005). In *Saccharomyces cerevisiae*, the homologous Tem1p Ras-like GTPase activates the mitotic exit network (MEN) signaling cascade and ultimately mitotic exit (Tan *et al.*, 2005). Thus, *U. maydis* *ras3* mutants appear normal morphologically but undergo mitosis in which the NE remains largely intact (Straube *et al.*, 2005). Taken together, observations of NE breakdown during mitosis in *U. maydis* suggest a mechanism more closely resembling the "open" mitosis described in animal cells. But rather than NE fragmentation, the envelope in *U. maydis* is stripped off and left in the mother cell for recycling in telophase (Straube *et al.*, 2005).

E. Connecting the signaling nodes

Given the earlier topics in this chapter, we may now ask how the cell cycle engine and cytoskeletal components are connected to the signaling pathways that regulate morphogenesis. A link between the cAMP pathway and an activator of the anaphase-promoting complex (APC) has been established in *U. maydis* (Castillo-Lluva *et al.*, 2004). The APC mediates cell cycle stage-specific degradation of cyclins such as Clb2 and other regulatory proteins in *Saccharomyces cerevisiae*, thus exerting cell cycle control (Tan *et al.*, 2005). In *U. maydis*, cAMP signaling regulates the expression of *cru1*, an APC that mediates cell cycle stage-specific protein degradation and thus maintains the cell cycle (Castillo-Lluva *et al.*, 2004). *cru1* is also required for appropriate control of budding cell length pheromone (*mfa1*) expression and dikaryotic filament formation. Compatible *cru1* deletion strains are capable of colonizing the plant but exhibit a drastic reduction in virulence, producing few small tumors that contain irregularly shaped teliospores. These defects indicate that Cru1-mediated APC activity has critical roles throughout the *U. maydis* developmental program and further support a connection between morphogenesis, virulence, and the cell cycle (Castillo-Lluva *et al.*, 2004). Additionally, a link between calcium homeostasis and the modulation of cytoskeletal functioning has been established in *U. maydis*. For example, Eca1, an endoplasmic reticulum-resident Ca^{2+}ATPase, likely cooperates in the same pathway with dynein, affecting Ca^{2+}/Calmodulin-dependent kinase activity (Adamikova *et al.*, 2004). In turn, this regulates MT dynamics and thus explains underlying temperature-dependent morphological defects in *eca1* mutants, including growth at both poles and the formation of multiple septa (Adamikova *et al.*, 2004).

V. PATHOGENESIS

A. Penetration and colonization

In *Magnaporthe grisea* and *Colletotrichum* species, involvement of the cAMP/PKA and MAP kinase pathways has been demonstrated in the early stages of plant infection (Deising *et al.*, 2000; Xu and Hamer, 1996). In *U. maydis*, tight regulation of the cAMP/PKA pathway is essential for the early stages of disease development. Mutants with low-PKA activity as a result of inactivation of genes in this pathway do not produce symptoms in inoculated plants. For example, compatible cAMP pathway mutant strains (*uac1*, *gpa3*, *adr1*) (Barrett *et al.*, 1993; Durrenberger *et al.*, 2001; Gold *et al.*, 1994a; Regenfelder *et al.*, 1997) produce no symptoms. Interestingly, inactivation of any of these genes in a wild-type haploid background leads to filamentous growth showing that filamentous morphology is necessary but not sufficient for pathogenesis. However, levels of PKA activity above normal do not appear to affect the penetration process as revealed by inoculation with strains with different degrees of PKA activation. For instance, *ubc1* (regulatory subunit of PKA) inactivated mutant strains infect and proliferate within plant tissue (Gold *et al.*, 1997; Kruger *et al.*, 2000). Also, dikaryons formed through mating of compatible activated *gpa3*$_{Q206L}$ strains, which exhibit elevated PKA activity, form appressoria and grow through the host epidermal layer normally (Kruger *et al.*, 2000). However, both the *ubc1* mutant and $gpa3_{Q206L}$-activated strains are affected in disease progression (Gold et al., 1997; Kruger et al., 2000), see below.

Plant pathogenic fungi, such as *Colletotrichum graminicola* and *Magnaporthe grisea*, produce appressoria, specialized infection structures with tough melanin-pigmented cell walls (Deising et al., 2000). Through appressoria, these fungi adhere tightly to plant surfaces and generate a localized turgor pressure that is sufficient for mechanical penetration (Deising et al., 2000). In *U. maydis*, appressorium-like structures are morphologically undifferentiated relative to appressoria formed by other pathogenic fungi and they are not melanized. Production of appressoria-like structures on the plant surface and the production of penetrating hyphae (Snetselaar and Mims, 1994) appear to be distinct steps in the infection process as evident in the analysis *b*-regulated MAP kinase Kpp6 mutant strains (Brachmann et al., 2003). Microscopic observations of plant surfaces after inoculation with compatible strains both carrying an inactivated mutant allele $kpp6^{T355A,Y357F}$ exhibited appressorium formation. However, only short filaments emerged from these appressoria that failed to penetrate plant cells (Brachmann et al., 2003).

In addition to mechanical forces, other mechanisms, such as production of lytic and other enzymes, are potentially involved in plant penetration and colonization. Little is known regarding the expression of lytic enzymes in *U. maydis* and their potential role in the infection process. In our laboratory we identified a gene encoding putative xylanases as highly expressed in budding wild-type cells and repressed in the constitutively filamentous *uac1* mutant (Garcia-Pedrajas and Gold, 2004). Strains deleted for this gene did not exhibit any reduction in virulence (M. D. Garcia-Pedrajas and S. E. Gold, unpublished data). Since deletion of single lytic enzymes has not had an effect on *U. maydis* pathogenicity, we are currently characterizing an orthologue of *snf1*, a general regulator of the expression of lytic enzymes and whose deletion has resulted in attenuated virulence in several fungal species such as *Cochliobolus carbonum* and *Fusarium oxysporum* (Ospina-Giraldo et al., 2003; Tonukari et al., 2000). Schauwecker et al. (1995) identified *egl1*, coding for a cellulase, which is not expressed in haploid cells but highly induced in the *b*-dependent filamentous form. However, mutants deleted for this gene were not affected in disease development. Cano-Canchola et al. (2000) obtained biochemical evidence for enzymatic activities in time course experiments, indicating pectate lyase in particular may have a role in the early steps of infection. Prior to and during colonization, cell wall function and biogenesis of *U. maydis* may be influenced via the production of hydrogen peroxide, which incidentally also may serve as a signaling molecule, either intracellularly or intercellularly with the host plant (Neill et al., 2002). Leuthner et al. (2005) used a restriction enzyme mediated integration (REMI)-based screen of solopathogenic haploids carrying an active b heterodimer to identify nonpathogenic strains. One of the identified nonpathogenic strains harbored the mutant *glo1* gene and exhibited a pleiotropic morphological defect including increased vacuolization and irregular branching. The *glo1* gene of *U. maydis* encodes a glyoxal oxidase that localizes to the plasma membrane and tips of budding cells. Compatible haploid cells deleted for *glo1* produce no symptoms although these strains are not completely sterile (Leuthner *et al.*, 2005). Null mutants of the related genes, *glo2* and *glo3*, had no affect on pathogenicity. Clearly, further studies to decipher the roles of enzymatic activities during *U. maydis* infection may prove valuable.

Once inside the host plant, how is growth and development of the filamentous dikaryon maintained? Since the filamentous dikaryon of *U. maydis* cannot be maintained in culture and is incapable of branching in culture, signals from the host are apparently necessary for the maintenance and proliferation of this structure. Additionally, certain mutant strains of *U. maydis* proliferate within plant tissue but the filamentous growth is altered (Abramovitch *et al.*, 2002; Gold *et al.*, 1997). For example, strains defective for *ukb1*, encoding a kinase that likely has a signaling role in the cAMP or a parallel pathway, exhibit an irregular branching pattern in proliferating hyphae *in planta* (Abramovitch *et al.*, 2002). Again, this suggests that signals from the host are necessary for appropriate development of the filamentous dikaryon. Yet the *in planta* signaling event(s) regulating hyphal maintenance remain unknown and their discovery remains a challenge for the future.

Another interesting aspect of the interaction between *U. maydis* and maize is that at the early stages of infection there is no clear evidence of a plant defense response at the ultrastructural level (Banuett and Herskowitz, 1996; Snetselaar and Mims, 1993). That has led researchers to speculate that *U. maydis* is able to escape plant recognition. Interestingly, deletion mutants of *gas1*, encoding an ER-localized catalytic subunit of a glucosidase II, may be altered in this ability; ultrastructural analyses indicate differences in the host–fungus interface at the early stages of infections between wild-type and *gas1* deletion strains (Schirawski *et al.*, 2005). Hyphae of *gas1* deletion strains are able to penetrate the plant but are arrested in the epidermal layer. Additionally, unlike the host–fungus interface observed in infections with wild-type strains, the host–fungus interface observed in *gas1* mutant infections is different, characterized by a wide, electron transparent zone, reflecting a different composition. Thus, the activity of Gas1 is required for the appropriate control of the host–fungus interface and may play a role in the suppression of host defense or in eliciting a host response, possibly for the acquisition of nutrients (Schirawski *et al.*, 2005).

B. Gall formation and *in planta* teliospore production

Infection of maize by *U. maydis* causes hypertrophy and hyperplasia of maize cells within the resultant galls (Banuett and Herskowitz, 1996; Callow and Ling,

1973; Snetselaar and Mims, 1994). The histology of these galls is consistent with the theory that an excess of plant hormones contributes to their development. In support of this idea, auxin or cytokinin levels are higher in infected maize tissues than in healthy tissues (Mills and Vanstaden, 1978; Moulton, 1942; Turian and Hamilton, 1960) and *U. maydis* itself makes auxin in culture (Martinez *et al.*, 1997; Wolf, 1952), but efforts to determine if the indole acetic acid made by *U. maydis* is essential for corn smut symptom development have been inconclusive (Basse *et al.*, 1996; Guevara-Lara *et al.*, 2000). Consequently, both the role and the origin of elevated hormones in this plant disease remain open questions, as does the mechanism by which *U. maydis* alters the morphology of maize cells.

U. *maydis* produces up to 6 billion teliospores/per cubic centimeter gall tissue (Christensen, 1963). These diploid, sexual spores are only produced when *U. maydis* is growing *in planta*, thus their development appears to be dependent on signals from the host plant. It may be that host-derived molecules directly stimulate teliospore production (Ruiz-Herrera *et al.*, 1999), or that the *in planta* environment may induce *U. maydis* to produce a sporulation factor or signal, which in turn promotes teliospore production once a critical concentration has been reached. The dependency of this stage of the *U. maydis* life cycle on the host plant, combined with the species' genetic tractability, makes *U. maydis* a particularly valuable model for predicting how other strictly obligate plant pathogens, such as the rust fungi, rely on their host plants.

The morphological events that are characteristic of teliospore development in *U. maydis* have been described in detail (Banuett and Herskowitz, 1996; Snetselaar and Mims, 1994). As corn smut galls develop, the hyphae within the galls proliferate to form large aggregates and they branch extensively, especially near their tips (Banuett and Herskowitz, 1996; Snetselaar and Mims, 1994). They then fragment within a mucilaginous matrix (Banuett and Herskowitz, 1996). The resulting cells become round and develop echinulate, darkly pigmented spore walls within the hyphae, which become gelatinous and deliquescent as the spores mature (Banuett and Herskowitz, 1996; Snetselaar and Mims, 1994). The host-derived triggers of teliosporogenesis have not been characterized nor is it known how *U. maydis* detects these signals.

Several *U. maydis* proteins that influence gall or teliospore formation by acting downstream of the initial signal perception steps have been identified. Some of these proteins are predicted to be regulators, either of signal transduction pathways or of transcriptional networks, which control whether these developmental programs are executed, while others are of unknown biochemical function. The following discussion emphasizes how study of these proteins has provided insight into the order of their activity during pathogenic and sexual development in *U. maydis* (Fig. 1.5), and why some of these proteins are predicted to regulate or interact with each other (Fig. 1.6). Note that proteins

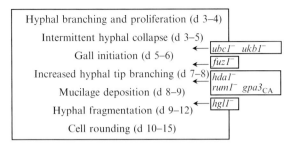

Hyphal branching and proliferation (d 3–4)
Intermittent hyphal collapse (d 3–5)
Gall initiation (d 5–6)   ← $ubc1^-$ $ukb1^-$
Increased hyphal tip branching (d 7–8) ←$fuz1^-$
Mucilage deposition (d 8–9) $hda1^-$
                                $rum1^-$ $gpa3_{CA}$
Hyphal fragmentation (d 9–12) ←$hgl1^-$
Cell rounding (d 10–15)

Figure 1.5. Steps in *U. maydis* gall formation and teliosporogenesis as described by Banuett and Herskowitz (1996). Approximate timing of events is shown in days (d) postinoculation. Developmental transitions at which different mutants are blocked are indicated with arrows. Only strains for which *in planta* morphological data have been collected are shown here. $gpa3_{CA}$ denotes a constitutively active allele of *gpa3*. Loss-of-function mutants are indicated as described in the text.

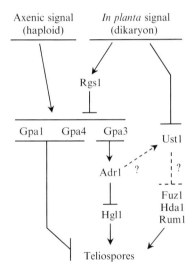

Figure 1.6. Model illustrating the regulatory pathways controlling teliosporogenesis in *U. maydis*. In haploid cells, Gpa3 is hypothesized to suppress sporulation via cAMP + PKA (Adr1). Activation of Ust1 and Ust1 is hypothesized to repress the transcription of at least three genes that are essential for various steps in teliospore development (dotted lines). In the interest of simplicity, the additional functions of Gpa1, Gpa3, and Gpa4 are not shown (e.g., regulation of mating and/or filamentous growth).

required for filamentation (e.g., many members of the MAPK pathway) typically are required for *U. maydis* proliferation *in planta* and for subsequent gall formation, but they are not emphasized here because they are discussed in detail elsewhere in this chapter.

Several *U. maydis* signal transduction proteins play important roles in the induction of galls or teliospores. Loss of Ubc1 (the regulatory subunit of cAMP-dependent PKA) or Ukb1 (a protein kinase B-related enzyme) activity does not entirely block hyphal proliferation, or the formation of short lateral branches *in planta*, but both proteins are needed for gall induction and the aggregation of the hyphal masses typically found in galls (Abramovitch *et al.*, 2002; Gold *et al.*, 1997). These phenotypes plus other shared phenotypes between *ukb1*⁻ and *ubc1*⁻ strains suggest that Ukb1 is somehow connected to cAMP signaling (Abramovitch *et al.*, 2002). Inactivation of Gpa3, a heterotrimeric G-protein α-subunit that activates the cAMP-signaling pathway in *U. maydis*, blocks tumor formation, while constitutive activation of Gpa3 suppresses teliospore formation (Kruger *et al.*, 1998, 2000; Regenfelder *et al.*, 1997). This implies that Gpa3 and the rest of the cAMP pathway act as repressors of teliospore development to ensure that spores are not made prematurely, that is, during axenic growth or mating (Fig. 1.6). Mutant strains in which Gpa3 is constitutively activated arrest before hyphal fragmentation (Kruger *et al.*, 2000) but after gall initiation, the step at which *ubc1*⁻ strains are blocked (Fig. 1.5). Because both of these mutants contain elevated PKA activity, their contrasting phenotypes suggest that slight differences in the level of cAMP signaling can influence downstream events significantly (Kruger *et al.*, 2000).

fuz1 mutants produce hyphae that branch and proliferate *in planta* like wild-type strains, but they do not produce the extracellular mucilage made by normal cells (Banuett and Herskowitz, 1996) (Fig. 1.5). *fuz1*⁻ hyphae also rupture and swell instead of fragmenting in an orderly fashion, suggesting that the organized dissolution of the cell wall at the septa is somehow dependent on the presence of the mucilage (Banuett and Herskowitz, 1996). The *fuz1* gene has not been characterized and the function of its encoded protein is unknown.

Hda1 functions as a histone deacetylase and thus is likely to influence levels of gene transcription by altering nucleosome positioning on the chromatin (Reichmann *et al.*, 2002). It is a transcriptional repressor in haploid cells of selected genes that are activated in dikaryons by the bE/bW heterodimer (Reichmann *et al.*, 2002). However, Hda1 also is required for bE/bW-dependent induction of at least one gene in dikaryons (Reichmann *et al.*, 2002). *In planta*, *hda1* loss-of-function mutants produce hyphae with normally branched tips, but the patches of these hyphae are smaller in volume than those made by wild-type strains (Reichmann *et al.*, 2002). The *hda1*⁻ hyphae are embedded within the usual mucilage, but they do not round up or develop into teliospores (Reichmann *et al.*, 2002) (Fig. 1.5).

Like Hda1, Rum1 represses a subset of genes in haploid cells that are induced by bE/bW in dikaryons (Quadbeck-Seeger *et al.*, 2000). Furthermore, the specific set of genes regulated by Rum1 is similar to that regulated by Hda1 (Reichmann *et al.*, 2002). The *rum1* gene encodes a protein with several conserved domains, including a DNA-binding domain and a histone deacetylase-interaction domain (Quadbeck-Seeger *et al.*, 2000). Therefore, Rum1 is proposed to form a complex with Hda1 in which specificity for certain promoters is conferred by Rum1 and regulation of transcription is conferred by Hda1 via its influence over nucleosome positioning (Reichmann *et al.*, 2002). *rum1* partial deletion mutants develop normally *in planta* up to the point of mucilage deposition, but they appear to be blocked slightly later in development than *hda1⁻* strains because they produce large, normal-sized patches of mucilage-embedded hyphae (Reichmann *et al.*, 2002) (Fig. 1.5).

Hgl1, a protein of unknown function that is required for teliospore formation *in planta* (Durrenberger *et al.*, 2001), is likely to be one of the teliospore-related proteins suppressed by activated Gpa3 in haploid cells. The *hgl1* gene was isolated as a genetic suppressor of a mutation in the catalytic subunit of PKA (Adr1), which is downstream of Gpa3 (Durrenberger *et al.*, 2001) (Fig. 1.6) and Ubc1 (Gold *et al.*, 1997) in the cAMP-signaling cascade. In addition, the Hgl1 protein is phosphorylated *in vitro* by Adr1 (Durrenberger *et al.*, 2001). *hgl1* partial deletion mutants develop normally *in planta* up to the point of hyphal fragmentation, but they do not make rounded cells or mature teliospores (Durrenberger *et al.*, 2001) (Fig. 1.5).

In recent work in our laboratories (Covert and Gold), four additional *U. maydis* proteins have been found to regulate aspects of teliospore development in *U. maydis*. Rgs1, a regulator of G-protein signaling that is predicted to accelerate the deactivation of G-protein α-subunits, appears to be an important molecular switch that reverses G-protein suppression of teliospore formation (Baker *et al.*, submitted for publication). In haploid cells, Rgs1 interacts genetically with Gpa3, as well as with Gpa1 and Gpa4, which are newly recognized as suppressors of teliospore formation *in planta* (Baker *et al.*, submitted for publication). Ust1, an APSES domain transcription factor, is predicted to be a global repressor of teliosporogenesis; in haploid cells, deletion of *ust1* leads to the *in vitro* production of copious, pigmented cells that strongly resemble teliospores (M. D. Garcia-Pedrajas and S. E. Gold, unpublished data). Potential locations for these four proteins in the networks regulating teliospore development are indicated in Fig. 1.6.

VI. GENOME-WIDE APPROACHES FOR THE STUDY OF *U. MAYDIS*

A. Genome structure

The *U. maydis* genome has been sequenced independently on three occasions. The first two sequencing projects were carried out with the resources of private companies (Bayer CropSciences and Excelixis, Inc.). Later, with public funding, the Whitehead Institute (now called the Broad Institute) sequenced and annotated the wild-type strain 521. The Broad Institute of MIT and Harvard consolidated the company data and released the information to the public (http://www.broad.mit. edu/annotation/fungi/ustilago_maydis/). This was accomplished as part of the Whitehead Institute's Fungal Genome Initiative (http://www.broad.mit.edu/anno-tation/fungi/fgi/). An additional important resource available to the public is the Munich Information Center for Protein Sequences (MIPS) MUMDB (MIPS *Ustilago maydis* Database, http://mips.gsf.de/projects/fungi/ustilago) at which manual annotation of the genome is currently well advanced.

The *U. maydis* genome is ~20 Mb in length and contains about 6500 genes on its 23 chromosomes. This is approximately the number of ORFs found in the *Saccharomyces cerevisiae* genome (6602; see SGD, http://www.yeastgenome. org/cache/genomeSnapshot.html). It is rather unusual for a filamentous fungus in that it contains an average of less than one intron per gene (ca. 4900). This is in comparison to sequenced ascomycete plant pathogens whose genomes are roughly twice the size and have approximately twice the number of predicted genes, each with an average of about two introns. Additionally, the relatively low number of introns in *U. maydis* differs markedly from the average number of 5.3 introns per protein-encoding gene predicted for *Cryptococcus neoformans* (Loftus *et al*., 2005).

The *U. maydis* genome possesses all necessary genes for biotrophic plant pathogenesis even though it has only half the coding capacity of other plant pathogens sequenced to date. For many gene families there seems to be a general reduction in the number of members (e.g., cutinase: *U. maydis* 1, *Magnaporthe grisea* 9; fungal specific transcription factor domain: *U. maydis* 20, *Magnaporthe grisea* 51; bZIP transcription factor: *U. maydis* 3, *Magnaporthe grisea* 7). On the other hand, there are numerous families that have roughly the same number of genes (ABC transporters: *U. maydis* 38, *Magnaporthe grisea* 35; pKinase domain-containing proteins: *U. maydis* 84, *Magnaporthe grisea* 85). A genome-wide comparison in search of "core" fungal genes found that *U. maydis* had almost as many *Saccharomyces cerevisiae* homologues as the ascomycetous fungi (Hsiang and Baillie, 2005). In particular, there is a high level of conservation of major signaling pathway constituents between yeast (and other ascomycetes) and *U. maydis* (Fig. 1.7). However, as discussed below, at least two adaptor proteins have evolved in the basidiomycetes differently than those found in the ascomy-cetes. We know that the Ubc2 domains lacking in the ascomycete counterparts are critical for signaling functionality in disease development. Thus, while there is wide conservation between the signaling pathways of the ascomycetes and basi-diomycetes as apparent in Fig. 1.7, there are important subtle differences that warrant exploration.

Another fascinating feature which has recently come to light through a combination of microarray experiments and genome analysis carried out in Germany is the presence of more than 10 gene islands encoding secreted

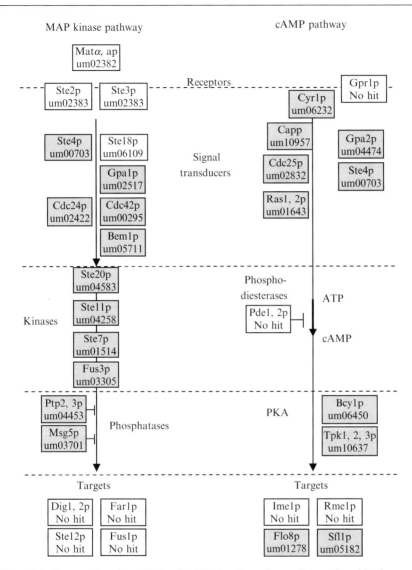

Figure 1.7. Conservation of the cAMP and MAPK-signaling pathways of U. *maydis* and *Saccharomyces cerevisiae*. *Saccharomyces cerevisiae* proteins used in homology searches are named. The annotation (um) number below the yeast protein name indicates a corresponding hit in strain 521 (*a1b1*, haploid) from the MUMDB or the Broad Institute databases. Boxes that are either shaded or white indicate similarities of the yeast and U. *maydis* proteins of e-values $<1 \times 10^{-10}$ and $>1 \times 10^{-10}$, respectively.

proteins that are induced *in planta*. Several of these clusters when mutated were found to dramatically affect pathogenicity (Kamper *et al.*, 2006). Overall the genome sequence combined with other tools developed has provided researchers of this fascinating fungus a critical tool for accelerated progress in the exploration of the function of its rather small set of genes.

B. An expanding toolbox for the study of *U. maydis*

The first reported attempt to produce protoplasts (spheroplasts) from sporidia of *U. maydis* dates to work of Waard (1976). It was not until 1983 when Banks reported the first protoplast-mediated transformation system for *U. maydis* by selecting for resistance to the antibiotic G418. However, the transformation frequency reported was very low, around 10 colonies per microgram of DNA. Wang *et al.* (1988) described a more efficient method using as a selectable marker, a bacterial HygB phosphotransferase gene fused to a *U. maydis hsp70* gene promoter, which confers resistance to hygromycin in *U. maydis*. The development of this efficient transformation method made possible cloning of genes by complementation of mutants and opened the doors to the genetic manipulation of *U. maydis*. Shortly thereafter, Tsukuda *et al.* (1988) cloned autonomously replicating sequences (ARSs) from *U. maydis* that when inserted into the integrative vector developed by Wang *et al.* (1988) increased the frequency of transformation to $\sim 10^4$ per microgram of input DNA. A year later, Fotheringham and Holloman (1989) and Kronstad *et al.* (1989) demonstrated that one-step gene disruption mediated by homologous recombination was possible in *U. maydis*.

In recent years, there has been a continuous improvement in transformation systems for *U. maydis*. These improvements include the addition of new selectable markers that make the system more versatile. For example, Keon *et al.* (1991) isolated a gene conferring resistance to the systemic fungicide carboxin (Cbx) and this is now widely used as a selectable marker for *U. maydis*. Currently, there are several dominant selectable markers for *U. maydis* transformation including those conferring resistance to hygromyin, Cbx, phleomycin, benomyl, geneticin, and nourseothricin (Gold *et al.*, 1994b; Kojic and Holloman, 2000). Expression vectors harboring constitutive and inducible promoters have also been developed for this fungal species. Among the former are vectors containing the glyceraldehyde-3-phosphate dehydrogenase (*gap*) promoter (Kinal *et al.*, 1993; Kojic and Holloman, 2000) and the *o2tef* promoter (Spellig *et al.*, 1996); and among the latter are *crg1* and *nar1* whose expressions are induced by arabinose and nitrate, respectively (Bottin *et al.*, 1996; Brachmann *et al.*, 2001). These inducible promoters have had interesting applications in the production of genetically engineered strains in which the expression of an active b heterodimer can be induced *in vitro* (Brachmann *et al.*, 2001). Spellig *et al.* (1996) described for the first time the successful expression of a green fluorescent protein (GFP) in *U. maydis*. Since then, GFP has become a very valuable tool to study gene expression, developmental processes, and host–pathogen interactions.

Researchers also have focused on the development of faster methods for the generation of gene deletion constructs. Since Fotheringham and Holloman (1989) and Kronstad *et al.* (1989) demonstrated that the frequency of homologous recombination in *U. maydis* is high enough to allow targeted gene replacement, this has become a central tool to study gene function. However, traditional methods to produce gene deletion constructs are labor intensive. Kamper (2004) has reported a method that uses polymerase chain reaction (PCR), digestion with *Sfi*I, a restriction enzyme from *Streptomyces fimbriatus* that allows directional cloning, and ligation to rapidly produce precise deletion constructs. In this approach, *Sfi*I sites are introduced in the PCR product during the amplification. Brachmann *et al.* (2004) have extended the use of *Sfi*I sites by employing a number of previously characterized resistance markers, reporter genes, and promoters to create a versatile battery of vectors with a choice of 32 different insertion cassettes all with the same basic structure: an optional reporter gene, a resistance cassette, and an optional promoter cassette flanked by two *Sfi*I recognition sites. These vectors allow the rapid generation of deletion constructs, insertion of reporter genes, expression under heterologous promoters, or combinations of the last two. In our laboratory (Gold), we have developed a method which combines PCR and Gateway-cloning technology from Invitrogen with use of the I-*Sce*I homing endonuclease to generate precise deletion constructs in a very simple, universal, and robust manner. We have called this method DelsGate, and in its final simplified version, it requires only the amplification of the gene flanks, followed by Gateway cloning (Invitrogen) and the generation of a final circular deletion construct *in vivo* by transformation of *E. coli*. We have modified a Gateway vector from Invitrogen to be used directly for *U. maydis* transformation (Garcia-Pedrajas *et al.*, submitted for publication). Clearly, the development of new methods and tools for manipulation of *U. maydis* holds promise in facilitating and enhancing the rate at which genetic analyses can be performed.

C. Differential screens

Since regulating the level of transcript produced in a particular condition is critical to gene function, it is expected that a subset of the genes whose expression patterns change at different stages of the life cycle of *U. maydis* play important roles in morphogenetic processes and pathogenicities. Differential screens have been performed to identify genes specifically induced or repressed in the different morphologies of the fungus grown *in vitro* (filamentous, budding, germinating teliospores), in specific nuclear conditions (haploid, diploid, and dikaryon), during pathogenic development; and also to identify genes regulated by specific transcription factors (i.e., the *b* heterodimer). Because of the critical role of the bE/bW heterodimer in regulating the switch to filamentous growth and pathogenic development, genes under the regulation of this transcription factor are of particular interest. Brachmann *et al*. (2001) combined an RNA-fingerprinting method with the use of inducible promoters to replace those native to *b* to identify genes regulated by the b heterodimer. They employed promoters that allowed the silencing or activation of the *b* genes in culture simply by modifying the nutritional components of the medium. This approach allowed the identification of both positively and negatively *b* transcriptionally regulated genes (Brachmann *et al*., 2001). Deletion of a number of the genes identified in this screen did not produce any discernible effect on morphology or pathogenicity, indicating that the ones characterized do not individually play a major role in these processes. However, in later work using the same approach, a novel MAP kinase upregulated by the bW/bE heterodimer and critical for plant tissue penetration was identified (Brachmann *et al*., 2003).

The dikaryon is the primary pathogenic form of *U. maydis*. However, in the laboratory, 1–3% of teliospore germinations result in diploid cultures (Holliday, 1961; Kojic *et al*., 2002) and production of diploid solopathogens in the field has been reported (Christensen, 1931). They can also be forced in the laboratory by mating compatible haploids carrying complementary auxotrophic mutations (Banuett and Herskowitz, 1989; Holliday, 1974). Although diploids are solopathogens, they are less pathogenic than dikaryons even when harboring exactly the same genetic background. This suggests that communication between the paired nuclei in the dikaryon may induce responses not stimulated in the diploid that are important for virulence. Nugent *et al*. (2004) constructed a cDNA library from a forced diploid culture of *U. maydis* growing as filaments and generated 7455 ESTs (expressed sequence tags) that are assembled into 3074 contiguous sequences. Babu *et al*. (2005) extended this work, studying upregulation of genes in diploid cells and filamentous dikaryons grown axenically as compared to haploid budding nonpathogenic cells. For this analysis, *U. maydis* ESTlibraries previously created (Nugent *et al*., 2004; Sacadura and Saville, 2003) were combined for comparative genomics analyses (Austin *et al*., 2004) and a subset of the analyzed cDNAs were selected for the construction of microarrays. Hybridization of these microarrays with probes prepared from the different material showed that more genes are upregulated in the filamentous dikaryon than in the filamentous diploid, relative to nonpathogenic-budding cells. The largest category of genes differentially expressed in the different morphologies and nuclear conditions was that of genes involved in metabolism (Babu *et al*., 2005). This is in agreement with our data in the identification of genes differentially expressed in two morphologies, budding haploid cells and the constitutively filamentous *uac1* mutant. In our laboratory (Gold), to gain insight into the role and significance of the budding phase of growth in the life and disease cycles of U. *maydis* in nature, we used suppression subtractive hybridization (SSH) to identify genes upregulated in budding cells. For this subtractive approach we took advantage of a *uac1* mutant strain, which is constitutively filamentous, to remove sequences common to both growth morphologies. We identified 37 genes specifically upregulated at least twofold in budding cells, 14 of which showed similarity with genes involved in metabolism, representing the largest single functional category (Garcia-Pedrajas and Gold, 2004). These studies together indicate that metabolism and morphology are closely linked in U. *maydis*.

Phytopathogenic fungi vary in strategies employed to survive between rounds of plant infection. U. *maydis* relies on the production of teliospores, resistant diploid cells with thick cell walls produced in mature galls. Teliospores of U. *maydis* are believed to overwinter in the soil and plant residues (Christensen, 1963). Meiosis occurs in germinating teliospores (O'Donnell and McLaughlin, 1984), leading to the production of haploid sporidia and thus may be considered the first step of infection. The morphological changes that take place during teliospore germination have been described in detail (O'Donnell and McLaughlin, 1984). However, little is known about the genetic regulation of this process. Sacadura and Saville (2003) initiated the identification of gene expression associated with teliospore germination on a large scale. They sequenced 2871 ESTs from a cDNA library prepared from germinating teliospores and performed Northern blots on selected cDNA clones. Zahiri *et al.* (2005) performed global gene expression studies during teliospore germination by production and hybridization of a cDNA library containing a set of 3918 nonredundant cDNAs. This work represents the first attempt to identify genes involved in teliospore germination on a large scale.

An area of particular interest is the identification of U. *maydis* genes specifically induced on maize infection. As discussed earlier, developmental events in the proliferating hyphae may be triggered on cue in response to multiple plant signals. However, little is known about such signals and the genes involved in inducing developmental responses in the fungus. But differential screens have been applied to the identification of those U. *maydis* genes induced during pathogenic development. An added problem in these screens is the need to deal with mixed samples containing both plant and fungal genes. Basse *et al.* (2000) used differential display to identify genes induced in U. *maydis* during maize infection. First, a maize-induced gene (*mig1*), which encodes a secreted protein, was identified. *In planta* observations of *mig1* expression with GFP in a reporter system revealed its strong upregulation following penetration until the development of sporogenic hyphae (Basse *et al.*, 2000). However, deletion of *mig1* did not compromise pathogenicity. Five additional *mig* genes similar to *mig1* were later characterized (Basse *et al.*, 2002). These genes (*mig2–1*, *mig2–2*, *mig2–3*, *mig2–4*, and *mig2–5*) are arranged as direct repeats in a 7.1-kb cluster. Deletion of the entire cluster did not have an effect on pathogenicity. Interestingly, features of the *mig* genes such as secretion, plant-inducible expression, and an even number of cysteines are reminiscent of fungal *avr* genes (Basse *et al.*, 2002).

Aichinger *et al.* (2003) combined REMI mutagenesis with enhancer trapping by using the gene for GFP as a reporter for detection to identify additional *in planta*-induced genes. These authors found three plant-induced genes (*pigs*) among a collection of 2350 insertion strains. *pig1* was found to be *mfa1*, and *pig2* showed similarity to a disulfide isomerase. The third integration event occurred in a locus that was designated the *p*-locus. It contains 11 genes and spans 24 kb. Five of the genes in the *p*-locus showed a plant-regulated expression pattern. Although deletion of both the *mig* and the *pig* gene clusters had no effect on pathogenicity, the analysis of the promoter region of the *mig2* gene cluster has led to the identification of a *cis*-acting element responsible for their high level of induction during pathogenic development (Farfsing *et al.*, 2005). It is expected that further characterization of plant-induced fungal genes will provide insight into the developmental program of U. *maydis* during pathogenesis. With the entire genome sequence available, new possibilities are open to study global expression patterns in the conditions of interest using microarrays representing all putative genes.

D. Adapted for pathogenicity

Multiple lines of evidence have implicated the adaptor protein Ubc2 as a regulator of the pheromone-responsive MAP kinase pathway. *ubc2* was identified by the complementation of a *uac1*-budding suppressor mutant to filamentous growth, implicating it in a pathway regulating filamentous growth (Mayorga and Gold, 1998). A multicopy plasmid encoding the Ubc4 MAPKK kinase partially complemented a *uac1 ubc2*[ts] mutant to filamentous growth (Mayorga and Gold, 1998), indicating a genetic and possibly a physical interaction between Ubc4 and Ubc2 and the specific involvement of Ubc2 in the regulation of the pheromone-responsive MAP kinase cascade. In support of this, *ubc2* mutant strains are morphogenetically unresponsive to mating pheromone (Mayorga and Gold, 2001). Additionally, deletion of an ortholog of *ubc2*, *cbc2*, in *Cryptococcus neoformans*, results in strains that are essentially incapable of mating or haploid filamentation (S. J. Klosterman, C. B. Nichols, J. Heitman, and S. E. Gold, unpublished data). Finally, *ubc2* encodes a protein that bears three of the hallmark domains present in adaptor proteins that have a well-documented role in the regulation of morphogenesis and mating in other fungi (Ramezani-Rad, 2003).

Among the *ubc* genes identified, the structure of *ubc2* is particularly intriguing, encoding an adaptor protein that has a basidiomycete-specific structure.

Related ascomycete proteins include the Ste50p and Mst50 adaptor proteins of *Saccharomyces cerevisiae* and *Magnaporthe grisea* (Ramezani-Rad, 2003; Zhao *et al.*, 2005), respectively. Analogous to the structure of Ste50p and related ascomycete proteins, Ubc2 possesses a sterile-alpha-motif (SAM) and an RA domain in a similar arrangement at the N-terminal half of the protein. The SAM domain of Ubc2 was previously found to have a role in supporting filamentous growth of *U. maydis* by complementation of a haploid mutant strain (Mayorga and Gold, 2001). But unlike the yeast Ste50p and other similar ascomycete adaptor proteins, Ubc2 contains two SH3 protein–protein interaction domains in the C-terminal half of the predicted protein. The presence of SH3 domains in the Ubc2-like proteins of bsidiomycetes, but not in the related proteins of the ascomycetes or *Rhizopus oryzae*, suggests that the SH3 domains of these proteins originated in basidiomycetes following the divergence from a common ancestor.

We have begun to elucidate the role of Ubc2 in regulating morphogenesis and pathogenicity in *U. maydis*. Like many of the deletion mutant strains of the Ras2/MAPK pathway in *U. maydis*, *ubc2* deletion mutants are also nonpathogenic (Mayorga and Gold, 2001). In support of the genetic interaction previously described between *ubc4* and *ubc2* (Mayorga and Gold, 1998) (Fig. 1.3), two-hybrid assays revealed that Ubc2 interacts physically with Ubc4 specifically via the SAM domains present at the N-termini of each of these proteins (S. J. Klosterman, A. D. Martinez-Espinoza, D. L. Andrews, J. R. Seay, and S. E. Gold, unpublished data). Site-directed mutagenesis of conserved amino acids in the putative functional domains of *ubc2*, followed by complementation of a budding *uac1ubc2* double mutant strain, revealed that both the SAM and RA domains are required for filamentous growth. Results of these complementation studies indicated that the C-terminal SH3 domains are unnecessary to support filamentous growth. Thus, the role of the SH3 domains was further investigated by deleting the entire region encoding the C-terminal half of *ubc2* in compatible strains of *U. maydis* and conducting mating and pathogenicity assays. Interestingly, analyses of the C-terminal SH3 deletion mutant strains and specific SH3 domain mutant strains of *ubc2* indicate these domains are not essential for mating but both are required for pathogenicity. In *U. maydis* in particular, these SH3 domains may represent a specific adaptation regulating pathogenicity (S. J. Klosterman, L. Esposito, J. R. Seay, D. L. Andrews, and S. E. Gold, unpublished data).

Library screens, using intact Ubc2 and portions thereof as bait, have revealed potential interacting proteins that may play roles in signaling, morphogenesis, and pathogenesis (S. J. Klosterman, A. D. Martinez-Espinoza, D. L. Andrews, J. R. Seay, and S. E. Gold, unpublished data). These experiments also revealed the efficacy of such an approach. That is, based on homology with proteins of the annotated *U. maydis* database and of proteins with known or predicted functions in the databases, few unknown proteins were obtained in these screens. One interesting example is an intersectin-like protein. U. *maydis* and other basidiomycetes possess a multiple SH3 domain-containing intersectin-type GEF protein. An orthologous protein regulates mitogenic signaling and endocytic machinery in animal cells (Adams *et al.*, 2000) and is apparently absent in *Saccharomyces cerevisiae* and other ascomycetes. Thus, the presence of unique adaptor-type proteins among the basidiomycetes may reflect differences in signal transduction via the traditional conserved signaling pathways. Other examples of potential Ubc2-interacting proteins include those displaying homology with a Cdc24-like GEF that interacts with Cdc42 GTPase in establishing cell polarity in *Saccharomyces cerevisiae* (Ziman and Johnson, 1994), a Sis1-like heat shock factor with a role in nuclear migration during mitosis in *Saccharomyces cerevisiae* (Luke *et al.*, 1991), and an unknown pleckstrin homology domain-containing protein (S. J. Klosterman, L. Esposito, J. R. Seay, D. L. Andrews, and S. E. Gold, unpublished data). Further characterization of the roles of these proteins is anticipated to provide insight into the mechanisms of Ubc2-mediated signaling in mating and pathogenic development.

VII. THE PLANT SIDE OF THE DISEASE EQUATION

The specialized interaction between U. *maydis* and its host obviously suggests cross-species signaling; the fungus induces significant developmental alterations in the plant, leading to the formation of galls. The fungus may achieve this on infection by sending signals to the host to modify its gene expression patterns. Since tumor formation is associated with cell enlargement and proliferation (Callow and Ling, 1973; Snetselaar and Mims, 1994), genes whose expression patterns are altered during gall formation likely include those involved in developmental processes. Thus, the study of up- and downregulation of maize genes in response to fungal infection is crucial to test these hypotheses and to gain insight into the pathogenic development of U. *maydis*. Basse (2005) published the first report of maize genes induced or repressed during U. *maydis* infection. These genes were identified in a differential display screen comparing galled tissue with noninfected tissue. In this way, 10 genes upregulated in galls and 2 genes downregulated in galls were identified. While expression of several genes dropped in more mature tissue in noninfected plants, they were highly induced in infected tissue of the same age. These results appear to substantiate the hypothesis that U. *maydis* is able to interfere with developmental processes in the plant, extending the undifferentiated state of immature expanding tissue which leads to enhanced cell proliferation and tumor formation. Additionally, the induction in galls of several genes associated with secondary metabolism and defense but the lack of effect of U. *maydis* infection on the expression of the defense-related maize gene *PR-1* led Basse (2005) to suggest that *U. maydis* is able to suppress classic defense mechanisms *in planta* while nonconventional responses are observed.

In our laboratory (Gold), we have identified genes up- and downregulated in galls by analyzing two subtractive libraries produced by SSH. In this differential screen, we used the wild-type gall producing *U. maydis* dikaryon and the contrasting disease symptoms caused by a *ubc1* mutant dikaryon (Gold *et al.*, 1997), able to colonize maize but unable to induce gall formation. In this way, we aimed to identify genes induced or repressed specifically at the stage of gall formation. The sequencing of SSH clones, shown to be differentially expressed in galls by reserve Northern blots, and comparison of these sequences with ESTs at ZmDB are allowing us to identify genes of maize origin whose expression patterns change on gall formation (M. D. Garcia-Pedrajas and S. E. Gold, unpublished data). The most abundant gene in the library enriched for genes upregulated in galls was a Bowman-Birk protease inhibitor. Northern blot analysis confirmed strong upregulation of this gene in galled tissue (M. D. Garcia-Pedrajas and S. E. Gold, unpublished data). Interestingly, seven proteases were identified as induced in the nongalled symptomatic tissue and repressed in galls. Among the genes we have identified as specifically induced in galls is an EST with similarity to a gene encoding an ETA subunit of the CCT chaperonin. This provides an interesting example because of its potential involvement in cell cycle regulation as it is essential for proper folding of cyclin E (Won *et al.*, 1998). Strong induction of this gene in galls has been confirmed by Northern blot analysis (M. Nadal, M. D. Garcia-Pedrajas and S. E. Gold, unpublished data). We postulate that this chaperonin could be involved in cell cycle deregulation leading to hyperplasia. We also identified differentially expressed genes with similarity to transcription factors or containing DNA-binding motifs (M. D. Garcia-Pedrajas and S. E. Gold, unpublished data). Characterization of these genes could give us clues as to possible signaling pathways altered during gall formation.

Through identification of the above-mentioned genes and other genes whose expression patterns are modified during *U. maydis* infection and gall induction, we have begun to unravel the genetics of the maize response. The increasing availability of genetic and genomic tools to work on maize is expected to accelerate the process of understanding the maize–*U. maydis* interaction. Under the Maize Gene Discovery Project, an NSF-funded plant genome initiative that started in 1998, several laboratories collaborate with the common goal of discovering new maize genes and developing tools for the phenotypic characterization of maize mutants (Lawrence *et al.*, 2004; Lunde *et al.*, 2003). This initiative includes EST sequencing, cDNA microarray production, and tools to study gene function by production of insertional mutants in maize using a recombinant Mu1 transposon (RescueMu) (Raizada, 2003). These and other resources are becoming available for public use and their description is available online at ZmDB (http://www.maizegdb.org/). With these tools, it is becoming feasible not only to study global expression patterns in the desired conditions and speculate on the possible mechanisms by which U. *maydis* induces morphological alterations in the plant, but also to test those hypotheses by analyzing maize lines mutated or overexpressing those genes.

VIII. CONCLUSION

U. *maydis* gained notoriety as a model system for the study of recombination and repair (Holliday, 2004), and as a model for studies involving mating, signaling, and pathogenesis (Feldbrugge *et al.*, 2004; Martinez-Espinoza *et al.*, 2002). Although the recent pace of advances in the study of U. *maydis* has been remarkable, as we have witnessed in this chapter, there remains substantial gaps in our understanding of how U. *maydis* signals and integrates various functions. Thus, there are numerous fascinating topics of study. One of the most intriguing and challenging of these is the genetics of the U. *maydis*–maize interaction. Moreover, the prospect of connecting the nodes of the various signaling pathways with other cellular processes, such as cytoskeletal regulation and the execution of the cell cycle, presents myriad challenges. Addressing these challenges and related questions promises to reveal a wealth of insight into the functions of important animal and plant pathogens, especially among related basidiomycetes. Ultimately, these insights may be applied for the development of target-specific control measures. The recent availability of the genome of U. *maydis* and the expanding toolbox for genetic manipulation of U. *maydis* will more thoroughly enable and accelerate large-scale approaches. Thus, the description of a more unified genetic model of U. *maydis* is likely in the near future.

References

Abramovitch, R. B., Yang, G., and Kronstad, J. W. (2002). The *ukb1* gene encodes a putative protein kinase required for bud site selection and pathogenicity in *Ustilago maydis*. *Fungal Genet. Biol.* **37**, 98–108.

Adamikova, L., Straube, A., Schulz, I., and Steinberg, G. (2004). Calcium signaling is involved in dynein-dependent microtubule organization. *Mol. Biol. Cell* **15**, 1969–1980.

Adams, A., Thorn, J. M., Yamabhai, M., Kay, B. K., and O'Bryan, J. P. (2000). Intersectin, an adaptor protein involved in clathrin-mediated endocytosis, activates mitogenic signaling pathways. *J. Biol. Chem.* **275**, 27414–27420.

Agrios, G. N. (1997). "Plant Pathology," 4th edn. Academic Press, San Diego.

Aichinger, C., Hansson, K., Eichhorn, H., Lessing, F., Mannhaupt, G., Mewes, W., and Kahmann, R. (2003). Identification of plant-regulated genes in *Ustilago maydis* by enhancer-trapping mutagenesis. *Mol. Genet. Genomics* **270**, 303–314.

Alexopolous, C. J., Mims, C. W., and Blackwell, M. (1996). "Introductory Mycology," 4th edn. John Wiley and Sons, New York.

Andrews, D. L., Egan, J. D., Mayorga, M. E., and Gold, S. E. (2000). The *Ustilago maydis* ubc4 and ubc5 genes encode members of a MAP kinase cascade required for filamentous growth. *Mol. Plant Microbe Interact.* **13,** 781–786.

Austin, R., Provart, N. J., Sacadura, N. T., Nugent, K. G., Babu, M., and Saville, B. J. (2004). A comparative genomic analysis of ESTs from *Ustilago maydis*. *Funct. Integr. Genomics* **4,** 207–218.

Babu, M. R., Choffe, K., and Saville, B. J. (2005). Differential gene expression in filamentous cells of *Ustilago maydis*. *Curr. Genet.* **47,** 316–333.

Baker, L. G., Gold, S. E., Sarah, F., and Covert S. F. A regulator of G-protein signaling in *Ustilago maydis* promotes sporulation *in planta* and regulates filamentous growth. *Mol. Microbiol.* (submitted for publication).

Bakkeren, G., Gibbard, B., Yee, A., Froeliger, E., Leong, S., and Kronstad, J. (1992). The a and b loci of *Ustilago maydis* hybridize with DNA sequences from other smut fungi. *Mol. Plant Microbe Interact.* **5,** 347–355.

Banks, G. (1983). Transformation of *Ustilago maydis* by a plasmid containing yeast 2-micron DNA. *Curr. Genet.* **7,** 73–77.

Banuett, F. (1995). Genetics of *Ustilago maydis*, a fungal pathogen that induces tumors in maize. *Annu. Rev. Genet.* **29,** 179–208.

Banuett, F., and Herskowitz, I. (1989). Different a alleles of *Ustilago maydis* are necessary for maintenance of filamentous growth but not for meiosis. *Proc. Natl. Acad. Sci. USA* **86,** 5878–5882.

Banuett, F., and Herskowitz, I. (1994). Identification of fuz7, a *Ustilago maydis* MEK/MAPKK homolog required for a-locus-dependent and -independent steps in the fungal life cycle. *Genes Dev.* **8,** 1367–1378.

Banuett, F., and Herskowitz, I. (1996). Discrete developmental stages during teliospore formation in the corn smut fungus, *Ustilago maydis*. *Development* **122,** 2965–2976.

Banuett, F., and Herskowitz, I. (2002). Bud morphogenesis and the actin and microtubule cytoskeletons during budding in the corn smut fungus, *Ustilago maydis*. *Fungal Genet. Biol.* **37,** 149–170.

Barrett, K. J., Gold, S. E., and Kronstad, J. W. (1993). Identification and complementation of a mutation to constitutive filamentous growth in *Ustilago maydis*. *Mol. Plant Microbe Interact.* **6,** 274–283.

Basse, C. W. (2005). Dissecting defense-related and developmental transcriptional responses of maize during *Ustilago maydis* infection and subsequent tumor formation. *Plant Physiol.* **138,** 1774–1784.

Basse, C. W., and Steinberg, G. (2004). *Ustilago maydis*, model system for analysis of the molecular basis of fungal pathogenicity. *Mol. Plant Pathol.* **5,** 83–92.

Basse, C. W., Lottspeich, F., Steglich, W., and Kahmann, R. (1996). Two potential indole-3-acetaldehyde dehydrogenases in the phytopathogenic fungus *Ustilago maydis*. *Eur. J. Biochem.* **242,** 648–656.

Basse, C. W., Stumpferl, S., and Kahmann, R. (2000). Characterization of a *Ustilago maydis* gene specifically induced during the biotrophic phase: Evidence for negative as well as positive regulation. *Mol. Cell. Biol.* **20,** 329–339.

Basse, C. W., Kolb, S., and Kahmann, R. (2002). A maize-specifically expressed gene cluster in *Ustilago maydis*. *Mol. Microbiol.* **43,** 75–93.

Biswas, K., and Morschhauser, J. (2005). The Mep2p ammonium permease controls nitrogen starvation-induced filamentous growth in *Candida albicans*. *Mol. Microbiol.* **56,** 649–669.

Bolker, M. (2001). *Ustilago maydis*—a valuable model system for the study of fungal dimorphism and virulence. *Microbiology* **147,** 1395–1401.

Bortfeld, M., Auffarth, K., Kahmann, R., and Basse, C. W. (2004). The *Ustilago maydis* a2 mating-type locus genes *lga2* and *rga2* compromise pathogenicity in the absence of the mitochondrial p32 family protein Mrb1. *Plant Cell* **16,** 2233–2248.

Bottin, A., Kamper, J., and Kahmann, R. (1996). Isolation of a carbon source-regulated gene from *Ustilago maydis*. *Mol. Gen. Genet.* **253,** 342–352.

Brachmann, A., Weinzierl, G., Kamper, J., and Kahmann, R. (2001). Identification of genes in the bW/bE regulatory cascade in *Ustilago maydis*. *Mol. Microbiol.* **42,** 1047–1063.

Brachmann, A., Schirawski, J., Muller, P., and Kahmann, R. (2003). An unusual MAP kinase is required for efficient penetration of the plant surface by *Ustilago maydis*. *EMBO J.* **22,** 2199–2210.

Brachmann, A., Konig, J., Julius, C., and Feldbrugge, M. (2004). A reverse genetic approach for generating gene replacement mutants in *Ustilago maydis*. *Mol. Genet. Genomics* **272,** 216–226.

Brefort, T., Muller, P., and Kahmann, R. (2005). The high-mobility-group domain transcription factor Rop1 is a direct regulator of *prf1* in *Ustilago maydis*. *Eukaryot. Cell* **4,** 379–391.

Callow, J. A., and Ling, I. T. (1973). Histology of neoplasms and chlorotic lesions in maize seedlings following injection of sporidia of *Ustilago maydis* (Dc) Corda. *Physiol. Plant Pathol.* **3,** 489.

Cano-Canchola, C., Acevedo, L., Ponce-Noyola, P., Flores-Martinez, A., Flores-Carreon, A., and Leal-Morales, C. A. (2000). Induction of lytic enzymes by the interaction of *Ustilago maydis* with *Zea mays* tissues. *Fungal Genet. Biol.* **29,** 145–151.

Casselton, L. A., and Olesnicky, N. S. (1998). Molecular genetics of mating recognition in basidiomycete fungi. *Microbiol. Mol. Biol. Rev.* **62,** 55–70.

Castillo-Lluva, S., Garcia-Muse, T., and Perez-Martin, J. (2004). A member of the Fizzy-related family of APC activators is regulated by cAMP and is required at different stages of plant infection by *Ustilago maydis*. *J. Cell Sci.* **117,** 4143–4156.

Christensen, J. J. (1931). Studies on the genetics of *Ustilago maydis*. *Phytopath. Z.* **4,** 129–188.

Christensen, J. J. (1963). Corn smut caused by *Ustilago maydis*. Monograph No. 2, American Phytopathological Society Press, Saint Paul.

Cooper, T. G. (2002). Transmitting the signal of excess nitrogen in *Saccharomyces cerevisiae* from the Tor proteins to the GATA factors: Connecting the dots. *FEMS Microbiol. Rev.* **26,** 223–238.

Deising, H. B., Werner, S., and Wernitz, M. (2000). The role of fungal appressoria in plant infection. *Microbes Infect.* **2,** 1631–1641.

Durrenberger, F., Wong, K., and Kronstad, J. W. (1998). Identification of a cAMP-dependent protein kinase catalytic subunit required for virulence and morphogenesis in *Ustilago maydis*. *Proc. Natl. Acad. Sci. USA* **95,** 5684–5689.

Durrenberger, F., Laidlaw, R. D., and Kronstad, J. W. (2001). The *hgl1* gene is required for dimorphism and teliospore formation in the fungal pathogen *Ustilago maydis*. *Mol. Microbiol.* **41,** 337–348.

Elliot, M. A., and Talbot, N. J. (2004). Building filaments in the air: Aerial morphogenesis in bacteria and fungi. *Curr. Opin. Microbiol.* **7,** 594–601.

Farfsing, J. W., Auffarth, K., and Basse, C. W. (2005). Identification of *cis*-active elements in *Ustilago maydis* mig2 promoters conferring high-level activity during pathogenic growth in maize. *Mol. Plant Microbe Interact.* **18,** 75–87.

Feldbrugge, M., Kamper, J., Steinberg, G., and Kahmann, R. (2004). Regulation of mating and pathogenic development in *Ustilago maydis*. *Curr. Opin. Microbiol.* **7,** 666–672.

Fischer, J. A., McCann, M. P., and Snetselaar, K. M. (2001). Methylation is involved in the *Ustilago maydis* mating response. *Fungal Genet. Biol.* **34,** 21–35.

Fotheringham, S., and Holloman, W. K. (1989). Cloning and disruption of *Ustilago maydis* genes. *Mol. Cell. Biol.* **9,** 4052–4055.

Fuchs, U., Manns, I., and Steinberg, G. (2005). Microtubules are dispensable for the initial pathogenic development but required for long-distance hyphal growth in the corn smut fungus *Ustilago maydis*. *Mol. Biol. Cell* **16,** 2746–2758.

Garcia-Muse, T., Steinberg, G., and Perez-Martin, J. (2003). Pheromone-induced G2 arrest in the phytopathogenic fungus *Ustilago maydis*. *Eukaryot. Cell* **2,** 494–500.

Garcia-Muse, T., Steinberg, G., and Perez-Martin, J. (2004). Characterization of B-type cyclins in the smut fungus *Ustilago maydis*: Roles in morphogenesis and pathogenicity. *J. Cell Sci.* **117,** 487–506.

Garcia-Pedrajas, M. D., and Gold, S. E. (2004). Fungal dimorphism regulated gene expression in *Ustilago maydis*: II. Filament downregulated genes. *Mol. Plant Pathol.* **5,** 295–308.

Garcia-Pedrajas, M. D., Kapa, L. B., Perlin, M. H., and Gold, S. E. DelsGate, a robust and rapid gene deletion construction method for functional genomics. *Nucleic Acids Res.* (submitted for publication).

Garrido, E., and Perez-Martin, J. (2003). The *crk1* gene encodes an Ime2-related protein that is required for morphogenesis in the plant pathogen *Ustilago maydis*. *Mol. Microbiol.* **47,** 729–743.

Garrido, E., Voss, U., Muller, P., Castillo-Lluva, S., Kahmann, R., and Perez-Martin, J. (2004). The induction of sexual development and virulence in the smut fungus *Ustilago maydis* depends on Crk1, a novel MAPK protein. *Genes Dev.* **18,** 3117–3130.

Gillissen, B., Bergemann, J., Sandmann, C., Schroeer, B., Bolker, M., and Kahmann, R. (1992). A two-component regulatory system for self/non-self recognition in *Ustilago maydis*. *Cell* **68,** 647–657.

Gold, S. E., and Kronstad, J. W. (1994). Disruption of two genes for chitin synthase in the phytopathogenic fungus *Ustilago maydis*. *Mol. Microbiol.* **11,** 897–902.

Gold, S., Duncan, G., Barrett, K., and Kronstad, J. (1994a). cAMP regulates morphogenesis in the fungal pathogen *Ustilago maydis*. *Genes Dev.* **8,** 2805–2816.

Gold, S. E., Bakkeren, G., Davies, J. E., and Kronstad, J. W. (1994b). Three selectable markers for transformation of *Ustilago maydis*. *Gene* **142,** 225–230.

Gold, S. E., Brogdon, S. M., Mayorga, M. E., and Kronstad, J. W. (1997). The *Ustilago maydis* regulatory subunit of a cAMP-dependent protein kinase is required for gall formation in maize. *Plant Cell* **9,** 1585–1594.

Guevara-Lara, F., Valverde, M. E., and Paredes-Lopez, O. (2000). Is pathogenicity of *Ustilago maydis* (huitlacoche) strains on maize related to *in vitro* production of indole-3-acetic acid? *World J. Microbiol. Biotechol.* **16,** 481–490.

Guevara-Olvera, L., Xoconostle-Cazares, B., and Ruiz-Herrera, J. (1997). Cloning and disruption of the ornithine decarboxylase gene of *Ustilago maydis*: Evidence for a role of polyamines in its dimorphic transition. *Microbiology* **143,** 2237–2245.

Gupta, M. (2006). Filamentous growth in *Ustilago maydis*: Bringing together ump mutants, cAMP concentrations, ammonium levels, and pH. M.S. Thesis, University of Louisville, Lovisville, KY.

Hartmann, H. A., Kahmann, R., and Bolker, M. (1996). The pheromone response factor coordinates filamentous growth and pathogenicity in *Ustilago maydis*. *EMBO J.* **15,** 1632–1641.

Hartmann, H. A., Kruger, J., Lottspeich, F., and Kahmann, R. (1999). Environmental signals controlling sexual development of the corn smut fungus *Ustilago maydis* through the transcriptional regulator Prf1. *Plant Cell* **11,** 1293–1306.

Hayman, M. L., Miller, M. M., Chandler, D. M., Goulah, C. C., and Read, L. K. (2001). The trypanosome homolog of human p32 interacts with RBP16 and stimulates its gRNA binding activity. *Nucleic Acids Res.* **29,** 5216–5225.

Hiscock, S. J., and Kues, U. (1999). Cellular and molecular mechanisms of sexual incompatibility in plants and fungi. *Int. Rev. Cytol.* **193,** 165–295.

Holliday, R. (1961). The genetics of *Ustilago maydis*. *Genet. Res.* **2,** 204–230.

Holliday, R. (1974). *Ustilago maydis*. In "Handbook of Genetics" (R. C. King, ed.), pp. 575–595. Plenum Press, New York.

Holliday, R. (2004). Early studies on recombination and DNA repair in *Ustilago maydis*. *DNA Repair* **3**, 671–682.

Hood, M. E., and Antonovics, J. (2000). Intratetrad mating, heterozygosity, and the maintenance of deleterious alleles in *Microbotryum violaceum* (=*Ustilago violacea*). *Heredity* **85**, 231–241.

Hsiang, T., and Baillie, D. L. (2005). Comparison of the yeast proteome to other fungal genomes to find core fungal genes. *J. Mol. Evol.* **60**, 475–483.

Kaffarnik, F., Muller, P., Leibundgut, M., Kahmann, R., and Feldbrugge, M. (2003). PKA and MAPK phosphorylation of Prf1 allows promoter discrimination in *Ustilago maydis*. *EMBO J.* **22**, 5817–5826.

Kahmann, R., and Kamper, J. (2004). *Ustilago maydis*: How its biology relates to pathogenic development. *New Phytol.* **164**, 31–42.

Kamper, J. (2004). A PCR-based system for highly efficient generation of gene replacement mutants in *Ustilago maydis*. *Mol. Genet. Genomics* **271**, 103–110.

Kamper, J., Kahmann, R., Bolker, M., Ma, L., Brefort, T., Saville, B. J., Banuett, F., Krenstad, J. W., Gold, S. E., Muller, O., Perlin, M. H., Wosten, H. A., *et al.* (2006). Insights from the genome of the biotrophic fungal plant pathogen *Ustilago maydis*. *Nature* **444**, 97–101.

Kamper, J., Reichmann, M., Romeis, T., Bolker, M., and Kahmann, R. (1995). Multiallelic recognition: Nonself-dependent dimerization of the bE and bW homeodomain proteins in *Ustilago maydis*. *Cell* **81**, 73–83.

Keon, J. P., White, G. A., and Hargreaves, J. A. (1991). Isolation, characterization and sequence of a gene conferring resistance to the systemic fungicide carboxin from the maize smut pathogen, *Ustilago maydis*. *Curr. Genet.* **19**, 475–481.

Khademi, S., O'Connell, J., III, Remis, J., Robles-Colmenares, Y., Miercke, L. J., and Stroud, R. M. (2004). Mechanism of ammonia transport by Amt/MEP/Rh: Structure of AmtB at 1.35 A. *Science* **305**, 1587–1594.

Kinal, H., Park, C. M., and Bruenn, J. A. (1993). A family of *Ustilago maydis* expression vectors: New selectable markers and promoters. *Gene* **127**, 151–152.

Klose, J., de Sa, M. M., and Kronstad, J. W. (2004). Lipid-induced filamentous growth in *Ustilago maydis*. *Mol. Microbiol.* **52**, 823–835.

Kojic, M., and Holloman, W. K. (2000). Shuttle vectors for genetic manipulations in *Ustilago maydis*. *Can. J. Microbiol.* **46**, 333–338.

Kojic, M., Kostrub, C. F., Buchman, A. R., and Holloman, W. K. (2002). BRCA2 homolog required for proficiency in DNA repair, recombination, and genome stability in *Ustilago maydis*. *Mol. Cell* **10**, 683–691.

Krainer, A. R., Mayeda, A., Kozak, D., and Binns, G. (1991). Functional expression of cloned human splicing factor SF2: Homology to RNA-binding proteins, U1 70K, and *Drosophila* splicing regulators. *Cell* **66**, 383–394.

Kronstad, J. W., and Leong, S. A. (1989). Isolation of two alleles of the *b* locus of *Ustilago maydis*. *Proc. Natl. Acad. Sci. USA* **86**, 978–982.

Kronstad, J. W., and Staben, C. (1997). Mating type in filamentous fungi. *Annu. Rev. Genet.* **31**, 245–276.

Kronstad, J. W., Wang, J., Covert, S. F., Holden, D. W., McKnight, G. L., and Leong, S. A. (1989). Isolation of metabolic genes and demonstration of gene disruption in the phytopathogenic fungus *Ustilago maydis*. *Gene* **79**, 97–106.

Kruger, J., Loubradou, G., Regenfelder, E., Hartmann, A., and Kahmann, R. (1998). Crosstalk between cAMP and pheromone signalling pathways in *Ustilago maydis*. *Mol. Gen. Genet.* **260**, 193–198.

Kruger, J., Loubradou, G., Wanner, G., Regenfelder, E., Feldbrugge, M., and Kahmann, R. (2000). Activation of the cAMP pathway in *Ustilago maydis* reduces fungal proliferation and teliospore formation in plant tumors. *Mol. Plant Microbe Interact.* **13,** 1034–1040.

Kuruvilla, F. G., Shamji, A. F., and Schreiber, S. L. (2001). Carbon- and nitrogen-quality signaling to translation are mediated by distinct GATA-type transcription factors. *Proc. Natl. Acad. Sci. USA* **98,** 7283–7288.

Lawrence, C. J., Dong, Q., Polacco, M. L., Seigfried, T. E., and Brendel, V. (2004). MaizeGDB, the community database for maize genetics and genomics. *Nucleic Acids Res.* **32,** 393–397.

Lee, N., and Kronstad, J. W. (2002). *ras2* controls morphogenesis, pheromone response, and pathogenicity in the fungal pathogen *Ustilago maydis*. *Eukaryot. Cell* **1,** 954–966.

Lee, N., D'Souza, C. A., and Kronstad, J. W. (2003). Of smuts, blasts, mildews, and blights: cAMP signaling in phytopathogenic fungi. *Annu. Rev. Phytopathol.* **41,** 399–427.

Lehmler, C., Steinberg, G., Snetselaar, K. M., Schliwa, M., Kahmann, R., and Bolker, M. (1997). Identification of a motor protein required for filamentous growth in *Ustilago maydis*. *EMBO J.* **16,** 3464–3473.

Leuthner, B., Aichinger, C., Oehmen, E., Koopmann, E., Muller, O., Muller, P., Kahmann, R., Bolker, M., and Schreier, P. H. (2005). A H2O2-producing glyoxal oxidase is required for filamentous growth and pathogenicity in *Ustilago maydis*. *Mol. Genet. Genomics* **272,** 639–650.

Leveleki, L., Mahlert, M., Sandrock, B., and Bolker, M. (2004). The PAK family kinase Cla4 is required for budding and morphogenesis in *Ustilago maydis*. *Mol. Microbiol.* **54,** 396–406.

Li, T., Stark, M. R., Johnson, A. D., and Wolberger, C. (1995). Crystal structure of the MATa1/MAT alpha 2 homeodomain heterodimer bound to DNA. *Science* **270,** 262–269.

Loftus, B. J., Fung, E., Roncaglia, P., Rowley, D., Amedeo, P., Bruno, D., Vamathevan, J., Miranda, M., Anderson, I. J., Fraser, J. A., Allen, J. E., Bosdet, I. E., *et al.* (2005). The genome of the basidiomycetous yeast and human pathogen *Cryptococcus neoformans*. *Science* **307,** 1321–1324.

Lorenz, M. C., and Heitman, J. (1998). The MEP2 ammonium permease regulates pseudohyphal differentiation in *Saccharomyces cerevisiae*. *EMBO J.* **17,** 1236–1247.

Loubradou, G., Brachmann, A., Feldbrugge, M., and Kahmann, R. (2001). A homologue of the transcriptional repressor Ssn6p antagonizes cAMP signalling in *Ustilago maydis*. *Mol. Microbiol.* **40,** 719–730.

Luke, M. M., Sutton, A., and Arndt, K. T. (1991). Characterization of SIS1, a *Saccharomyces cerevisiae* homologue of bacterial dnaJ proteins. *J. Cell Biol.* **114,** 623–638.

Lunde, C. F., Morrow, D. J., Roy, L. M., and Walbot, V. (2003). Progress in maize gene discovery: A project update. *Funct. Integr. Genomics* **3,** 25–32.

Magasanik, B., and Kaiser, C. A. (2002). Nitrogen regulation in *Saccharomyces cerevisiae*. *Gene* **290,** 1–18.

Marini, A. M., and Andre, B. (2000). In vivo N-glycosylation of the mep2 high-affinity ammonium transporter of *Saccharomyces cerevisiae* reveals an extracytosolic N-terminus. *Mol. Microbiol.* **38,** 552–564.

Marini, A. M., Soussi-Boudekou, S., Vissers, S., and Andre, B. (1997). A family of ammonium transporters in *Saccharomyces cerevisiae*. *Mol. Cell. Biol.* **17,** 4282–4293.

Martinez, V. M., Osuna, J., Paredes-Lopez, O., and Guevara, F. (1997). Production of indole-3-acetic acid by several wild-type strains of *Ustilago maydis*. *World J. Microbiol. Biotechnol.* **13,** 295–298.

Martinez-Espinoza, A. D., Garcia-Pedrajas, M. D., and Gold, S. E. (2002). The Ustilaginales as plant pests and model systems. *Fungal Genet. Biol.* **35,** 1–20.

Martinez-Espinoza, A. D., Ruiz-Herrera, J., Leon-Ramirez, C. G., and Gold, S. E. (2004). MAP kinase and cAMP signaling pathways modulate the pH-induced yeast-to-mycelium dimorphic transition in the corn smut fungus *Ustilago maydis*. *Curr. Microbiol.* **49,** 274–281.

Mayorga, M. E., and Gold, S. E. (1998). Characterization and molecular genetic complementation of mutants affecting dimorphism in the fungus *Ustilago maydis*. *Fungal Genet. Biol.* **24,** 364–376.

Mayorga, M. E., and Gold, S. E. (1999). A MAP kinase encoded by the *ubc3* gene of *Ustilago maydis* is required for filamentous growth and full virulence. *Mol. Microbiol.* **34,** 485–497.

Mayorga, M. E., and Gold, S. E. (2001). The *ubc2* gene of *Ustilago maydis* encodes a putative novel adaptor protein required for filamentous growth, pheromone response and virulence. *Mol. Microbiol.* **41,** 1365–1379.

Mills, L. J., and Vanstaden, J. (1978). Extraction of cytokinins from maize, smut tumors of maize and *Ustilago maydis* cultures. *Physiol. Plant Pathol.* **13,** 73–80.

Mosch, H. U., Kubler, E., Krappmann, S., Fink, G. R., and Braus, G. H. (1999). Crosstalk between the Ras2p-controlled mitogen-activated protein kinase and cAMP pathways during invasive growth of *Saccharomyces cerevisiae*. *Mol. Biol. Cell* **10,** 1325–1335.

Moulton, J. (1942). Extraction of auxin from maize, from smut tumors of maize, and from *Ustilago zeae*. *Bot. Gazette* **103,** 725–739.

Muller, P., Aichinger, C., Feldbrugge, M., and Kahmann, R. (1999). The MAP kinase kpp2 regulates mating and pathogenic development in *Ustilago maydis*. *Mol. Microbiol.* **34,** 1007–1017.

Muller, P., Katzenberger, J. D., Loubradou, G., and Kahmann, R. (2003a). Guanyl nucleotide exchange factor Sql2 and Ras2 regulate filamentous growth in *Ustilago maydis*. *Eukaryot. Cell* **2,** 609–617.

Muller, P., Weinzierl, G., Brachmann, A., Feldbrugge, M., and Kahmann, R. (2003b). Mating and pathogenic development of the smut fungus *Ustilago maydis* are regulated by one mitogen-activated protein kinase cascade. *Eukaryot. Cell* **2,** 1187–1199.

Muller, P., Leibbrandt, A., Teunissen, H., Cubasch, S., Aichinger, C., and Kahmann, R. (2004). The Gbeta-subunit-encoding gene *bpp1* controls cyclic-AMP signaling in *Ustilago maydis*. *Eukaryot. Cell* **3,** 806–814.

Neill, S., Desikan, R., and Hancock, J. (2002). Hydrogen peroxide signalling. *Curr. Opin. Plant Biol.* **5,** 388–395.

Nugent, K. G., Choffe, K., and Saville, B. J. (2004). Gene expression during *Ustilago maydis* diploid filamentous growth: EST library creation and analyses. *Fungal Genet. Biol.* **41,** 349–360.

O'Donnell, K. L., and McLaughlin, D. J. (1984). Ultrastructure of meiosis in *Ustilago maydis*. *Mycologia* **76,** 468–485.

O'Farrell, P. H. (2001). Triggering the all-or-nothing switch into mitosis. *Trends Cell Biol.* **11,** 512 519.

Ospina-Giraldo, M. D., Mullins, E., and Kang, S. (2003). Loss of function of the *Fusarium oxysporum SNF1* gene reduces virulence on cabbage and Arabidopsis. *Curr. Genet.* **44,** 49–57.

Quadbeck-Seeger, C., Wanner, G., Huber, S., Kahmann, R., and Kamper, J. (2000). A protein with similarity to the human retinoblastoma binding protein 2 acts specifically as a repressor for genes regulated by the *b* mating type locus in *Ustilago maydis*. *Mol. Microbiol.* **38,** 154–166.

Raizada, M. N. (2003). RescueMu protocols for maize functional genomics. *Methods Mol. Biol.* **236,** 37–58.

Ramezani-Rad, M. (2003). The role of adaptor protein Ste50-dependent regulation of the MAPKKK Ste11 in multiple signalling pathways of yeast. *Curr. Genet.* **43,** 161–170.

Regenfelder, E., Spellig, T., Hartmann, A., Lauenstein, S., Bolker, M., and Kahmann, R. (1997). G proteins in *Ustilago maydis*: Transmission of multiple signals? *EMBO J.* **16,** 1934–1942.

Reichmann, M., Jamnischek, A., Weinzierl, G., Ladendorf, O., Huber, S., Kahmann, R., and Kamper, J. (2002). The histone deacetylase Hda1 from *Ustilago maydis* is essential for teliospore development. *Mol. Microbiol.* **46,** 1169–1182.

Romeis, T., Brachmann, A., Kahmann, R., and Kamper, J. (2000). Identification of a target gene for the bE-bW homeodomain protein complex in *Ustilago maydis*. *Mol. Microbiol.* **37,** 54–66.

Ruiz-Herrera, J., Leon, C. G., Guevara-Olvera, L., and Carabez-Trejo, A. (1995). Yeast-mycelial dimorphism of haploid and diploid strains of *Ustilago maydis*. *Microbiology* **141,** 695–703.

Ruiz-Herrera, J., Leon-Ramirez, C., Cabrera-Ponce, J. L., Martinez-Espinoza, A. D., and Herrera-Estrella, L. (1999). Completion of the sexual cycle and demonstration of genetic recombination in *Ustilago maydis in vitro*. *Mol. Gen. Genet.* **262,** 468–472.

Sacadura, N. T., and Saville, B. J. (2003). Gene expression and EST analyses of *Ustilago maydis* germinating teliospores. *Fungal Genet. Biol.* **40,** 47–64.

Sanchez-Martinez, C., and Perez-Martin, J. (2001). Dimorphism in fungal pathogens: *Candida albicans* and *Ustilago maydis*—similar inputs, different outputs. *Curr. Opin. Microbiol.* **4,** 214–221.

Schauwecker, F., Wanner, G., and Kahmann, R. (1995). Filament-specific expression of a cellulase gene in the dimorphic fungus *Ustilago maydis*. *Biol. Chem. Hoppe Seyler* **376,** 617–625.

Schirawski, J., Heinze, B., Wagenknecht, M., and Kahmann, R. (2005). Mating type loci of *Sporisorium reilianum*: Novel pattern with three *a* and multiple *b* specificities. *Eukaryot. Cell* **4,** 1317–1327.

Schulz, B., Banuett, F., Dahl, M., Schlesinger, R., Schafer, W., Martin, T., Herskowitz, I., and Kahmann, R. (1990). The *b* alleles of *U. maydis*, whose combinations program pathogenic development, code for polypeptides containing a homeodomain-related motif. *Cell* **60,** 295–306.

Sgarlata, C., and Perez-Martin, J. (2005a). Inhibitory phosphorylation of a mitotic cyclin-dependent kinase regulates the morphogenesis, cell size and virulence of the smut fungus *Ustilago maydis*. *J. Cell Sci.* **118,** 3607–3622.

Sgarlata, C., and Perez-Martin, J. (2005b). The cdc25 phosphatase is essential for the G2/M phase transition in the basidiomycete yeast *Ustilago maydis*. *Mol. Microbiol.* **58,** 1482–1496.

Smith, D. G., Garcia-Pedrajas, M. D., Gold, S. E., and Perlin, M. H. (2003). Isolation and characterization from pathogenic fungi of genes encoding ammonium permeases and their roles in dimorphism. *Mol. Microbiol.* **50,** 259–275.

Smith, D. G., Garcia-Pedrajas, M. D., Hong, W., Yu, Z., Gold, S. E., and Perlin, M. H. (2004). An *ste20* homologue in *Ustilago maydis* plays a role in mating and pathogenicity. *Eukaryot. Cell* **3,** 180–189.

Snetselaar, K. M., and Mims, C. W. (1993). Infection of maize stigmas by *Ustilago maydis*: Light and electron-microscopy. *Phytopathology* **83,** 843–850.

Snetselaar, K. M., and Mims, C. W. (1994). Light and electron-microscopy of *Ustilago maydis* hyphae in maize. *Mycol. Res.* **98,** 347–355.

Soll, D. (1985). *Candida albicans*. *In* "Fungal Dimorphism with Emphasis on Fungi Pathogenic for Humans" (P. J. Szaniszlo, ed.), pp. 167–195. Plenum Press, New York.

Spellig, T., Bottin, A., and Kahmann, R. (1996). Green fluorescent protein (GFP) as a new vital marker in the phytopathogenic fungus *Ustilago maydis*. *Mol. Gen. Genet.* **252,** 503–509.

Steinberg, G., and Fuchs, U. (2004). The role of microtubules in cellular organization and endocytosis in the plant pathogen *Ustilago maydis*. *J. Microsc.* **214,** 114–123.

Steinberg, G., Schliwa, M., Lehmler, C., Bolker, M., Kahmann, R., and McIntosh, J. R. (1998). Kinesin from the plant pathogenic fungus *Ustilago maydis* is involved in vacuole formation and cytoplasmic migration. *J. Cell Sci.* **111,** 2235–2246.

Steinberg, G., Wedlich-Soldner, R., Brill, M., and Schulz, I. (2001). Microtubules in the fungal pathogen *Ustilago maydis* are highly dynamic and determine cell polarity. *J. Cell Sci.* **114,** 609–622.

Straube, A., Enard, W., Berner, A., Wedlich-Soldner, R., Kahmann, R., and Steinberg, G. (2001). A split motor domain in a cytoplasmic dynein. *EMBO J.* **20,** 5091–5100.

Straube, A., Brill, M., Oakley, B. R., Horio, T., and Steinberg, G. (2003). Microtubule organization requires cell cycle-dependent nucleation at dispersed cytoplasmic sites: Polar and perinuclear microtubule organizing centers in the plant pathogen *Ustilago maydis*. *Mol. Biol. Cell* **14,** 642–657.

Straube, A., Weber, I., and Steinberg, G. (2005). A novel mechanism of nuclear envelope breakdown in a fungus: Nuclear migration strips off the envelope. *EMBO J.* **24,** 1674–1685.

Szabo, Z., Tonnis, M., Kessler, H., and Feldbrugge, M. (2002). Structure-function analysis of lipopeptide pheromones from the plant pathogen *Ustilago maydis. Mol. Genet. Genomics* **268**, 362–370.

Tan, A. L. C., Rida, P. C. G., and Surana, U. (2005). Essential tension and constructive destruction: The spindle checkpoint and its regulatory links with mitotic exit. *Biochem. J.* **386**, 1–13.

Tonukari, N. J., Scott-Craig, J. S., and Walton, J. D. (2000). The *Cochliobolus carbonum* SNF1 gene is required for cell wall-degrading enzyme expression and virulence on maize. *Plant Cell* **12**, 237–248.

Tsukuda, T., Carleton, S., Fotheringham, S., and Holloman, W. K. (1988). Isolation and characterization of an autonomously replicating sequence from *Ustilago maydis. Mol. Cell. Biol.* **8**, 3703–3709.

Turian, G., and Hamilton, R. H. (1960). Chemical detection of 3-indolylacetic acid in *Ustilago zeae* tumors. *Biochim. Biophy. Acta* **41**, 148–150.

Waard, M. A. (1976). Formation of protoplasts from *Ustilago maydis. Antonie Van Leeuwenhoek* **42**, 211–216.

Walter, J. M. (1935). Factors affecting the development of corn smut, *Ustilago maydis. Minnesota Agric. Exp. Stn. Tech. Bulletin.* **111**, 1–67.

Wang, J., Holden, D. W., and Leong, S. A. (1988). Gene transfer system for the phytopathogenic fungus *Ustilago maydis. Proc. Natl. Acad. Sci. USA* **85**, 865–869.

Wang, P., Perfect, J. R., and Heitman, J. (2000). The G-protein beta subunit GPB1 is required for mating and haploid fruiting in *Cryptococcus neoformans. Mol. Cell. Biol.* **20**, 352–362.

Weber, I., Gruber, C., and Steinberg, G. (2003). A class-V myosin required for mating, hyphal growth, and pathogenicity in the dimorphic plant pathogen *Ustilago maydis. Plant Cell* **15**, 2826–2842.

Weber, I., Assmann, D., Thines, E., and Steinberg, G. (2006). Polar localizing class V myosin chitin synthases are essential during early plant infection in the plant pathogenic fungus *Ustilago maydis. Plant Cell* **18**, 225–242.

Wedlich-Soldner, R., Bolker, M., Kahmann, R., and Steinberg, G. (2000). A putative endosomal t-SNARE links exo- and endocytosis in the phytopathogenic fungus *Ustilago maydis. EMBO J.* **19**, 1974–1986.

Wedlich-Soldner, R., Straube, A., Friedrich, M. W., and Steinberg, G. (2002). A balance of KIF1A-like kinesin and dynein organizes early endosomes in the fungus *Ustilago maydis. EMBO J.* **21**, 2946–2957.

Weinzierl, G., Leveleki, L., Hassel, A., Kost, G., Wanner, G., and Bolker, M. (2002). Regulation of cell separation in the dimorphic fungus *Ustilago maydis. Mol. Microbiol.* **45**, 219–231.

Whiteway, M., Hougan, L., Dignard, D., Thomas, D., Bell, L., Saari, G., Grant, F., O'Hara, P., and MacKay, V. (1989). The STE4 and STE18 genes of yeast encode potential beta and gamma subunits of the mating factor receptor-coupled G protein. *Cell* **56**, 467–477.

Wolf, F. T. (1952). The production of indole acetic acid by *Ustilago zeae*, and its possible significance in tumor formation. *Proc. Natl. Acad. Sci. USA* **38**, 106–111.

Won, K. A., Schumacher, R. J., Farr, G. W., Horwich, A. L., and Reed, S. I. (1998). Maturation of human cyclin E requires the function of eukaryotic chaperonin CCT. *Mol. Cell. Biol.* **18**, 7584–7589.

Wosten, H. A., and Willey, J. M. (2000). Surface-active proteins enable microbial aerial hyphae to grow into the air. *Microbiology* **146**, 767–773.

Wosten, H. A., Bohlmann, R., Eckerskorn, C., Lottspeich, F., Bolker, M., and Kahmann, R. (1996). A novel class of small amphipathic peptides affect aerial hyphal growth and surface hydrophobicity in *Ustilago maydis. EMBO J.* **15**, 4274–4281.

Xiang, X., and Plamann, M. (2003). Cytoskeleton and motor proteins in filamentous fungi. *Curr. Opin. Microbiol.* **6**, 628–633.

Xoconostle-Cazares, B., Leon-Ramirez, C., and Ruiz-Herrera, J. (1996). Two chitin synthase genes from *Ustilago maydis*. *Microbiology* **142,** 377–387.

Xoconostle-Cazares, B., Specht, C. A., Robbins, P. W., Liu, Y., Leon, C., and Ruiz-Herrera, J. (1997). Umchs5, a gene coding for a class IV chitin synthase in *Ustilago maydis*. *Fungal Genet. Biol.* **22,** 199–208.

Xu, J. R., and Hamer, J. E. (1996). MAP kinase and cAMP signaling regulate infection structure formation and pathogenic growth in the rice blast fungus *Magnaporthe grisea*. *Genes Dev.* **10,** 2696–2706.

Yee, A. R., and Kronstad, J. W. (1993). Construction of chimeric alleles with altered specificity at the *b* incompatibility locus of *Ustilago maydis*. *Proc. Natl. Acad. Sci. USA* **90,** 664–668.

Zahiri, A. R., Babu, M. R., and Saville, B. J. (2005). Differential gene expression during teliospore germination in *Ustilago maydis*. *Mol. Genet. Genomics* **273,** 394–403.

Zhao, X., Kim, Y., Park, G., and Xu, J. R. (2005). A mitogen-activated protein kinase cascade regulating infection-related morphogenesis in *Magnaporthe grisea*. *Plant Cell* **17,** 1317–1329.

Zheng, L., Kostrewa, D., Berneche, S., Winkler, F. K., and Li, X. D. (2004). The mechanism of ammonia transport based on the crystal structure of AmtB of *Escherichia coli*. *Proc. Natl. Acad. Sci. USA* **101,** 17090–17095.

Ziman, M., and Johnson, D. I. (1994). Genetic evidence for a functional interaction between *Saccharomyces cerevisiae* CDC24 and CDC42. *Yeast* **10,** 463–474.

2 Enabling a Community to Dissect an Organism: Overview of the Neurospora Functional Genomics Project

Jay C. Dunlap,* Katherine A. Borkovich,[†] Matthew R. Henn,[‡]
Gloria E. Turner,[§] Matthew S. Sachs,[¶] N. Louise Glass,**
Kevin McCluskey,[††] Michael Plamann,[††] James E. Galagan,[‡]
Bruce W. Birren,[‡] Richard L. Weiss,[§] Jeffrey P. Townsend,[‡‡]
Jennifer J. Loros,* Mary Anne Nelson,[§§] Randy Lambreghts,*
Hildur V. Colot,* Gyungsoon Park,[†] Patrick Collopy,*
Carol Ringelberg,* Christopher Crew,[†] Liubov Litvinkova,[†]
Dave DeCaprio,[‡] Heather M. Hood,[¶] Susan Curilla,* Mi Shi,*
Matthew Crawford,[‡] Michael Koerhsen,[‡] Phil Montgomery,[‡]
Lisa Larson,[‡] Matthew Pearson,[‡] Takao Kasuga,**
Chaoguang Tian,** Meray Baştürkmen,[¶] Lorena Altamirano,[†]
and Junhuan Xu[§§]

*Department of Genetics, Dartmouth Medical School, Hanover
New Hampshire 03755
[†]Department of Plant Pathology, University of California
Riverside, California 92521
[‡]Broad Institute of MIT & Harvard, Cambridge, Massachusetts 02142
[§]Department of Chemistry and Biochemistry, University of California
Los Angeles, California 90095
[¶]Oregon Health & Science University, Beaverton, Oregon 97006
**Department of Plant and Microbial Biology, University of California
Berkeley, California 94720
[††]Fungal Genetics Stock Center, School of Biological Sciences
University of Missouri, Kansas City, Missouri 64110
[‡‡]Department of Ecology and Evolutionary Biology, Yale University
New Haven, Connecticut 06520
[§§]Department of Biology, University of New Mexico, Albuquerque
New Mexico 87131

Advances in Genetics, Vol. 57
0065-2660/07 $35.00
DOI: 10.1016/S0065-2660(06)57002-6

ABSTRACT

A consortium of investigators is engaged in a functional genomics project centered on the filamentous fungus Neurospora, with an eye to opening up the functional genomic analysis of all the filamentous fungi. The overall goal of the four interdependent projects in this effort is to acccomplish functional genomics, annotation, and expression analyses of *Neurospora crassa*, a filamentous fungus that is an established model for the assemblage of over 250,000 species of nonyeast fungi. Building from the completely sequenced 43-Mb Neurospora genome, Project 1 is pursuing the systematic disruption of genes through targeted gene replacements, phenotypic analysis of mutant strains, and their distribution to the scientific community at large. Project 2, through a primary focus in Annotation and Bioinformatics, has developed a platform for electronically capturing community feedback and data about the existing annotation, while building and maintaining a database to capture and display information about phenotypes. Oligonucleotide-based microarrays created in Project 3 are being used to collect baseline expression data for the nearly 11,000 distinguishable transcripts in Neurospora under various conditions of growth and development, and eventually to begin to analyze the global effects of loss of novel genes in strains created by Project 1. cDNA libraries generated in Project 4 document the overall complexity of expressed sequences in Neurospora, including alternative splicing alternative promoters and antisense transcripts. In addition, these studies have driven the assembly of an SNP map presently populated by nearly 300 markers that will greatly accelerate the positional cloning of genes.   © 2007, Elsevier Inc.

I. INTRODUCTION

The availability of whole genomic sequences has vastly accelerated the pace of research in eukaryotic model systems. However, to exploit this resource, research communities must (1) annotate the genome to extract the relevant information, (2) systematically disrupt the function of the identified genes, (3) examine the regulation of the genes in different biological contexts, and finally (4) communicate this information to the scientific community at large, particularly to those studying similar problems in other systems. We may then integrate all these aspects of phenotype and regulation into a comprehensive portrait describing the biology of organisms. It is a tautology to state that the simplest organisms are the easiest to dissect but may reveal the least, and that the most complicated organisms, while the most information rich, may be beyond the scope of current efforts. Yet it is apparent from the extant genomic comparisons that conservation of important biological processes is the rule, and that simple models can inform more complex systems. The desirability of rich biology coupled with the realistic need for approachable genetics recommends the filamentous fungi in general, and prominent among the established model organisms are *Neurospora* and *Aspergillus*.

A. Why study fungi?

Fungi, plants, and animals represent the three phylogenetic kingdoms within the eukaryotes. Within the ~250,000 different species of fungi, about 75% belong to the Ascomycetes (90% being filamentous fungi, the remainder being yeasts) and 25% are Basidiomycetes (those that form fruiting bodies such as mushrooms). The fungi have an enormous impact on the United States and world economies. Because fungi constitute the Kingdom most closely related to Animalia, and yet are exceptionally tractable experimentally, fungi are universally used as model organisms for understanding all aspects of basic cellular regulation. These regulatory networks include cell cycle progression, gene expression, circadian timing, light sensing, recombination, secretion, and multicellular development. Additionally, mycorrhizal fungi (those that grow interdependently with the roots of plants) are essential symbionts without which most trees and many grasses cannot live; fungi also carry out most biomass turnover. The filamentous fungi also include serious plant and human pathogens; the latter have particular impact on the health of those that are immunocompromised. In addition, because they are eukaryotes, treatment of opportunistic infections by fungi poses special risks and challenges not encountered with bacteria. Filamentous fungi are furthermore metabolically gifted with the ability to produce secondary metabolites, many of which are now recognized as both important pharmaceuticals and salient environmental toxins. Pharmaceutical manufacture using fungi constitutes a multi-billion dollar per year industry. Penicillin and similar β-lactams, all produced by fungi, are the world's largest-selling antibiotics. At the same time estimates are that 10–35% of the world's food supply is lost each year due to fungal contamination, a loss of over $200 billion per year. Contamination of grains with aflatoxin, produced by *Aspergillus flavus*, is considered the chief cause of cancers in developing countries. Within the United States alone, fungicide sales constitute more than a $1 billion per year industry. Industrial production of chemicals by filamentous fungi constitutes a greater than $35 billion per year industry. The United States is a net importer of some of these chemicals (such as citrate), representing in excess of a $1 billion annually. Industrial production of enzymes, largely by filamentous fungi, constitutes a $1.5 billion per year industry.

B. Why Neurospora?

Neurospora crassa has been studied for decades and is perhaps the best understood filamentous fungus. It is a saprophyte that displays both asexual and sexual life cycles. Neurospora exists vegetatively as an incompletely septate syncytium, growing equally well in simple liquid or on solid media of known composition. It is nonpathogenic, although very closely phylogenetically related to pathogens. Both asexual development and sexual differentiation are highly influenced by environmental factors including nutrient, light, and temperature. Neurospora is typically haploid undergoing only a very transient diploid stage immediately prior to meiosis. Unlike the yeasts, *N. crassa* elaborates at least 28 distinct cell types that contribute to a wonderfully complex life cycle (Fig. 2.1) (Bistis *et al.*, 2003; Borkovich *et al.*, 2004). Because of its interesting and diverse biology, ease of culture, facile genetics, and rapid growth rate, Neurospora remains a widely used model that sustains a wide community.

The reference-quality genomic sequence (ca. 16-fold coverage) is complete and automated annotation (http://www-genome.wi.mit.edu/annotation/fungi/neurospora/index.html) predicts about 10,000 genes (Galagan *et al.*, 2003).

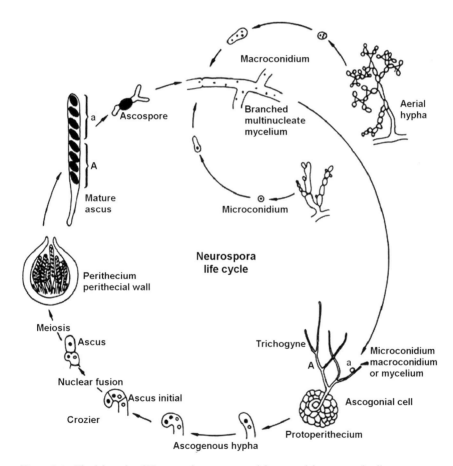

Figure 2.1. The life cycle of *N. crassa* showing some of the many life stages and cell types seen in the sexual and asexual cycles. Adapted from {www.fgsc.net/Neurospora/sectionB2.htm}.

The genetics of Neurospora is unparalleled within the filamentous fungi—most identified genes, densest and most accurate genetic map—a legacy of 70 years effort that began concurrently with Drosophila and now supports a large research community. The field of biochemical genetics arose from work in Neurospora (Beadle and Tatum, 1945) and the ease of culture, rapid growth (mass doubling time of 140 min), and ease of harvest continue to aid research. Crosses are technically trivial and the genetic generation time (from progeny to progeny) is about 3 weeks. Sexual spores are stable for years at 4°C and asexual spores or mycelia can be stored for decades. Molecular tools are comfortably advanced and being steadily improved as is typical in a vibrant research community. Neurospora was the first filamentous fungus to be transformed (soon after yeast; Davis, 2000), and transformation is routine at frequencies up to many thousands of transformants per microgram of DNA. A variety of selectable markers exist, and several regulatable promoters are routinely used to control expression of transgenes (Aronson *et al.*, 1994a). As in animal cells, transformation in completely wild-type cells is typically the result of ectopic insertion, although targeted disruption by reciprocal homologous integration is widely used to generate knockouts of known sequences via insertion of selectable markers (Aronson *et al.*, 1994b) at frequencies now routinely approaching 100% of transformants depending on the recipient strain, construct, and the locus (Colot *et al.*, 2006; Ninomiya *et al.*, 2004). Alternative methods for knockouts or knockdowns are also used with success: first is RIP (repeat-induced point mutation; Selker, 1990), a rapid method whereby duplicated genes are detected in a parental strain during a sexual cross and mutated prior to meiosis. "Knockdowns" are also made through quelling, a form of RNAi/cosuppression found in Neurospora (Cogoni and Macino, 1997).

Several independent estimates have suggested a surprising degree of genetic novelty in the Neurospora genome—repeated estimates suggest that over a third of Neurospora genes have no homologues or orthologues in GenBank (perhaps because of the 13 million sequences in GenBank, only a small fraction comes from Ascomycetes). These are either novel genes with novel functions or novel genes representing different ways of carrying out known functions; in either case, they will be of great interest. The complexity of the Neurospora genome approaches that of Drosophila. This degree of complexity—approximately twice that of yeasts—appears typical for filamentous fungi. Moreover, because of mechanisms in Neurospora that target and eliminate duplications (e.g., RIP), there are few gene families so that nearly all of the added sequence complexity reflects actual diversity. Neurospora shares gene sequences with a variety of taxonomic groups, so information from Neurospora informs projects covering the breadth of biology, not just within the fungi. Although both nonpathogenic and easy to manipulate, Neurospora is phylogenetically very closely allied and genetically syntenic, both with important animal and plant pathogens and with agriculturally and industrially important production strains (such as Cochliobolus, Fusarium, Magnaporthe, and *Trichoderma reesii*). Strong parallels have been noted in signaling pathways, photobiology, developmental regulation, and many aspects of metabolism including secondary metabolism, to name a few. Information from Neurospora, especially in terms of functional genomics and regulation, is readily transferable. An understanding of the functional genomics of Neurospora provides a gateway to the genomes of the fungi.

C. Overview of the functional genomics effort

The overall effort to develop the functional genomics of Neurospora has been divided into four subprojects, each of which is partially dependent on and informs the others.

The first project is concerned with the systematic mutation of genes in the organism through targeted knockouts and is centered at Dartmouth and UC Riverside. The initial phenotypic characterization of knockout strains generated at these two sites is coordinated by investigators at UC Los Angeles (UCLA). This group shares a major ongoing commitment to minority undergraduate education, so this effort has enlisted significant involvement of minority under-graduates in the MBRS/IMSD, MARC, and CAMP programs at UCLA. In addition, they have seeded smaller efforts elsewhere to characterize additional genes. In addition to the phenotypic characterization, strains are archived at the Fungal Genetics Stock Center (FGSC) at the University of Missouri from which they are distributed to the research community at large. The generation of knockout constructs in Project 1 is informed by the annotation completed in Project 2 using the Expressed Sequence Tag (EST) sequences generated in Project 4, and knockouts created in Project 1 are being analyzed via microarrays in Project 3.

Annotation and Genomics Core work has been the focus of Project 2 centered at the Broad Institute, Massachusetts Institute of Technology, and Harvard (formerly the Whitehead Institute Center for Genome Research; WICGR) and at Oregon Health & Science University (OHSU). The sequence of the Neurospora genome was completed at WICGR (Galagan *et al.*, 2003), and automated annotation is ongoing there. OHSU has provided liaison with the Neurospora community in terms of linking the genomics with the genetic map and the existing Neurospora compendium of genes and phenotypes. Aims of Project 2 have been to (1) build a platform for electronically capturing commu-nity feedback and data about the existing annotation (gene calls), (2) build and maintain a database to capture and display information about phenotypes resulting from gene knockouts and disruptions, and (3) utilize data from ongoing EST analyses in Project 4 (as described below) to refine the gene structures.

Project 3, transcriptional profiling, began with the creation of DNA microarrays corresponding to the ~10,000 uniquely distinguishable transcripts in Neurospora. Centered at UC Berkeley, this project has provided a baseline analysis of gene expression for *N. crassa* under a variety of growth conditions and environmental stresses, and has begun to analyze the global effects of loss of novel genes in strains created by Project 1. These results provide the foundation for the effects of mutations on particular pathways of gene expression, and the availability of affordable microarrays has nucleated this technology in the community.

Progress in Project 4 at Dartmouth and the University of New Mexico has focused on EST analyses and creation of a Single Nucleotide Polymorphism (SNP) map, which has greatly facilitated the cloning of genes and analysis of complex genetic traits. To make maximum use of the products of this EST-sequencing effort, the Mauriceville strain of Neurospora is being used for construction of the cDNA libraries at OHSU as well as at Dartmouth and the University of New Mexico. This is an alternative wild-type strain that is fully interfertile with the Oak Ridge strain (sequenced to 16-fold coverage by the Broad), but that displays a sufficient number of nucleotide variants such that it has been used as the crossing strain for an existing *N. crassa* restriction fragment length polymorphism (RFLP) map (Metzenberg and Grotelueschen, 1995). The SNPs arising from comparison of EST sequences from Mauriceville and Oak Ridge have provided the basis for constructing the SNP map. Additionally, unlike the apparent case in yeasts, alternative splicing and use of alternative promoters appear to contribute widely to the overall complexity of expressed sequences in Neurospora (Colot *et al.*, 2005), and there are known cases of long antisense transcripts (Kramer *et al.*, 2003). The Broad Institute has carried out limited EST sequencing of cDNA libraries and begun to use these data to establish the prevalence of alternative splicing and antisense transcripts. Project 4 has relied on resources from Project 2 and provided data essential for the annotation in Project 2 which, in turn, informed the creation of knockouts in Project 1. Additionally, EST sequences generated in Project 4 continue to inform the choice of probes used for the creation of microarrays in Project 3.

The interplay and interrelatedness among projects are shown schematically in Fig. 2.2. This serves as a useful scaffold for information flow as the efforts and progress of each project are considered in more detail below.

II. PROJECT 1: SYSTEMATIC GENE KNOCKOUTS

The goal of this project is straightforward. Neurospora has ~10,000 genes, a number that continues to fluctuate as annotation continues. Phenotypes are now associated with about 15% of these, and roughly a third of the genes have no

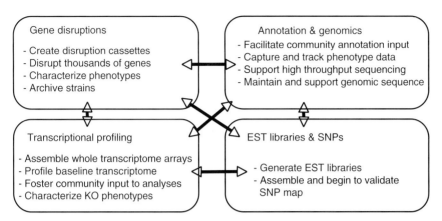

Gene disruptions

- Create disruption cassettes
- Disrupt thousands of genes
- Characterize phenotypes
- Archive strains

Annotation & genomics

- Facilitate community annotation input
- Capture and track phenotype data
- Support high throughput sequencing
- Maintain and support genomic sequence

Transcriptional profiling

- Assemble whole transcriptome arrays
- Profile baseline transcriptome
- Foster community input to analyses
- Characterize KO phenotypes

EST libraries & SNPs

- Generate EST libraries
- Assemble and begin to validate SNP map

Figure 2.2. Goals within and interactions among the four major research groups in the Neurospora functional genomics project.

strong sequence homologues in other organisms; thus, there are no clues to their functions. Furthermore, it is likely that functions of some genes sharing sequence similarity with genes in other organisms will be modified slightly or greatly, reflecting the novel biology of filamentous fungi. Our long-term goal is to create gene knockouts in all of these genes as a first step to determining each gene's function(s), and the immediate goal of Project 1 is to facilitate this. Following on the remarkable success and utility of the complete set of knockouts made in the yeast *Saccharomyces cerevisiae* (Martin and Drubin, 2003; Ooi et al., 2006; Winzeler et al., 1999), this approach scarcely requires defense.

There are several differences between this effort and the one originally carried out in yeast. First, this work follows on and benefits from that in yeast, and we tried to make good use of rationales, approaches, and lessons learned from that project. Second, many Neurospora genes are also found in yeast where they likely perform similar functions, so there may be less pressing need to examine them all. There are, however, many more genes in filamentous fungi as compared to yeasts, and in some cases the biology of a filamentous fungus may influence the functional significance of even known genes in unknown ways, so the magnitude of the endeavor here is somewhat larger than that faced by the yeast effort. Third, there exist several possible different ways in which gene function can be eliminated in Neurospora. One is RIP, which utilizes a Neurospora-specific phenomenon in which duplicated sequences, when passed through meiosis, undergo frequent C to T transitions thus usually resulting in loss of function (Selker, 1997). Additionally, genes can be targeted for "knockdowns" through RNAi or through "Quelling" (Cogoni and Macino, 1997). Finally and importantly, gene disruption yielding unambiguous nulls can be achieved by replacement through standard double homologous recombination. In this regard, unengineered Neurospora is more typically eukaryotic than yeast in that the frequency of homologous recombination rarely approaches 100% but lies instead between a few percent and 30% depending on the gene and the length of the homologous DNA flanking the gene in the construct.

A. Creation of gene knockouts in Neurospora

We chose to knock out genes in Neurospora by targeted gene replacement through homologous recombination. To achieve this, we needed to make a knockout cassette carrying a selectable marker for each gene to be replaced, to transform these cassettes into Neurospora, and to identify among the selected transformants the ones carrying the replacement but not other extraneous copies of the selectable marker. The knockout project can thus be thought of in three parts: construction of the cassettes, transformation, and examination of the transformants. At the outset of this effort we have devoted considerable time and energy to optimizing steps in this process, and in this section go over some of the rationale behind these choices as well as describing the final design of the work. This provides both greater detail and more insight into the process than is available in the initial description of this work (Colot *et al.*, 2006).

1. Creation of the knockout cassettes

a. The selectable marker

We chose the hygromycin phosphotransferase gene (Gritz and Davies, 1983) encoding resistance to hygromycin because (1) it is a very widely used marker in the filamentous fungal community; (2) it can be selected in *Escherichia coli*, yeast, and Neurospora; and (3) it is a dominant selectable marker, so the recessive auxotrophic mutations that are commonly used in genetic analysis do not have to be sacrificed. It is best when this bacterial gene is driven by a promoter from another fungus that contains no Neurospora sequences that could be mutated by RIP (Selker, 1997). We tried two different promoters for use in driving expression of the selectable marker: (1) the Ashbya *translation elongation factor* promoter p-TEF [driving *hph* (Goldstein and McCusker, 1999), total size of cassette is ∼2 kb]; this is a weaker promoter but one that works in *E. coli*, yeast, and Neurospora; and (2) the *A. nidulans trpC* promoter (Pandit and Russo, 1992). In all cases the flanks of the cassette end in the recognition sequence for the restriction enzyme Mme1 (TCCpuAC) for reasons that will be explained further below.

The knockout cassettes are generated by assembly of PCR-generated DNA fragments in yeast (Oldenburg *et al.*, 1997; Raymond *et al.*, 1999)

(Fig. 2.3). The entire procedure through production of the final Neurospora knockout cassette was designed for high throughput to facilitate use of a 96-well format. Homologous recombination in yeast was, obviously, a cornerstone of the methodologies used for the yeast knockout project, where it was noted that there is a trade-off between the efficiency of recombination and the length of the

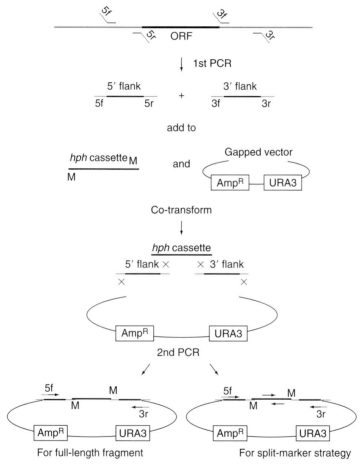

Figure 2.3. Strategy for creating deletion cassettes. 3′ and 5′ flank fragments are PCR'd separately from genomic DNA using primers 3f + 3r and 5f + 5r; primers 3f and 5r incorporate MmeI sites and have 5′ tails matching the *hph* cassette whereas 3r and 5f match the vector. The two flanks, *hph* cassette and gapped vector are cotransformed into yeast where homologous recombination recreates the circular construct. The final linear deletion cassette is PCR'd from the pooled yeast DNA using primers 3r and 5f. The cassette is constructed so that *hph* is transcribed antisense relative to the target gene.

oligonucleotide-derived overlap. Regarding the length of oligonucleotides used for gene replacement constructs, 45 base pairs (bp) of homology yielded 80% correct disruptants whereas 30 bp yielded only about 50% (Mark Johnston, personal communication). As a rule they used 45-mers. However, we used the overlap only to assemble fragments and determined that lower efficiency of recombination resulted in only a marginally lower yield of the correctly assembled fragments. In addition, we ultimately amplified the final linear knockout construct from crude yeast DNA preparations using PCR, thus eliminating any background due to incorrect or incomplete plasmids in the DNA used to transform Neurospora. Following this logic, we settled on overlapping homologous regions of 29 bp in the DNA fragments comprising the initial gene replacement "knockout cassette" used to transform yeast. This afforded considerable financial savings in the oligonucleotide syntheses necessary for the genome-wide knockout strategy.

b. The DNA template for PCR generating the fragments and design of primers

Initially, we had believed that it would be necessary to use cosmid or BAC DNA as the template for the PCRs generating the to-be-assembled fragments. However, we determined that genomic DNA worked fine: 49-bp primers (having 20 bp of overlap with the appropriate genome sequence) and Neurospora genomic DNA plus polymerase in the appropriate buffer yield products of sufficient mass that they can be directly transformed into yeast with no cleanup. This results in a considerable savings in time and supplies.

For primers to be gene-specific, the left and right flanks for each cassette must be tailored to individual genes. Thus, the choice of primers is a computationally intense problem. We used a custom-built application (written by John Jones, UCR) to retrieve regions adjacent to each Open Reading Frame (ORF) and pass them to Primer3 (http://frodo.wi.mit.edu/cgi-bin/primer3/primer3_www.cgi), which would automatically select a list of candidate primers with the following parameters: "PRIMER_GC_CLAMP = 2," "PRIMER_OPT_TM = 56.0," "PRIMER_ MIN_TM = 50.0," "PRIMER_MAX_TM = 63.0," "PRIMER_MIN_GC = 50.0," "PRIMER_MAX_GC = 65.0." Each selected primer pair was checked for uniqueness in the genome.

c. Molecular "bar coding" to individually identify knockout strains

In the yeast knockout project, each gene knockout was constructed so that it was marked by unique 20-bp sequences that provided novel sequence tags ("bar codes") that could be used to identify the mutant. These short sequences were incorporated into the knockout cassettes in separate PCR steps, and thereafter into the genome of each mutant; they can be detected by PCR or by hybridization to a membrane containing the sequence of the tag. The construction to include bar codes required an extra PCR step (half again more primers and more labor) or longer primers, but because the cost of retrofitting such tags would be prohibitive they were included even though no experiments were planned for their use at that time. Subsequently, of course, they have proven to be quite useful. They provide unambiguous identifiers for each strain, and allow large numbers of deletion stains to be pooled and analyzed in parallel following selective growth. In yeast they can greatly reduce the cost and increase the efficiency of identifying phenotypes associated with deletion strains.

Generally, the utility of the tags in yeast has often hinged on the ability of yeast to grow without fusing, whereas Neurospora hyphae in culture rapidly anastomose to form local heterokaryons, so that rare recessive mutations are quickly complemented; this suggested that they might not be useful in Neurospora. However, even in yeast the present uses were not anticipated or even fully appreciated at the time the tags were first incorporated, but later developed as a result of their availability. We expected the same would be true in Neurospora, but the cost of the extra primers and PCR step to incorporate the yeast sequence tags was prohibitive. As a work-around, we developed what we hope will be an economical way in which to incorporate such molecular bar codes. The restriction enzyme Mme1 has found use chiefly for SAGE analysis in which use is made of the fact that Mme1 has a 5-bp recognition sequence but makes a 2-bp staggered cut 20-bp 3′ from the recognition sequence. We incorporated Mme1 sites at the right- and left-hand ends of the *hph* cassette (Figs. 2.3 and 2.4). Oligo 3f primes from the left-hand side of the rightward piece of genomic DNA flanking the *hph* gene in the knockout cassette and oligo 5r from the right side of the left flank. In the construction of knockout cassettes for each new gene, both oligos 3f and 5r must be individually synthesized since they must contain the gene-specific 20 bp as well as the portion overlapping the *hph* cassette; that is, they are gene specific and unique in the genome and thus fulfilled the prime requirement for a bar code and no extra cost (Fig. 2.4). When the knockout cassette is fully assembled, the Mme1 sites will lie next to the "bar code." When genomic DNA is digested with Mme1, many fragments will be generated, but only two will contain the right- and left-hand margins of the *hph* gene, respectively, and it should be possible to amplify the bar codes through ligation-mediated PCR, just as if they were SAGE tags (Fig. 2.4).

d. The construction method

In this first series of PCR reactions, three fragments are produced: (1) The "selectable marker" (*hph*) was amplified using a simple pair of primers, generating a generic fragment that can be used for all knockouts. (2) Depending on the desired construct, 1–3 kb of sequence immediately upstream of the target ORF was amplified by a pair of 49-mers that use 20 bp to prime on the genomic

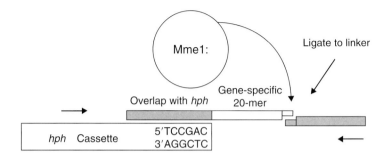

$$\text{Mme1:} \quad \begin{array}{l} 5'\text{TCCPuAC(N)}_{20}{}^{\wedge} \\ 3'\text{AGGPyTC(N)}_{18}{}^{\wedge} \end{array}$$

Figure 2.4. Detailed view of a primer (3f or 5r) used for assembly of the knockout cassette and one end of the *hph* cassette used for gene deletions, showing the location of the bar code. The circle represents the restriction enzyme Mme1 binding to its recognition sequence, and the attached curved arrow points to the position where it will cut genomic DNA containing a disrupted gene. After cutting with Mme1, the "gene-specific 20-mer" comprises a molecular bar code that may be detected through ligation of a linker (shown as a gray bar on the right) followed by PCR using primers (horizontal thick black arrows) specific for the linker and the *hph* cassette.

template and add 29 bp of DNA homologous to left side of the vector gap (5f) and the left portion of the *hph* cassette (5r), respectively. (3) Similarly, the downstream-flanking sequence was amplified by a pair of 49-mers that use 20 bp to prime on the genomic template and add 29 bp of DNA homologous to right side of the vector gap (3r), and the right margin of the *hph* cassette (3f). By design, the last (5′) 6 bp of the 29-bp *hph*-related part of the 49-mers were the Mme1 sites used as a part of our molecular bar-coding scheme.

    In the next step, these fragments were transformed into yeast along with the gapped yeast shuttle vector. As originally implemented, the vector contained a cycloheximide sensitivity gene (*CYH2*) that provided a positive selection for recombination by being displaced on recombination of the left side fragment. However, our selection for circle closure and replication serves the same purpose and we found both selections redundant. We used lithium acetate transformation of yeast with success. After brief recovery, yeast cells were cultured in medium lacking uracil to stationary phase. The product of this step is a yeast strain harboring a shuttle plasmid containing the three fragments recombined in the proper order: left (5′) flank, *hph* cassette, and right (3′) flank. When driving *hph* with TEF, the resulting plasmid confers hygromycin resistance to the recipient yeast, but this was not the case using the *trpC* promoter.

This method of assembly is extremely robust and has only rarely failed, having been successfully used by five different investigators in two different laboratories. In later experiments, there seemed no reason to continue to monitor individual plasmids so PCR was simply used to generate the correctly assembled cassette from the mixed yeast on plates, and in these cases the correctly assembled cassette was recoverable by PCR in nearly all of the constructions.

PCR using LA Taq (TaKaRa), a genomic DNA template, and the primers as described above consistently gave very high yields of product such that a few microliters (ca. 0.5 μg) of each PCR reaction can be combined directly with 0.5 μg of the standard *hph* fragment and 100 ng of the gapped vector in the PEG/lithium acetate transformation. DNA from the transformed yeast culture was prepared by a streamlined protocol (Colot *et al.*, 2006) and was then used directly for the production of the final Neurospora knockout cassettes. The knockout cassettes were amplified using LA Taq and the same 5f and 3r primers that flank the ends of the Neurospora DNA in the yeast vector. The resulting linear DNA fragments were subjected to a PCR cleanup protocol and then frozen until ready for transformation into Neurospora (see below). The production of knockout cassettes was successfully automated using a Biomek NX robot (Beckman Coulter, Fullerton, CA) so that production was between 400 and 600 per week, and the cassettes for nearly the entire genome were completed within a year.

2. Transformation of Neurospora

Although we had limited data for transformation using circular DNA, it appeared that linearization of the gene replacement vector and/or gel purification of the insert prior to transformation reduced the incidence of ectopic integration events in transformants with the desired homologous recombination event. Also, higher rates of homologous recombination seemed generally correlated with larger flanks; for wild-type strains 3-kb flanks are the desired size. The logic of the full size cassette was the same as in the original yeast knockout procedure. A selectable marker (here *hph*) is flanked by DNA homologous to chromosomal DNA flanking a gene-to-be-deleted. On transformation, reciprocal homologous recombination on both sides of the targeted gene serves to replace the targeted gene with the selectable marker. Replacement transformants in which the targeted gene was knocked out were identified based on the selectable marker, and confirmed by Southern analysis. The method worked dependably in that there has never been a gene that we could not delete in this way. However, it has the drawback that the entire selectable marker is incorporated into the knockout cassette, so for organisms in which ectopic insertion happens in a finite proportion of the total transformations (such as Neurospora), ectopic insertion events constitute a background to the desired single gene replacements. This problem is

exacerbated by the fact that both an ectopic and a targeted replacement in a
strain might provide as much as twice the drug resistance of a single replacement,
so there is a mild selection in favor of ectopic events. One solution to this
problem is to split the selectable marker into pieces (the split-marker technique),
and another solution on which we settled was to use the stronger A. *nidulans trpC*
promoter to drive hygromycin resistance so that there would be less marginal
selection for ectopic transformations.

In split-marker transformation, the whole disruption cassette is supplied
in the transformation cocktail as two separate pieces of DNA; neither piece on
its own encodes *hph* but recombination recreates the complete gene. A successful gene replacement, then, requires recombination not only in the flanks but
also within the overlapping central part of the selectable marker. This has the
effect of lowering transformation efficiency. However, there is no longer any
selective advantage to ectopic insertions, so that a higher proportion of the
hygromycin-resistant strains contain only the desired replacements. The split-
marker technique has been used in yeast (Fairhead *et al.*, 1996) and in
Cochliobolus (Catlett *et al.*, 2003), and in our hands yielded gene replacements
in 44% of transformants for which 68% were free from ectopic insertions (Colot
et al., 2006). These results were clearly adequate, and we were prepared to use a
split-marker strategy to replace all the genes until a significant improvement
appeared.

Work in yeast (Ooi *et al.*, 2001) has suggested that the Ku70 and Ku80
proteins are important for the nonhomologous end-joining process that gives
rise to ectopic insertions. On the basis of this we originally proposed in our
application for funding to delete the corresponding genes in Neurospora in
hopes of improving the efficiency of homologous replacements; however, non-
filamentous fungal-based members of the study section recommended that this
aspect should be deleted. Fortunately, Inoue and colleagues independently
came up with the idea, performed the experiments in Neurospora, and found
that the resulting mutant strains were remarkably efficient at gene replacement
(Ninomiya *et al.*, 2004). Because Inoue and colleagues used *hph* (the selectable
marker in our knockout cassettes) to generate the Δ*mus-51* and Δ*mus-52*
(homologues of *ku70* and *ku80*) mutants, we reengineered these mutants using
bar as a selectable marker (confers resistance to phosphinothricin) (Colot *et al.*,
2006). In our hands, with the analysis of over 600 independent transformants by
Southern analysis, we found that over 98% showed clean gene replacements
inserting the knockout cassette in place of the resident gene with no accom-
panying ectopic insertion (Colot *et al.*, 2006). The great success of this method
also meant that the relatively long 3-kb flanks were no longer needed and gave
rise to the final plan using 1-kb flanks in the knockout constructs. This led to
more efficient amplification of the shorter fragments using PCR, thus further
streamlining our protocol.

3. High-throughput production of gene replacements in Neurospora to generate unambiguous functional knockouts

Knockouts are being generated in the Oak Ridge wild-type genetic background that was used for the 16-fold coverage genome sequencing at the Broad Institute (Galagan *et al.*, 2003). Since Neurospora naturally and rapidly forms hetero-karyons, and because the conidia that will serve as transformation recipients have, on average, 2.5 nuclei per spore, primary transformants are heterokaryons that serve to shelter loss of essential genes in the transformed nucleus. We carry out the transformation using the Neurospora knockout cassettes described above, and select for resistance to hygromycin. Primary transformants are then crossed to a closely related but opposite mating type Oak Ridge wild-type strain. Phosphinothricin-sensitive progeny (lacking the *bar* marker and the corresponding *mus* mutation) are then screened from the hygromycin-resistant progeny to select for strains in which a single gene knockout resides in an otherwise wild-type genetic background.

Final molecular verification of the knockout is provided by Southern analysis of hygromycin-resistant and phosphinothricin-sensitive progeny. We use a custom software program (written by John Jones) to both predict the best restriction enzymes to use for Southern analysis of a given gene and to provide the sizes of hybridizing fragments from the knockout and wild-type alleles. All Southerns are probed using the full-length knockout cassette, thus allowing detection of any rare ectopic integrations. In addition to confirming the presence of knockout mutations and absence of ectopic integrations, this step verifies strain integrity. In practice, the software program has performed very well, greatly facilitating an extremely tedious manual procedure.

As soon as the confirmed homokaryons are generated in one or both mating types, they are shipped to the FGSC for distribution to the research community at large. Since some genes being disrupted prove to be essential to life, it is not always possible to recover hygromycin-resistant homokaryotic knockout strains from crosses. In such cases, we assume the gene is essential (either for life or for meiosis/spore viability) and the original heterokaryotic transformant is checked using Southern analysis (should contain both wild-type and knockout-hybridizing fragments) and then sent to the FGSC. To date nearly 2300 strains have been deposited and the fulfillment of requests to the FGSC makes the products of the knockout collection a heavily used resource within the FGSC's operations.

An up-to-date listing of knockout strains produced or in progress as well as protocols and methods for making or requesting knockouts can be found at [http://www.dartmouth.edu/~neurosporagenome/1_s1.html]. All strains, procedures, and reagents used for knockouts have been deposited with the FGSC and are available there.

a. Schedule and throughput

In production mode, we try to adhere to a schedule wherein a new plate of 96 strains begins every 3 weeks at both Dartmouth and UCR. Because it takes closer to 10–12 weeks to complete a full knockout construction and verification by Southern, this means that an investigator will have several overlapping cohorts ongoing at one time at varying stages of completion. To monitor progress, track strains, and house the primer design and Southern restriction enzyme software, we have implemented a Laboratory Information Management System (LIMS; written by John Jones; http://www.borkovichlims.ucr.edu). The current LIMS tracks the more than 30 steps in the entire knockout procedure. Plates and products at various stages of the protocol are labeled with physical bar codes that can be scanned by a hand-held bar code scanner. The progress of a strain or set of 96 strains can then be tracked through the entire procedure, culminating in assignment of an FGSC number to the final homokaryotic or heterokaryotic mutant strains. These bar codes can be recognized by personnel at the FGSC and UCLA using their own bar code scanners by logging in to the UCR database.

By maintaining overlapping cohorts of strains, one person can handle between eight and nine batches of knockouts ($\sim$8.5 × 96 = 816 genes) per year. In this scenario, two laboratories each with two people at the bench will produce about 3200 knockouts per year. If 65% of these are clean knockouts that are easy to characterize, this yields 2100 genes per year, with a total of just over 7000 by 2009. If 95% are clean knockouts, 3100 genes can be mutated each year, with just over 10,000 completed by 2009.

These perspectives raise the issue of what can and does go wrong. So far cassette construction (as noted above) has worked extremely well, with the only problems being rare operator errors, mistakes with the oligo supplier, and a very few problem sequences that may not be compatible with the yeast system. Clean ectopic-free gene replacements occur at a dependable rate of about 98%, so there are always a few strains that need to be redone, but this is not onerous. Verification by Southern that strains do indeed contain gene replacements and are free of ectopic insertions has been the greatest remaining bottleneck. This step requires growth of the strain, preparation of sufficiently pure genomic DNA to allow restriction digestion, followed by electrophoresis, blotting, probing, and interpretation of the results. As mentioned above, our custom-written software for choosing an appropriate enzyme and predicting the correct Southern pattern has immensely helped in this endeavor. We also have a robust nonradioactive method for generating probes and detecting hybridizing fragments. However, isolation of genomic DNA from recalcitrant strains that grow poorly and/or are hard to lyse has presented technical hurdles and required making modifications to the standard protocol in some instances. Other than this, another challenge is the continual process of genome annotation that periodically adds or deletes many hundreds of genes based on new molecular evidence from ESTs or comparative studies. Each change immediately sends that gene back to square one, and these will undoubtedly comprise the major source of uncompleted genes when the project winds down.

b. Choosing genes to disrupt and coordinating the effort

The most recent annotation version 3 of assembly 7 of the Neurospora genome has revealed 9826 potential protein-coding "genes," although this number continues to fluctuate as a result of the manual annotation in Project 2 and EST analyses ongoing in Project 4. Although in the best of all possible worlds it may be possible to knock out most of these genes, there must be priorities assigned to which ones to do first. Our initial effort went into a proof-of-principle experiment where just over 100 genes encoding transcription factors were deleted and the phenotypes analyzed (Colot *et al.*, 2006). Subsequent efforts have been organized on the principle of generating the strains that will be of the greatest utility to the community. These have been chosen in two different ways.

First anyone can request that knockout strains be constructed and such requests immediately go into the queue. Such requests are made to a knockout-specific email address (knockouts@dartmouth.edu) and care is taken to be sure that all orders are handled anonymously so that a particular request is never associated with the name of the requestor. To date we have generated several hundred strains requested by the community at large, including some sizable blocks of genes such as >100 products associated with hyphal growth.

Second, we have chosen for disruption groups of genes that can form the basis of research projects or that are commonly among those that are needed for dissection of biological processes. These groups have included, for instance, the genes encoding the remaining ~100 transcription factors, genes encoding protein kinases and phosphatases, and chromatin-remodeling enzymes.

More than a third of the genes in Neurospora have no homologues outside of the filamentous fungi, so the elucidation of the phenotypes of such genes promises to inform much ongoing work among these organisms. With this in mind, as we move beyond the analysis of known genes, some general rules have been established:

1. Is it novel? If yes do it; if not, go on to 2.
2. Is it already associated with a phenotype deriving from a known mutation in Neurospora? If yes, skip it; if no, go on.
3. Is it in *Saccharomyces cerevisiae* or *Schizosaccharomyces pombe*? If no, do it; if yes, go on.
4. Is it essential in either or both yeasts? If yes, skip it; if no, go on.
5. Is it associated with an obvious housekeeping function in yeast, such as intermediary metabolism, such that the phenotype of a null is easily predictable? If yes, skip it; if no, do it.

The products of this high-throughput knockout effort are strains bearing a molecularly verified gene replacement marked by hygromycin resistance. Each tube is physically bar coded, this time with a bar code corresponding to the FGSC number. Each laboratory generating the knockouts retains a tube for each mutant, and a replicate bar-coded tube is sent to the FGSC for distribution to the community and, if needed, to UCLA for phenotypic analysis.

B. Basic phenotypic characterization of mutants

It is not feasible to carry out a comprehensive analysis of every phenotype caused by each knockout mutation. However, a preliminary characterization of each knockout strain—to provide enough basic phenotypic information to enable other researchers to productively use the set of mutants—has enormous added value. Because the techniques associated with this work are relatively straightforward, and because Neurospora is absolutely nonpathogenic, we determined early on that it would be desirable to use the characterization of these strains as a vehicle to introduce undergraduate students, and in particular where possible underrepresented minority students, to Neurospora and microbiology in general. This was the origin of the very successful Neurospora Genetics and Genomics Summer Research Institute (NGGSRI) at UCLA where phenotypic analysis protocols for beginning science students have been developed. These methodologies are now being implemented at several other sites worldwide.

Novel knockout strains arrive from the FGSC, Dartmouth, or UCR. Two replicates are generated and serve as a backup and a student set. Using the student set, each student generates a separate stock for their individual knockout mutant analysis. Multiple students perform the analyses on an individual strain so that collectively each strain is examined in quadruplicate. This redundancy, along with supervision of the final data set, ensures quality control of the data.

Five assays are performed for each mutant analysis, which examine morphology, asexual development, growth, and sexual development using wild-type strains as a reference.

First, analysis of colony growth and morphology is performed on solid Vogel's minimal (VM) medium and on minimal medium plus yeast extract (VM + YE) at both 25 and 37 °C. Petri dishes inoculated at their centers are cultured under ambient light/dark conditions and digital images recorded at 24 and 48 h [Infinity Camera with a Navitar Zoom 7000 lens (Lumenera Scientific, Ottawa, Canada)]. For each strain, the hyphae at the edge of the colony are photographed after 24 and/or 48 h using an S8 Apo Stereo Zoom microscope mounted with a DFC 280 digital camera (Leica, Wetzlar, Germany). Second, the rate of extension of basal hyphae is measured on race tubes (Dunlap and Loros, 2005) containing 13 ml of VM agar medium at 25 °C in ambient light/dark conditions. Growth marks are recorded twice per day (morning and afternoon) over a 72-h period, the data graphed to verify linearity, and the growth rate expressed as mm/day. These procedures identify growth and morphological aberrations.

Third, the production of aerial hyphae and conidia is measured on slants containing VM medium grown at 25 °C for 3 days and then put at room temperature for 3–5 days. Conidiation, pigmentation, and aerial hyphae are scored using the wild-type strains as a reference. The extension of aerial hyphae is measured in 13×100 mm test tubes containing 2 ml of VM or VM + YE as standing liquid cultures at 25 °C. The top edge of the mycelial mat as seen after 24 h is marked on the tubes, the cultures incubated statically for a total of 72 h, and the difference in height in mm recorded. These results reveal mutants in asexual development.

Fourth, the sexual developmental pathway is analyzed in the following increments. Formation of protoperithecia (female sexual structures) are assessed following 7–8 days growth at room temperature (22–25 °C) on plates or tubes of synthetic crossing medium (Davis and deSerres, 1970) containing either 0.1% or 1% sucrose. Fifth, plates or tubes are then fertilized with 10^6 wild-type conidia of the opposite mating type, incubated at room temperature for another 7–8 days, and mature perithecial formation is scored using a stereomicroscope. Plates or tubes are then finally checked for ascospore development 2 weeks after fertilization.

The phenotype data are recorded and, along with the digital images, uploaded to the Broad Institute (Fig. 2.5). Students are given user names and passwords for the Broad and each establishes his or her own database. A summary tool developed by the Broad Institute allows access to all databases for the UCLA curator. The summary tool is organized by NCU No. and contains all images and data for every mutant analysis. Following curation and verification by the UCLA staff, the data are made public in the format that is shown in Project 2 (Fig. 2.8). To date nearly 400 strains have been characterized at UCLA and several additional phenotyping sites will come online in 2007 using the UCLA protocols.

C. Deposition of the strains in the Fungal Genetics Stock Center (FGSC) and their distribution to the scientific community at large

The FGSC receives bar-coded hetero- or homokaryotic strains from UCR or Dartmouth. The complete set of mutants as it is assembled is listed in the FGSC online catalog, and made available to the scientific community by the FGSC through its normal mechanisms. Protocols have now been developed whereby sets of 96 knockouts can be archived, replicated, and sent out in deep-well microtiter plates. Requests for individual knockout strains as well as for the entire emerging collection make this a heavily used resource within the FGSC's portfolio.

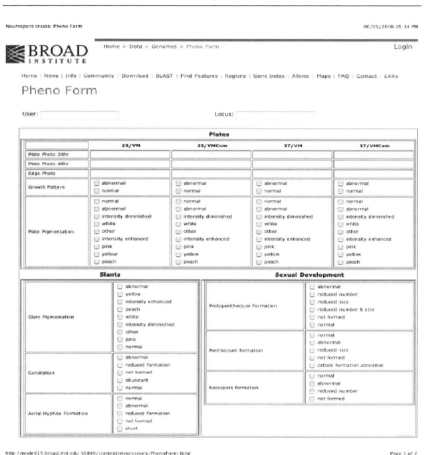

Figure 2.5. Screen shot of page 1 of the phenotype data entry form used by undergraduates for web-paged entry of basic phenotypic data.

D. Summary of gene knockouts

The first systematic mutation of a eukaryotic genome was that of Saccharomyces, and the multiple impacts of that work have by all accounts been spectacularly successful. The products of this approach, the assembled mutants potentially addressing the function of each gene in a genome, constitute one of the central cornerstones of a modern research model system. Given the central importance of the filamentous fungi to medicine, to agriculture, and to industry, and considering their surprising genomic complexity and novelty, the Neurospora knockout collection is proving to be an invaluable resource.

III. PROJECT 2: GENOME INFORMATICS AND FUNCTIONAL ANNOTATION STUDIES

A. Introduction

To increase the value of the *N. crassa* genome sequence we must continue to refine the genome annotation and also capture and integrate with the genome sequence the wealth of information within the research community about the genes and genetics of the organism. To this end, the Annotation and Genomics Group of the NIH Neurospora Program Project has the specific aims of (1) generating EST data and using them to both improve gene predictions in the genome and delineate single nucleotide polymorphisms (SNPs) between the Oak Ridge and Mauriceville strains, (2) building a platform for capturing and curating community input about the genome annotation, and (3) building and maintaining a database to capture and display information about phenotypes resulting from gene knockouts and disruptions.

The Annotation and Genomics Group of the Neurospora Program Project is centered at the Broad Institute of MIT and Harvard with an associated community-based annotation effort centered at OHSU. It uses the Broad's Calhoun system as the main infrastructure. Calhoun is an informatics system developed for whole genome annotation and analysis. The system is based on a modular and extensible architecture, and has been applied successfully to the genome annotation and analysis of a multitude of fungi, microbes, and other organisms sequenced through the Broad's Fungal Genome Initiative and Microbial Sequencing Center. The features of Calhoun include (1) an Oracle relational database for the storage and management of genome sequence data, (2) an extensive Application Programming Interface (API) for data insertion, retrieval, and manipulation, (3) a high-throughput sequence analysis pipeline, (4) extensive tools for data mining, and (5) sophisticated user interfaces and client tools. Calhoun's architecture is organized into data storage, data interface, analysis, and presentation layers.

B. Improving automated genome annotation

To realize the full potential of genomic sequence, important genetic features must be identified, and these features must be associated with biologically relevant functional information. High-quality gene annotation is the starting point for harnessing the power of genome sequence. For example, an accurate annotation of the location of genes is essential for the development of microarrays, and errors and revisions to the number, structure, and locations of genes have major impacts on the quality and cost of gene knockout efforts.

The task of identifying all genes in a genome and associating functional information with them is not a completely solved problem. Many tools exist for computationally predicting the location and structure of genes. These include *ab initio* gene prediction tools, homology search tools, and sequence alignment tools. However, these tools do not produce perfect predictions, and their ability to identify genes in a particular organism depends heavily on additional supporting data. Among the different forms of evidence used for gene calling, EST sequences are especially valuable both as training sets for *ab initio* gene predictors and as raw data for use by gene-building systems.

Neurospora EST data from previously characterized libraries (Bell-Pedersen *et al.*, 1996; Nelson *et al.*, 1997; Zhu *et al.*, 2001) have been valuable for training gene predictors, validating automated gene predictions, and constructing microarrays for functional studies. Although the Neurospora gene annotation is highly accurate in detecting gene loci, the fine structure (introns/exon boundaries, start and stop codons, untranslated regions, and alternative splicing) still requires refinement. Additional EST coverage is the surest method for improving gene structure prediction. Currently only ~1/3 of Neurospora genes have EST support. To expand this coverage, the Broad Institute is sequencing ESTs generated for the Mauriceville strain by the Neurospora Program Project and integrating these EST data into the automated annotation of Neurospora genes. The Broad has built and continues to refine an automated gene-calling pipeline that uses multiple gene prediction algorithms and selects the best gene call using multiple forms of evidence including ESTs. ESTs are the highest ranked form of evidence used to call a gene structure from the structures derived by the prediction algorithms. When EST coverage is available, gene predictions can be evaluated for (1) Spurious Predicted Introns: predicted gene structures with EST alignments completely spanning an intron (without corresponding gaps in the alignment), (2) Missing Exons: EST alignments not in predicted coding regions, and (3) Incorrect Splice Sites: predicted gene structures with EST alignments covering an intron and an adjacent exon. The gene predictions with the most agreements to the aligned ESTs for (1) through (3) are selected (Fig. 2.6A). In addition, the ESTs are used to define untranslated regions (UTR) and alternative transcripts (Fig. 2.6B). The Broad Institute is currently implementing an algorithm that directly use EST data for gene prediction versus only for gene calling.

In addition to using the new ESTs in gene calling, the Annotation and Genomics Core is using the ESTs to identify SNPs that are in turn being used to build a new SNP-based genetic map of Neurospora ESTs from the highly polymorphic Mauriceville strain are automatically aligned to the Oak Ridge genome sequence using the Calhoun infrastructure (Section V). To increase the specificity in our SNP identification, SNPs are defined using the neighborhood quality scoring algorithm (NQS) developed at the Broad Institute (Altshuler *et al.*, 2000). A polymorphism is called a SNP if there is a mismatch between

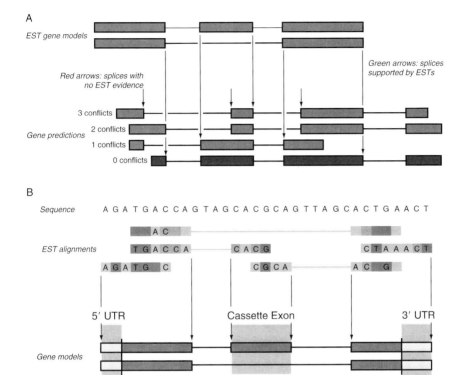

Figure 2.6. How EST sequences are used for (A) gene calling and (B) UTR and alternative splice prediction. (See Color Insert.)

two bases in the alignment, each base has a PHRED quality score of at least 25, and that polymorphism is in a neighborhood of bases (defined by five bases upstream and downstream of the mismatch) each with PHRED of at least 20. SNPs delineated using NQS are then validated using both PCR-based and restriction digest-based validation methods at participating laboratories in the Neurospora Program Project. Validated SNPs are displayed on the genome sequence and integrated with the Neurospora genetic map.

C. Community annotation

Computational methods alone are not adequate to produce the highest quality genome annotation; manual annotation and curation are also necessary. Automated gene prediction systems produce systematic as well as specific errors that are difficult to correct without manual intervention. Furthermore, automated methods remain incapable of tapping into the vast wealth of knowledge about the genes and genetics of an organism that are collectively contained within the research community.

One way to provide for manual annotation and to capture community knowledge is to create an infrastructure that enables researchers in the community to submit and curate gene annotations. Although variations on this theme exist, they are typically referred to as "Community Annotation Projects" (CAP). The Annotation and Genomics Core has developed a robust infrastructure called CAP that uses the automated annotation of the Neurospora genome as a scaffold onto which expert knowledge about genes is mapped to the genome. CAP is powered by Calhoun and is accessed through the Broad Institute's web interface for the Neurospora genome pages. CAP provides researchers the ability to improve the Neurospora annotation by integrating information including: (1) refined gene structures and alternative splices, (2) gene symbols, (3) gene synonyms, (4) gene product names, (5) functional information such as ontology terms, (6) associated literature, and other information directly onto the genome. Community annotations made using CAP are attached to the official annotation and CAP provides tools to make these community annotations part of the official annotation through a curation process (Fig. 2.7). CAP annotations are immediately searchable and retrievable by all users once saved into the Calhoun database. Until made official, CAP annotations are visualized in a summary format at the bottom of a "gene detail page" (Fig. 2.7); clicking on a CAP "ID" opens a detailed view of the CAP annotation. Community annotations can be viewed, edited, as well as commented on by other members of the research community.

More than 60 years of research on Neurospora constitutes a vast resource of information contained within the Neurospora community for interpreting the genome sequence. Capturing a portion of this information in the genome annotation will provide an invaluable resource for Neurospora researchers and the wider scientific community. CAP provides this opportunity. For example, the large amount of data in Alan Radford's electronic version of the Neurospora compendium (Perkins *et al.*, 2001) has been incorporated into CAP, as have the new gene names and phenotypes associated with the transcription factor knockouts generated in the initial work of Project 1 (Borkovich *et al.*, 2004; Colot *et al.*, 2006).

Finally, as part of the CAP project, a web-accessible Textpresso database of the Neurospora literature has been developed to enable rapid full-text searching of papers that refer to this organism. This resource (www.textpresso.ebs.ogi.edu) is useful for annotation and for providing access to specific information in this literature. The Textpresso application, originally created by researchers at Caltech for curation of the *Caenorhabditis elegans* literature (Muller *et al.*, 2004), has been used by us to build a Neurospora database. Textpresso accepts

Community
Annotation

↓

Attachment of
Annotation

↓

Manual Curation

↓

Incorporation into
Official
Annotation

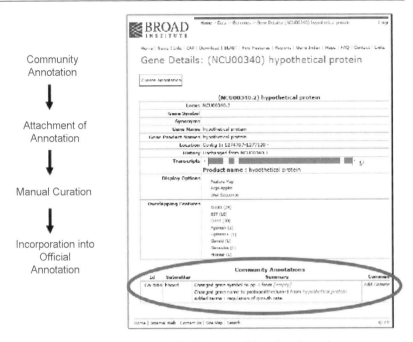

Figure 2.7. Community Annotation Project (CAP) process and Gene Detail page showing community annotation. CAP is an infrastructure developed by the Broad Institute in collaboration with members of the Neurospora Functional Genomics Project that uses the automated genome annotation as a scaffold on which the research community can build a refined manual annotation. CAP provides the research community the ability to annotate the structure of a gene as well as capture a wealth of functional information about a gene through a user-friendly web interface. Community annotations are attached to the official genome annotation, are immediately searchable, and are incorporated into the official annotation through a manual curation process (see http://www.broad.mit.edu/annotation/genome/neurospora/CASearchAnnotations.html/direct).

queries to search for specific words alone or in combination, as well as ontology relationships defined by specific categories. It responds by providing the queried terms in the original context of the research paper in which they are found, as well as links to PubMed. Thus, for example, it is possible to search the literature for instances in which specific alleles of genes are used by searching on the allele name, or to search for instances in which specific compounds are named, and rapidly find the specific contexts in which the alleles or compounds have been discussed.

D. Annotation of allele-specific phenotypes

The automated annotation of the Neurospora genome sequence has revealed 9826 genes, many of which have as yet no clear homologues in other organisms.

To provide functional annotation of these genes, the results of large-scale gene disruption analyses of these genes must be mapped to the genome and put in the context of what is already known about N. *crassa* mutant phenotypes. Mutations in N. *crassa* can be considered in terms of a variety of major phenotypic effects. The most fundamental is how they affect viability. Such observations date from the demonstration in the 1940s that single gene mutants in N. *crassa* caused inviability at high temperature, important evidence for showing that the one-gene one-enzyme theory did not apply only to dispensable genes such as those that caused auxotrophy. Major common types of phenotypic effects include alterations in morphology, physiology, and development. More specific effects include alterations to nutritional requirements, sexual fertility, the circadian clock, the capacity to form heterokaryons, and to posttranscriptional gene silencing. Integrating with the genome both existing phenotype data as well as new information obtained from ongoing knockout experiments that are part of the Neurospora Program Project will leverage both sequence and phenotype data to enable functional annotation.

To this end, the Annotation and Genomics Group developed, maintains and continues to refine the database that captures and displays information about alleles derived from the phenotypic analysis of mutants carried out at UCLA and other sites as described in Section II.B. This infrastructure is powered by Calhoun and is accessed through a web interface. Like CAP, the genome annotation is used as a scaffold onto which allele phenotypes are mapped. The system has the ability to track multiple alleles associated with a gene, capture defined phenotypic characterizations, and obtain such information for multiple assay conditions (Fig. 2.8). Researchers characterize knockout strains using a defined vocabulary and images of the mutant's morphology and then upload this information using a web-based data-entry form (Section II.B). To maintain consistency in the annotation of phenotypic data and provide a database that is fully searchable, a controlled vocabulary is used to define phenotypes. To ensure this annotation meets the scientific needs of the *Neurospora* community and is valuable to the wider fungal community, a substantial effort is dedicated to establishing and maintaining the vocabulary for describing abnormal phenotypes. This task is undertaken by an Annotation Steering Committee.

Once characterizations are expertly curated using tools developed by the Broad they are added as a feature in the gene annotation and can be searched and viewed through the Broad Neurospora webpages. Since allele summaries (Fig. 2.8) are stored as a feature of a gene they are accessed directly from a gene's detail page; alternatively they are retrievable by browsing a table of phenotype characterizations and associated genes on the Neurospora alleles homepage.

The Annotation and Informatics Group of the NIH Neurospora Program Project continues to enhance the structural and functional annotation

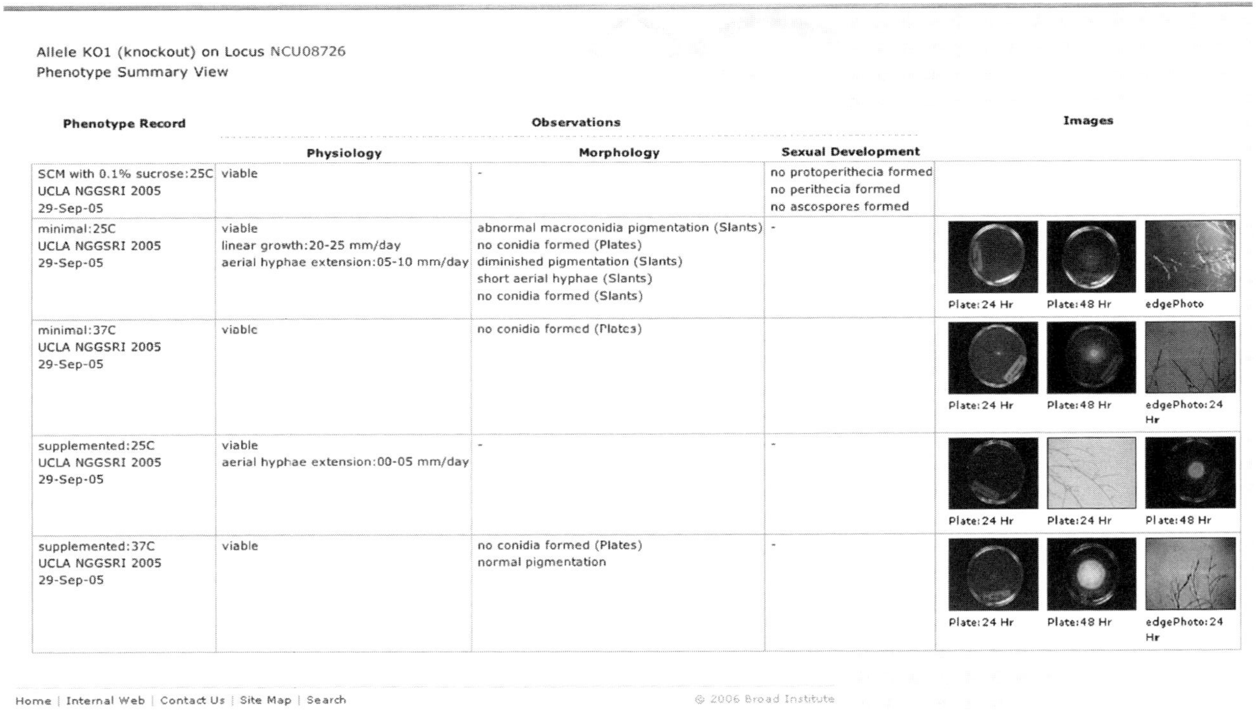

Allele KO1 (knockout) on Locus NCU08726
Phenotype Summary View

Phenotype Record	Observations			Images
	Physiology	Morphology	Sexual Development	
SCM with 0.1% sucrose:25C UCLA NGGSRI 2005 29-Sep-05	viable	-	no protoperithecia formed no perithecia formed no ascospores formed	
minimal:25C UCLA NGGSRI 2005 29-Sep-05	viable linear growth:20-25 mm/day aerial hyphae extension:05-10 mm/day	abnormal macroconidia pigmentation (Slants) no conidia formed (Plates) diminished pigmentation (Slants) short aerial hyphae (Slants) no conidia formed (Slants)	-	Plate:24 Hr Plate:48 Hr edgePhoto
minimal:37C UCLA NGGSRI 2005 29-Sep-05	viable	no conidia formed (Plates)		Plate:24 Hr Plate:48 Hr edgePhoto:24 Hr
supplemented:25C UCLA NGGSRI 2005 29-Sep-05	viable aerial hyphae extension:00-05 mm/day	-	-	Plate:24 Hr Plate:24 Hr Plate:48 Hr
supplemented:37C UCLA NGGSRI 2005 29-Sep-05	viable	no conidia formed (Plates) normal pigmentation	-	Plate:24 Hr Plate:48 Hr edgePhoto:24 Hr

Home | Internal Web | Contact Us | Site Map | Search

© 2006 Broad Institute

Figure 2.8. A typical "alleles phenotype characterization" summary page for a Neurospora gene, taken from http://www.broad.mit.edu/annotation/genome/neurospora/AlleleDetails.html.

of the Neurospora genome. This is accomplished through the development of community annotation and allele phenotype characterization infrastructures that for the first time provide the research community the ability to link information about genetic features with the *Neurospora* genome, to refine gene structures and to curate all entries in a searchable database. Gene models are also improved using EST sequence that moreover enable the generation of an SNP-based genetic map. All of these activities will provide the high-quality annotation of the Neurospora genome that will empower researchers to new discoveries in eukaryotic biology.

IV. PROJECT 3: PROFILING TRANSCRIPTION IN *NEUROSPORA*

The biology of filamentous fungi remains relatively unexplored, especially when compared to fungal model yeast species, such as *Saccharomyces cerevisiae* and *Schizosaccharomyces pombe*. As mentioned above, in contrast to yeasts, filamentous fungi are characterized by a complex life cycle, and there are at least 28 distinct cell types in *N. crassa* (Bistis *et al.*, 2003; Borkovich *et al.*, 2004). To help to provide a solid foundation for understanding this interesting and yet tractable level of complexity, and to enhance the technology for inference of gene expression levels in *Neurospora*, we are creating whole-genome-spotted oligonucleotide microarrays. Spotted microarrays allow the measurement of the abundance of transcripts from thousands of genes simultaneously by the competitive hybridization of labeled cDNA transcripts (targets) to immobilized probes cross-linked to the surface of amine-coated microscope slides (Derisi and Iyer, 1999; Eisen and Brown, 1999). Gene expression data generated by microarray technology can be used to expedite the annotation process of the *N. crassa* genome and will aid in assigning putative functions for unique genes.

A. Oligonucleotide design and synthesis

We designed 70-mer oligonucleotide-immobilized probes from 10,526 open-reading frames (ORFs) predicted from the *N. crassa* genome sequence (Broad Institute, http://www.broad.mit.edu/annotation/fungi/neurospora_crassa_7/index.html and MIPS, http://pedant.gsf.de/cgi-bin/wwwfly.pl?Set = Ncrassa_annotations&Page = index) using the bioinformatic tool ArrayOligoSelector (Bozdech *et al.*, 2003). Array OligoSelector identifies a unique 70-bp segment to represent each ORF, avoiding self-annealing structures and repetitive sequences. It preferentially chooses oligonucleotides that are located close to the 3′-terminal region of each gene and that conform to a narrow range of GC content. A total of 10,910 70-mer oligonucleotides were synthesized by Illumina, Inc., San Diego CA. These represent the 10,526 predicted genes as well as an additional 384 70-mer oligonucleotides designed for intergenic or telomeric regions. *Neurospora* full genome microarrays are printed onto γ-aminopropyl silane slides and are available at cost from the FGSC (http://www.fgsc.net/). Information on the oligonucleotide gene set is available at the *Neurospora* Functional Genomics Database (http://www.yale.edu/townsend/Links/ffdatabase/introduction.htm).

B. Experimental design for microarray experiments

Two samples of target cDNA are competitively hybridized against the immobilized probe in spotted DNA microarray hybridizations. Thus, primary analysis yields the ratios of gene expression between two samples. However, most experimental designs incorporate multiple developmental, genetic, and/or environmental states. Inference of gene expression level in multiple states from ratiometric measurements requires judicious experimental design. In closed circuit designs (Fig. 2.9), each strain is compared head-to-head with other strains, in a circular or multiple-pairwise fashion. The ability to detect differences is maximized because the comparisons are directly between individual strains or conditions of interest, and the problems of using a common standard (Townsend, 2003) are avoided. The sole disadvantage of this method is that immediate inference from the raw data is not easy because ratios observed across multiple pairwise comparisons do not all compare or contrast in an intuitive way. However, this is a problem that is rapidly solved by accessible methods of analysis (Kerr and Churchill, 2001; Townsend and Hartl, 2002; Wolfinger *et al.*, 2001). For instance, a Bayesian analysis of gene expression levels (Townsend and Hartl, 2002) uses all transitive and direct comparisons from any replicated, interconnected experimental design to infer relative gene expression levels and 95% confidence intervals. The results are reported with gene expression in the sample with lowest expression as one unit; the samples are scaled appropriately. Although the inferred expression levels are of arbitrary unit scale, this scaling has the intuitive appeal that all gene expression level measurements are positive, as they should be. Circuit designs of microarray comparisons have been strongly endorsed by statisticians (Kerr and Churchill, 2001; Yang and Speed, 2002) and have demonstrated dramatically improved resolution with regard to identifying differential gene regulation below the twofold level (Townsend and Hartl, 2002; Vinciotti *et al.*, 2005).

Subsequent hierarchical clustering of inferred levels of gene expression can greatly assist in understanding the function of previously uncharacterized genes as there is a tendency for sets of coexpressed genes to be involved in common cellular functions (Eisen *et al.*, 1998). Data can be analyzed by multiple computational methods, including self-organizing maps (SOMs), k-means clustering, and principal component analysis. Motif searches conducted, using programs such as BioProspector (McGuire *et al.*, 2000), MDscan (Liu *et al.*, 2002), and MEME (Bailey and Elkan, 1994), can be utilized to identify *cis*-regulatory elements attributable to the coregulation of gene clusters using profiling data from *N. crassa* (Kasuga *et al.*, 2005).

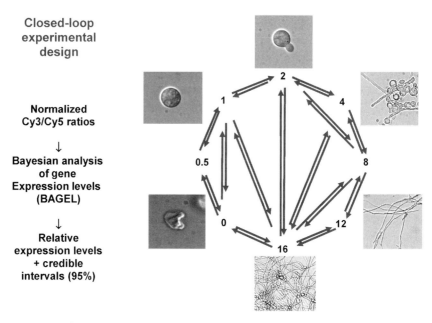

Figure 2.9. Schematic representation of the data comparisons executed in a tightly fashioned closed-loop design for microarray data collection. Images (not to scale) represent the shape of cells from the time conidia are inoculated into growth medium (time 0) through the first 16 h of growth. RNA samples collected at each time point are hybridized to multiple microarrays such that each pair of arrows represents one experiment of reference and control (Kasuga *et al.*, 2005).

C. Technical aspects to transcriptional profiling: RNA extraction, cDNA labeling, image acquisition, and normalization procedures

Protocols for RNA extraction and cDNA labeling are available at the *Neurospora* Functional Genomics Database (http://web.uconn.edu/townsend/Links/ ffdatabase/introduction.htm). Briefly, total RNA is extracted from samples using TRIzol reagent (Invitrogen Life Technologies, Burlington, ON). A 100-μg sample of total RNA from each sample is cleaned using RNeasy Mini Protocol for RNA cleanup (Qiagen, Valencia, CA). For cDNA synthesis, 20 μg of total RNA is mixed with 5 μg of an anchored 17-mer oligo dT and 3.3 ng of ArrayControl single RNA spike mixture (Ambion, Austin, TX). cDNA is synthesized in a final volume of 30 μl with 500 μM each of dATP, dCTP, and dGTP, 300 μM of dTTP, 200 μM of aminoallyl-dUTP, 10-mM DTT, and 100 units Stratascript reverse transcriptase (Stratagene, La Jolla, CA). For conjugation to fluorescent dyes, 10 μl of 0.05-M sodium bicarbonate is added to the monofunctional NHS-esters of Cy3 or Cy5 (CyDye Post-Labeling Reactive Dye,

Amersham Bioscience, Piscataway, NJ) and 5 μl of the dye solution is added to the cDNA solution. The labeled cDNA is purified with the CyScribe GFX Purification Kit (Amersham) and dried under vacuum.

Slides are prehybridized and labeled cDNA is resuspended in hybridization buffer and pipetted into the space between a microarray slide and a LifterSlip cover glass (Erie Scientific, Portsmouth, NH). An Axon Gene-Pix 4000B scanner (Axon Instruments, CA) is used to acquire images and GenePix Pro 4.1 software is used to quantify hybridization signals.

To evaluate the ratio of mRNA from comparative hybridizations for normalization purposes, control RNA spikes are used as internal standards. The control spikes consist of eight polyadenylated bacterial mRNAs at concentrations ranging from 50 to 1000 pg/μl, which are complementary to eight Array-Control Sense oligonucleotides (Ambion). The ArrayControl oligonucleotides are added as duplicate spots to the N. crassa oligonucleotide microarrays. A total of 3.3 ng of each control mRNA spike is added to each of the 20-μg total RNA samples for each time point.

D. *Neurospora* functional genomics microarray database

We constructed a *Neurospora* microarray database in the public standard minimal information about a microarray experiment (MIAME) format (Brazma, 2001) (http://web.uconn.edu/townsend/Links/ffdatabase/introduction.htm). MIAME is a reporting protocol that describes the minimum information required to ensure that microarray data can be easily interpreted and that results derived from its analysis can be independently verified. Using this reporting protocol will facilitate the database deposition into public repositories and enable the development and use of novel data analysis tools. The *Neurospora* microarray database stores raw and normalized expression data from microarray experiments and also provides detailed discussions and information on experimental design, data analysis methods, and web interfaces for scientists to retrieve, analyze, and visualize their data. In the future, this database will be integrated into the N. crassa database at the Broad Institute (http://www.broad.mit.edu/annotation/fungi/neurospora_crassa_7/index.html).

E. Proof-of-principle: Transcriptional profiling of conidial germination

Filamentous fungi undergo complex asexual and sexual developmental programs. In addition, their mycelial growth habit differs substantially from that of unicellular organisms such as *Saccharomyces cerevisiae*. However, since both N. crassa and *Saccharomyces cerevisiae* are fungi, and extensive profiling experiments have been performed on *Saccharomyces cerevisiae* (see Saccharomyces Genome Database; http://www.yeastgenome.org/), a comparison of transcriptional profiles of these two species in response to environmental and nutritional factors, DNA damage, and the cell cycle will be especially informative. Other aspects of the life cycle in *N. crassa* have no obvious counterpart in *Saccharomyces cerevisiae*, such as development of female reproductive structures, formation of asexual spores, creation of the interconnected mycelium, and asexual spore germination (Davis, 2000). As a proof-of-principle experiment for the development of oligonucleotide arrays, we performed transcriptional profiling of conidial germination in *N. crassa* on a partial genome array comprising 3366 immobilized probes for predicted genes (Kasuga *et al.*, 2005). Conidial germination in filamentous fungi is a highly regulated process that is triggered by environmental stimuli. Biochemical and morphological aspects associated with conidial germination have been well documented in *N. crassa* (for review, see Bonnen and Brambl, 1983; Denfert, 1997; Riquelme *et al.*, 1998; Roca *et al.*, 2005; Schmidt and Brody, 1976). However, fundamental genetic mechanisms, such as those that drive the germination process and underlie the timing of gene expression and metabolic pathway activation, remain obscure.

We chose a closed-circuit experimental design to determine the expression levels of genes relevant to conidial germination (Fig. 2.9) (Kasuga *et al.*, 2005). RNA was isolated from eight time points during the germination process: times 0 and 30 min, and times 1, 2, 4, 8, 12, and 16 hours post inoculation. In this circuit of experimental comparisons, each sample was compared head-to-head with other samples, in a circular, and in some cases, multiple-pairwise fashions. We obtained precise statistical estimates of expression levels for 1287 genes during the process of conidial germination. Estimates of gene expression levels were remarkably consistent with previous data assessing transcript levels of a number of genes and with biochemical processes that have been associated with conidial germination (Kasuga *et al.*, 2005; Sachs and Yanofsky, 1991; Schmidt and Brody, 1976).

Of the 1287 genes for which strong estimates of gene expression level were acquired, 473 have been described as hypothetical, conserved hypothetical, or putative genes. Gene expression data for the remaining 814 genes with functional annotation were evaluated by the MIPS functional catalogue (FunCat) (Kasuga *et al.*, 2005), which is an annotation scheme for the functional description of proteins (Ruepp *et al.*, 2004). Concordance was apparent between predicted function of transcriptionally regulated genes and previously identified biochemical processes associated with conidial germination (Fig. 2.10). For example, a large number of genes belonging to the FunCat category ribosomal biogenesis showed maximum expression between 1 and 4 h after conidia are induced to germinate by inoculation into growth medium, consistent with biochemical data indicating that RNA and protein syntheses are activated soon after the induction of germination (Schmidt and Brody, 1976). Many genes in the FunCat cell cycle and DNA processing showed maximum expression between 30 min and 4 h after inoculation, consistent with biochemical observations that the initiation of DNA replication occurs ~2 hours postinoculation (Schmidt and Brody, 1976).

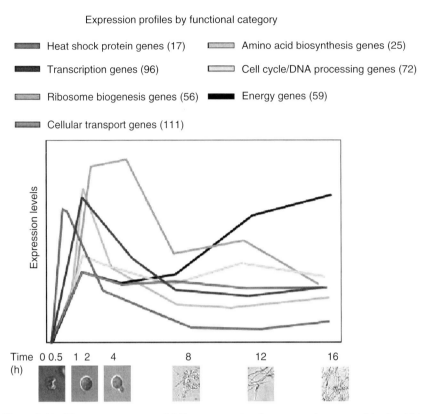

Figure 2.10. Changes in expression of different categories of genes over the course of the first 16-h growth in Neurospora as elucidated by microarray analysis. (See Color Insert.)

Transcriptional profiles for some of the 473 hypothetical, conserved hypothetical, or putative genes correlated well with biochemical processes associated with conidial germination. Thus, microarray data in N. *crassa* will guide future laboratory experiments with regard to functional annotation of hypothetical genes.

F. Future prospects

1. Baseline transcriptional profiling of filamentous fungal colonial growth

The mycelial colony of filamentous fungi is the hallmark of this group of organisms. We currently use full genome arrays (10,910 oligonucleotides) to obtain baseline transcriptional profiling data growth and reproduction in

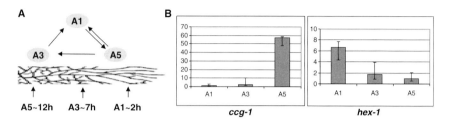

Figure 2.11. An example of detailed changes in gene expression associated with early development in Neurospora as revealed by microarray analysis.

N. crassa. For example, we profiled *N. crassa* vegetative growth by comparing the transcriptomes of ~2-, ~7-, and ~12-h old sections of a colony grown on solid medium (Fig. 2.11). Expression of over 7000 genes was detected at statistically significant levels; 72 genes showed statistically significant differences in expression level between 2- and 12-h old hyphae within a *Neurospora* colony. As an example, expression of *ccg-1* at 12 h was 30-fold higher than in 2-h old hyphal tips (Fig. 2.11B). In contrast, *hex-1* (encoding the structural element of the Woronin body) expression was fivefold higher in the colony periphery versus 12-h old hyphae (Fig. 2.11B). These results are consistent with published Northern RNA blot results for these genes (Tey *et al.*, 2005).

2. Deciphering the transcriptional regulatory network of Neurospora

The *N. crassa* genome encodes ~176 transcription factors with five DNA-binding motif families, that is, basic helix loop helix (HLH), bZIP, zinc finger C2H2 (zf-C2H2), GATA zinc finger, and Zn(2)-Cys(6) binuclear cluster (Zn2Cys6) (Borkovich *et al.*, 2004). We have performed phylogenetic analysis of predicted transcription factors of *N. crassa* as compared to other filamentous fungi, including *Saccharomyces cerevisiae.* Many of the predicted transcription factor genes have close orthologs in other filamentous fungi but not to *Saccharomyces cerevisiae.* The vast majority of these transcription factors are completely uncharacterized. Transcriptome analysis of three mutants (NCU00340, a *ste-12* ortholog; NCU03725, an NDT80 homologue; and NCU07392, a zinc binuclear cluster TF) have been conducted. As an example, the NCU00340 mutant showed 129 differentially expressed genes. A putative *cis*-element (CATCNTCAT) was enriched in the downregulated gene set ($p = 0.00017$, fisher test). The identified *cis*-element of NCU00340 putative target genes does not show similarity to the *Saccharomyces cerevisiae* Ste12p DNA-binding site (ATGAAAC) (Zeitlinger *et al.*, 2003). These data suggest that at least some transcriptional regulatory networks have diverged in fungi, as previously documented for Rpn4 orthologs (Gasch *et al.*, 2004). Further analysis of transcription factor mutants in *N. crassa* in comparison to wild type will reveal transcriptional regulatory networks that have been conserved among fungi and others that have diverged among filamentous fungi, yeast, and more distantly related eukaryotic species.

V. PROJECT 4: cDNA LIBRARIES AND THE GENERATION OF A HIGH-DENSITY SNP MAP

A. Introduction and rationale for the design of the project

SNPs are the most common genetic variants between individual genomes. Dense SNP maps can be used to precisely and rapidly map single gene mutations as well as polygenic traits and quantitative trait loci (QTL). This approach has been exploited for the identification of genes in the nematode (Wicks *et al.*, 2001), the fruit fly (Berger *et al.*, 2001), and yeast (Winzeler *et al.*, 1999), as well as many other organisms. An eventual future application would be the generation of oligonucleotide microarrays based on the identified SNPs. All SNPs in a strain could then be typed in parallel in a single experiment, by hybridizing under conditions that disallow the formation of stable hybrids if there is a single mismatch (or SNP).

We reasoned that by simply using a wild-collected *N. crassa* strain for the isolation of mRNA, one with many nucleotide differences from the standard laboratory (Oak Ridge) strain, one could generate the sequence information required to construct a detailed SNP map for Neurospora. Moreover, if the strain of *N. crassa* was genetically close enough to Oak Ridge, the ESTs generated would serve to bolster the gene calling and annotation aspects of this effort while still providing the genetic variability necessary for an SNP map. The wild-type Neurospora strain isolated in Mauriceville Texas is the strain used for RFLP mapping in Neurospora; it is fully interfertile with *N. crassa* Oak Ridge, but since it is sufficient for identification of RFLPs we reasoned that it would be distinct enough that SNPs could reliably be found in the same manner that RFLPs have been reliably found for the past 16 years (Metzenberg *et al.*, 1985). This was confirmed by comparing several noncoding regions of the Mauriceville and Oak Ridge genomes: SNPs were detected at a rate of ~2% in our hands, data comparable to published reports showing variable frequencies, from 0.2% to 2.1%, in three Neurospora ORFs (Table 2.1).

On the basis of averaged data from the whole genome (Section 2 and Galagan *et al.*, 2003), Neurospora genes appear about every 3.7 kb along the chromosomes. The average transcript is 1.3 kb in length, and preliminary data suggested that the frequency of SNPs within coding regions is, conservatively, >0.2%.

Table 2.1. SNP Frequency in Neurospora Open Reading Frames[a]

Gene	Number of polymorphisms in Mauriceville with respect to Oak Ridge	Nature of polymorphisms (all are bp substitutions)	References
cys-3	15 in 708-bp ORF	9 silent; 6 substitutions in nonessential regions	Coulter and Marzluf, 1998
mtr	3 in 1472-bp ORF	3 silent	Dillon and Stadler, 1994
nmr	6 in 1464-bp ORF	5 silent; 1 substitution truncates ORF by 3 codons	Young and Marzluf, 1991

[a]The percentage of SNPs in these three ORFs is 0.7%.

This suggested that the average cDNA would identify several SNPs. Since the rate of recombination in Neurospora is low compared to Saccharomyces, 1 map unit (1% recombination) corresponds to about 20 kbp in a typical region in the middle of a chromosome arm (McClung *et al.*, 1989). A 20-kbp region might on average, then, contains 5 genes and 10–40 SNPs at saturation. These data suggested that sequencing of genomic DNA would not be necessary to ensure an adequate distribution of SNPs across the genome unless there are areas that have a paucity of genes or are, mysteriously, devoid of SNPs while still encoding transcripts.

Analysis of EST sequences from Mauriceville should reveal a trove of SNPs that can be used to augment the genetic map and greatly improve the efficiency with which informative mutations can be mapped. An application of RFLP/SNPs using a similar strategy in C. *elegans* demonstrated mapping to chromosome arm, and subsequently to cosmid, within a day after initial progeny were scored for presence/absence of a marker to be mapped (Wicks *et al.*, 2001). Overall then, this project would win in two ways, both by supplying needed EST-based data to bolster gene calling for annotation in Project 2 and in providing the basis for an SNP map.

B. Construction of the map

1. Source of SNPs

A cDNA library was made from FGSC 2225 (*N. crassa* Mauriceville, A). Conidia were harvested and inoculated into liquid culture (2% glucose, $1\times$ Vogel's salts, 0.5% arginine, and 0.05-μg/ml biotin) and grown at 30°C in

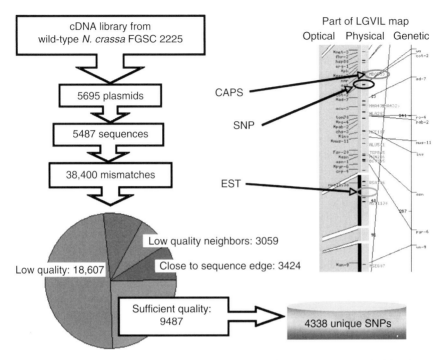

Figure 2.12. Flow chart for the preliminary SNP generation and mapping project. On the left is shown the progressive stages of data generation and screening, and on the right an example of the final product. This shows a fragment of chromosome VI L including parts of two contigs (green and black). Shown are the optical, physical, and genetic maps. Green dashes in the EST band indicate regions with EST coverage (at 3-kb resolution), black dashes are unconfirmed SNPs, and validated CAPS are identified by a three-letter enzyme designator and a number in red (see http://www.broad.mit.edu/annotation/genome/neurospora/maps/ViewMap.html?sp=5). (See Color Insert.)

light for 4 h. The germinated conidia were harvested, total RNA isolated and poly A + RNA from this. One microgram of this Mauriceville mRNA was used to synthesize cDNA (Bell-Pedersen *et al.*, 1996), which was sized on a Sepharose CL-2B column. Two cDNA fractions were ligated with the Uni-ZAP XR (Stratagene; precut with XhoI and EcoRI) vector, and the ligation products packaged *in vitro* (Stratagene Gigapack III kit), yielding a titer of about 3.3×10^4 pfu/ml. Following infection into XL1-blue MRF bacterial cells, phagemids were mass-excised and introduced into an SOLR strain to get a stable plasmid form. 5695 of these plasmids derived from germinated conidia of a wild-type strain of *N. crassa* from Mauriceville, Texas were then sequenced at the Broad Institute (Fig. 2.12A).

2. Identification of SNPs within ESTs

To be maximally useful, the SNP data must be integrated with the sequence and genetic maps of Neurospora; this integration will be achieved through collaboration with the Broad Institute (Section III.B). Briefly, traces from EST sequencing are aligned with the assembled genomic sequence (from the Oak Ridge wild-type strain) and the top-scoring hit identified as the corresponding gene from Mauriceville. Due to RIP and other duplication scanning mechanisms (Selker, 1997), few problems with identification of duplicated sequences in the Neurospora genome were anticipated, so Mauriceville ESTs would have one and only one corresponding gene in Oak Ridge; this has been the case so far. The sequences are aligned (or multiply aligned where more than one sequence is available due to multiple ESTs being sequenced or to overlap of the traces from each end of an EST) and the probability of a polymorphism determined. Since the error rate in the genomic sequence is well under 1/10,000 (Project 2) equivalent to a PHRED of 40, putative SNPs are called only if they fall within a region of otherwise high PHRED in the EST sequence. Likewise, for insertions and deletions, a SNP is inferred only if the PHRED value on either side of the gain/loss is high, and if the insertion/deletion does not occur within a region prone to compression artifacts. It is a truism that some SNPs will create RFLPs, so we will be able to independently verify our assignment criteria, in addition to amplifying the existing RFLP map and tying it directly to the assembled genomic sequence. Since the SNPs are being generated in the context of an assembled genome, they are already mapped with regard to each other, and can be used directly for fine mapping of novel genes.

Specifically then, sequences arising from the ESTs were aligned with the Oak Ridge genome using BLAT (Jim Kent, UCSC) for SNP detection. A mismatch that meets the NQS (Neighborhood Quality Standard) is called SNP; its PHRED quality score is higher than 25 and the five bases to either side display a PHRED score higher than 20 and are not mismatched themselves. This algorithm was adopted for detecting SNPs in human genome with an accuracy rate around 95% (Altshuler *et al.*, 2000). As predicted, SNPs have appeared among the sequences at a raw frequency of about 1 per kilobase. They are less common among coding sequences than in noncoding transcribed regions or intergenic regions, but there are still plenty of SNPs within genes for purposes of map construction.

3. Detection of SNPs

In all cases, SNPs are detected among cross progeny using PCR-based strategies, but two different methods have been used successfully in our hands. The first is differential amplification (http://ausubellab.mgh.harvard.edu) and consists basically of a sensitized PCR screen in which three oligonucleotides are needed to differentiate one SNP with certainty. Oligonucleotides are designed to have an exact match at the 3′ end with one or the other SNP, and each is paired with a third common oligonucleotide with which it can produce an amplified product (Drenkard et al., 2000). Ideally, each diagnostic oligo will only form a product with one of the two sequence variations of the SNP; however, in our hands the results have been mixed such that often several different oligo sequences must be tried before a good one can be identified, and even then weak-amplified bands sometimes appear from the "wrong" genotype. Due in part to these uncertainties, for generation of the SNP map we have settled more recently on another simple and efficient PCR-based method, CAPS (cleaved amplified polymorphic sequence), or "snip-SNPs." CAPS is based on the notion that some SNPs will make or destroy restriction enzyme-recognition sites, and this will facilitate their detection. With CAPS, a region around a (typically single nucleotide) polymorphism is amplified by PCR using two primers designed to invariant regions of sequence. The products are subjected to digestion with a restriction enzyme that acts on one version of the polymorphism but not on the other (Konieczny et al., 1991), thus creating an RFLP (Fig. 2.13).

The real power of the method lies in the ability to analyze groups of progeny for a single SNP in a single step. A single DNA preparation for each individual offspring can be used for screening hundreds of these "snip-SNPs" in parallel. As seen in Fig. 2.13, after a cross between a mutant isolated in an Oak Ridge background and the FGSC 2225 Mauriceville mapping strain, progenies are separated by phenotype and DNA pools made of mutant versus wild types. Aliquots of the pool are used for CAPS mapping with different markers; the PCR products are digested with the diagnostic enzyme and visualized on a gel. A measure of linkage to a given SNP can be estimated as the ratio of cut to uncut DNA for that marker, and unlinked markers ought to be equally represented in each DNA preparation. Going from gross to fine resolution within a chromosomal region is then just a matter of the numbers of SNPs used and their location, just as in classical transmission genetics where enhanced resolution is achieved through analysis of individual allele (here SNP) segregation in more and more individual segregants. In principal, once the SNP map is populated and using only common tools such as PCR and restriction digestion, it should be quite feasible to begin with isolation of DNA from bulked cultures of similar phenotype arising from 200 ascospores and then proceed to 1 map unit resolution (about 20 kb, within a cosmid) within a few days.

C. Validation of the SNP map

Of course, this approach is contingent on the availability of a sufficient number of CAPS markers well spaced over the entire genome. As seen in Fig. 2.12A, the

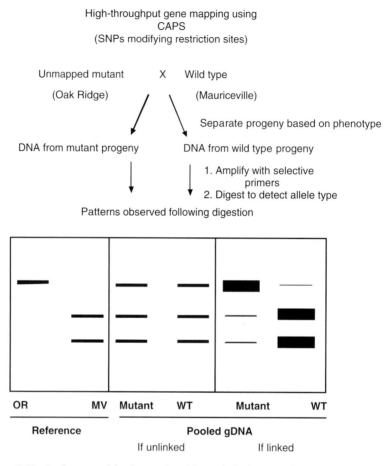

Figure 2.13. A schematic of the data produced from a bulked progeny SNP-mapping experiment.

sequence information from the 5695 plasmids yielded about 5500 sequences and nearly 40,000 putative mismatches (SNPs), but most of these were the result of sequencing ambiguities and were screened out immediately. This left about 9500, about half of which were duplicates, leaving 4338 unique SNPs. Of these, for the 12 most commonly used four-cutters that we have examined, 669 created an extra site in Mauriceville, 707 an extra site in Oak Ridge, and 45 extra sites in both.

These are all putative CAPS markers, but two additional criteria must be met for them to be useful. First, the DNA fragments resulting from digestion must be distinguishable from each other and from nonunique bands arising in the same digestion. For instance, creation of a new site in a sea of closely placed identical sites will not yield a useful SNP. Second, the markers must be well spaced through the genome. Two markers lying within a few kilobases of each other will be identifiable at the molecular level but redundant genetically.

A script was developed to select potential CAPS markers by these criteria, and both served to eliminate a number of useless markers. Figure 2.14

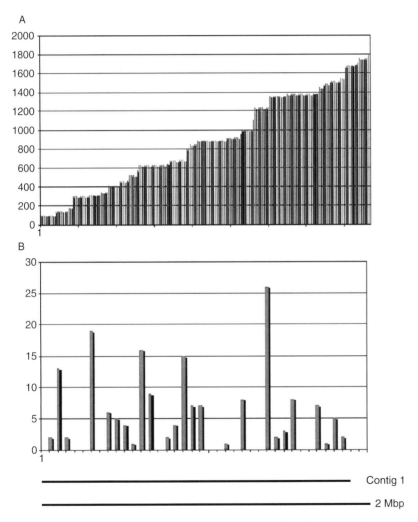

Figure 2.14. SNP Clustering. (A) Position of SNPs within the 1,800,000 bp of contig 1 in assembly 7 of the Neurospora genome. Note the long horizontal stretches that correspond to clusters. (B) Number of SNPs in each region of 50 kb along contig 1.

shows evidence for clustering of SNPs on contig 1. SNP positions tend to cluster closely together, with large gaps in between clusters (due to the fact that the length of sequenced ESTs is very small compared with their low frequency); hence a naïve-clustering method is sufficient to select among these: simply seed the first cluster in a given contig with the first SNP, and then for each consecutive SNP add it to the current cluster if it is closer to the previous SNP than a cutoff value "minsep," or create a new cluster to accommodate it if it is not. This gives very similar results for different values of minsep within a reasonable range (Table 2.2). Next, within the set of usable clusters, potential CAPS markers were identified based on whether an enzyme could be found that yielded a unique diagnostic fragment at least 50 bases different in size from any non-unique bands. Applying both criteria, the current set of EST sequences led to the selection and validation of close to 300 confirmed CAPS markers, spanning about 70% of the predefined SNP clusters (at a 10K resolution).

D. Implementation of the SNP map

As SNPS markers are validated they are placed into a database that will be maintained both at Dartmouth and integrated with the online physical maps at the Broad Institute. In this visual context, they can be associated with the maps of each linkage group as shown in Fig. 2.12B.

As the SNP map develops more, the goal is to have markers at least every 2 centimorgans (cM) along each chromosome; current resolution is about 7 cM on average. Searchable databases will list the identity of each CAP as well as its sequence context, the sequence of suitable primers for it amplification, and the restriction enzyme used for its detection.

Table 2.2. Clustering Results for Different Values of Minsep

Minsep	5K	10K	20K	50K	100K
Cluster width (average ± standard deviation)	0.6 ± 1.2	1.1 ± 2.5	3.2 ± 7.7	16 ± 28	43 ± 73
Maximal cluster width	10	16	64	188	360
Number of clusters	452	424	373	263	163
Number of nonsingleton clusters	309	289	252	179	89
Cluster gap (average ± standard deviation)	66 ± 60	71 ± 60	80 ± 59	110 ± 58	160 ± 55

Cluster width is defined as the difference between highest and lowest position within a cluster, and cluster gap as that between the highest in a cluster and the lowest in the next (both in kilobases).

VI. CONCLUSIONS

We have described here the ongoing progress on the functional genomic analysis of the Neurospora genome. As outlined in Section I there are a large number of genes that are unique to the filamentous fungi, and many of these organisms are of great importance as pathogens or for industrial manufacture but are experimentally less tractable than Neurospora. Abundant evidence from the comparisons of fungal genomes shows that the repertoire of genes in fungi is in part distinct from other organisms but is held in common among the filamentous fungi. There is thus every reason to believe that the functions that can be ascribed to genes in Neurospora will inform the study of a wide variety of related organisms. And this is, after all, the function of a model system.

Acknowledgments

This work was supported by grant P01 GM068087 from the National Institute of General Medical Sciences. We would like to thank John Jones for software design and LIMS implementation, and the following students who participated in the Neurospora Genetics and Genomics Summer Reasearch Institute (NGGSRI) at UCLA for phenotypic analysis: Cynthia Aguirre, Eliana Alcantar, Andrea Cahuantzi, Natalie Cornejo, Zachary W. Cue, Evelyn De Los Santos, Anthony Dualo, Thomas J. Dunehew, Mina El-Mastry, Jonathan Finley, Lizette C. Flores, Christopher Fonseca, Rukhsana A. Khan, Carolyn Kingsley, Juan Lupercio, Criseyda Martinez, Rosaura Ochoa, Olufisayo Oke, Cam M. Phu, Chloe Rivera, Michael Smith, Desree L. Tesada, Tuan D. Tran, and Jackelyn Valladares. The NGGSRI program was supported by NIH/NIGMS 5 R25 GM050067 and NIH/NIGMS 5 R25 GM055052.

References

Altshuler, D., Pollara, V. J., Cowles, C. R., Van Etten, W. J., Baldwin, J., Linton, L., and Lander, E. S. (2000). An SNP map of the human genome generated by reduced representation shotgun sequencing. *Nature* **407**(6803), 513–516.

Aronson, B. D., Johnson, K., Loros, J. J., and Dunlap, J. C. (1994a). Negative feedback defining a circadian clock: Autoregulation in the clock gene *frequency*. *Science* **263**, 1578–1584.

Aronson, B. D., Lindgren, K. M., Dunlap, J. C., and Loros, J. J. (1994b). An efficient method of gene disruption in *Neurospora crassa* with potential for other filamentous fungi. *Mol. Gen. Genet.* **242**, 490–494.

Bailey, T. L., and Elkan, C. (1994). Fitting a mixture model by expectation maximization to discover motifs in biopolymers. *In* "Proceedings of the Second International Conference on Intelligent Systems for Molecular Biology," pp. 28–36. AAAI Press, Menlo Park, CA.

Beadle, G. W., and Tatum, E. L. (1945). Neurospora II. Methods of producing and detecting mutations concerned with nutritional requirements. *Am. J. Bot.* **32**, 678–686.

Bell-Pedersen, D., Shinohara, M., Loros, J., and Dunlap, J. C. (1996). Circadian clock-controlled genes isolated from *Neurospora crassa* are late night to early morning specific. *Proc. Nat. Acad. Sci. USA* **93**, 13096–13101.

Berger, J., Suzuki, T., Senti, K. A., Stubbs, J., Schaffner, G., and Dickson, B. J. (2001). Genetic mapping with SNP markers in *Drosophila*. *Nat. Genet.* **29**(4), 475–481.

Bistis, G. N., Perkins, D. D., and Read, N. D. (2003). Different cell types in *Neurospora crassa*. *Fungal Genet. Newslett.* **50,** 17–19.

Bonnen, A., and Brambl, R. (1983). Germination physiology of *Neurospora crassa* conidia. *Exp. Mycol.* **7,** 197–207.

Borkovich, K., Alex, L., Yarden, O., Freitag, M., Turner, G., Read, N., Seiler, S., Bell-Pedersen, D., Paietta, J., Plesofsky, N., Plamann, M., Schulte, U., *et al.* (2004). Lessons from the genome sequence of *Neurospora crassa*: Tracing the path from genomic blueprint to multicellular organism. *Microbiol. Mol. Biol. Rev.* **68,** 1–108.

Bozdech, Z., Zhu, J., Joachimiak, M. P., Cohen, F. E., Pulliam, B., and DeRisi, J. L. (2003). Expression profiling of the schizont and trophozoite stages of *Plasmodium falciparum* with a long-oligonucleotide microarray. *Genome Biol.* **4**(2), R9.

Brazma, A. (2001). On the importance of standardisation in life sciences. *Bioinformatics* **17**(2), 113–114.

Catlett, N. L., Lee, B., Yoder, O. C., and Turgeon, B. G. (2003). Split-marker recombination for efficient targeted deletion of fungal genes. *Fungal Genet. Newslett.* **50,** 9–11.

Cogoni, C., and Macino, G. (1997). Isolation of quelling-defective (qde) mutants impaired in posttranscriptional transgene-induced gene silencing in *Neurospora crassa*. *Proc. Natl. Acad. Sci. USA* **94,** 10233–10238.

Colot, H., Park, G., Jones, J., Turner, G., Borkovich, K., and Dunlap, J. C. (2006). High throughput knockout of transcription factors in *Neurospora* reveals diverse phenotypes. *Proc. Natl. Acad. Sci. USA* **103,** 10352–10357.

Colot, H. V., Loros, J. J., and Dunlap, J. C. (2005). Temperature-modulated alternative splicing and promoter use in the Circadian clock gene frequency. *Mol. Biol. Cell* **16**(12), 5563–5571.

Coulter, K. R., and Marzluf, G. A. (1998). Functional analysis of different regions of the positive-acting CYS3 regulatory protein of *Neurospora crassa*. *Curr. Genet.* **33**(6), 395–405.

Davis, R. H. (2000). "Neurospora: Contributions of a Model Organism." Oxford University Press, Oxford, UK.

Davis, R. L., and deSerres, D. (1970). Genetic and microbial research techniques for *Neurospora crassa*. *Methods Enzymol.* **27A,** 79–143.

Denfert, C. (1997). Fungal spore germination-insights from the molecular genetics of *Aspergillus nidulans* and *Neurospora crassa*. *Fungal Genet. Biol.* **21**(2), 163–172.

Derisi, J. L., and Iyer, V. R. (1999). Genomics and array technology. *Curr. Opin Oncol.* **11**(1), 76–79.

Dillon, D., and Stadler, D. (1994). Spontaneous mutation at the mtr locus in *Neurospora*: The molecular spectrum in wild-type and a mutator strain. *Genetics* **138**(1), 61–74.

Drenkard, E., Richter, B. G., Rozen, S., Stutius, L. M., Angell, N. A., Mindrinos, M., Cho, R. J., Oefner, P. J., Davis, R. W., and Ausubel, F. M. (2000). A simple procedure for the analysis of single nucleotide polymorphisms facilitates map-based cloning in Arabidopsis. *Plant Physiol.* **124**(4), 1483–1492.

Dunlap, J. C., and Loros, J. J. (2005). Analysis of circadian rhythms in Neurospora: Overview of assays and genetic and molecular biological manipulation. *Methods Enzymol.* **393,** 3–22.

Eisen, M. B., and Brown, P. O. (1999). DNA arrays for analysis of gene expression. *Methods Enzymol.* **303,** 179–205.

Eisen, M. B., Spellman, P. T., Brown, P. O., and Botstein, D. (1998). Cluster analysis and display of genome-wide expression patterns. *Proc. Natl. Acad. Sci. USA* **95,** 14863–14868.

Fairhead, C., Llorente, B., Denis, F., Soler, M., and Dujon, B. (1996). New vectors for combinatorial deletions in yeast chromosomes and for gap-repair cloning using "split-marker" recombination. *Yeast* **12,** 1439–1457.

Galagan, J. E., Calvo, S. E., Borkovich, K. A., Selker, E. U., Read, N. D., Jaffe, D., FitzHugh, W., Ma, L. J., Smirnov, S., Purcell, S., Rehman, B., Elkins, T., *et al.* (2003). The genome sequence of the filamentous fungus *Neurospora crassa*. *Nature* **422,** 859–868.

Gasch, A. P., Moses, A. M., Chiang, D. Y., Fraser, H. B., Berardini, M., and Eisen, M. B. (2004). Conservation and evolution of cis-regulatory systems in ascomycete fungi. *PLoS Biol.* **2**(12), e398.

Goldstein, A. L., and McCusker, J. H. (1999). Three new dominant drug resistance cassettes for gene disruption in *Saccharomyces cerevisiae. Yeast* **15**, 1541–1543.

Gritz, L., and Davies, J. (1983). Plasmid-encoded hygromycin B resistance: The sequence of the hygromycin B phosphotransferase gene and its expression in *Escherichia coli* and *Saccharomyces cerevisiae. Gene* **25**, 179–188.

Kasuga, T., Townsend, J. P., Tian, C., Gilbert, L. B., Mannhaupt, G., Taylor, J. W., and Glass, N. L. (2005). Long-oligomer microarray profiling in *Neurospora crassa* reveals the transcriptional program underlying biochemical and physiological events of conidial germination. *Nucleic Acids Res.* **33**(20), 6469–6485.

Kerr, M. K., and Churchill, G. A. (2001). Statistical design and the analysis of gene expression microarray data. *Genet. Res.* **77**(2), 123–128.

Konieczny, A., Voytas, D. F., Cummings, M. P., and Ausubel, F. M. (1991). A superfamily of *Arabidopsis thaliana* retrotransposons. *Genetics* **127**(4), 801–809.

Kramer, C., Loros, J. J., Dunlap, J. C., and Crosthwaite, S. K. (2003). Role for antisense RNA in regulating circadian clock function in *Neurospora crassa. Nature* **421**, 948–952.

Liu, X. S., Brutlag, D. L., and Liu, J. S. (2002). An algorithm for finding protein-DNA binding sites with applications to chromatin-immunoprecipitation microarray experiments. *Nat. Biotechnol.* **20**(8), 835–839.

Martin, A. C., and Drubin, D. G. (2003). Impact of genome-wide functional analyses on cell biology research. *Curr. Opin. Cell Biol.* **15**, 6–13.

McClung, C. R., Fox, B. A., and Dunlap, J. C. (1989). The *Neurospora* clock gene *frequency* shares a sequence element with the *Drosophila* clock gene *period. Nature* **339**, 558–562.

McGuire, A. M., Hughes, J. D., and Church, G. M. (2000). Conservation of DNA regulatory motifs and discovery of new motifs in microbial genomes. *Genome Res.* **10**(6), 744–757.

Metzenberg, R. L., and Grotelueschen, J. (1995). Restriction polymorphism maps of *Neurospora crassa*: Update. *Fungal Genet. Newslett.* **42**, 82–90.

Metzenberg, R. L., Stevens, J. N., Selker, E. U., and Morzycka-Wroblewska, E. (1985). A restriction fragment length polymorphism map for *Neurospora crassa. Proc. Natl. Acad. Sci. USA* **82**, 2067–2071.

Muller, H. M., Kenny, E. E., and Sternberg, P. W. (2004). Textpresso: An ontology-based information retrieval and extraction system for biological literature. *PLoS Biol.* **2**, e309.

Nelson, M. A., Kang, S., Braun, E. L., Crawford, M. E., Dolan, P. L., Leonard, P. M., Mitchell, J., Armijo, A. M., Bean, L., Blueyes, E., Cushing, T., Errett, A., *et al.* (1997). Expressed sequences from conidial, mycelial, and sexual stages of *Neurospora. Fungal Genet. Biol.* **21**, 348–363.

Ninomiya, Y., Suzuki, K., Ishii, C., and Inoue, H. (2004). Highly efficient gene replacements in *Neurospora* strains deficient for nonhomologous end-joining. *Proc. Natl. Acad. Sci. USA* **101**(33), 12248–12253.

Oldenburg, K. R., Vo, K. T., Michaelis, S., and Paddon, C. (1997). Recombination-mediated PCR-directed plasmid construction *in vivo* in yeast. *Nucleic Acids Res.* **25**, 451–452.

Ooi, S. L., Shoemaker, D. D., and Boeke, J. D. (2001). A DNA microarray-based genetic screen for nonhomologous end-joining mutants in *Saccharomyces cerevisiae. Science* **294**, 2552–2556.

Ooi, S. L., Pan, X., Peyser, B. D., Ye, P., Meluh, P. B., Yuan, D. S., Irizarry, R. A., Bader, J. S., Spencer, F. A., and Boeke, J. D. (2006). Global synthetic-lethality analysis and yeast functional profiling. *Trends Genet.* **22**(1), 56–63.

Pandit, N. N., and Russo, V. E. (1992). Reversible inactivation of a foreign gene, *hph*, during the asexual cycle in *Neurospora crassa* transformants. *Mol. Gen. Genet.* **234**, 412–422.

Perkins, D. D., Radford, A., and Sachs, M. S. (2001). "The Neurospora Compendium." Academic Press, San Diego.

Raymond, C. K., Powder, T. A., and Sexson, S. L. (1999). General method for plasmid construction using homologous recombination. *Biotechniques* **26,** 134–141.

Riquelme, M., Reynaga-Peña, C. G., Gierz, G., and Bartnicki-García, S. (1998). What determines growth direction in fungal hyphae? *Fungal Genet. Biol.* **24**(1–2), 101–109.

Roca, M. G., Arlt, J., Jeffree, C. E., and Read, N. D. (2005). Cell biology of conidial anastomosis tubes in *Neurospora crassa*. *Eukaryot. Cell* **4,** 911–919.

Ruepp, A., Zollner, A., Maier, D., Albermann, K., Hani, J., Mokrejs, M., Tetko, I., Guldener, U., Mannhaupt, G., Munsterkotter, M., and Mewes, H. W. (2004). The FunCat, a functional annotation scheme for systematic classification of proteins from whole genomes. *Nucleic Acids Res.* **32**(18), 5539–5545.

Sachs, M. S., and Yanofsky, C. (1991). Development expression of genes involved in conidiation and amino acid biosynthesis in *Neurospora crassa*. *Dev. Biol.* **148**(1), 117–128.

Schmidt, J. C., and Brody, S. (1976). Biochemical genetics of *Neurospora crassa* conidial germination. *Bacteriol. Rev.* **40,** 1–41.

Selker, E. U. (1990). Premeiotic instability of repeated sequences in *Neurospora crassa*. *Annu. Rev. Genet.* **24,** 579–613.

Selker, E. U. (1997). Epigenetic phenomena in filamentous fungi. *Trends Genet.* **13,** 296–301.

Tey, W. K., North, A. J., Reyes, J. L., Lu, Y. F., and Jedd, G. (2005). Polarized gene expression determines Woronin body formation at the leading edge of the fungal colony. *Mol. Biol. Cell* **16**(6), 2651–2659.

Townsend, J. P. (2003). Multifactorial experimental design and the transitivity of ratios with spotted DNA microarrays. *BMC Genomics* **4**(1), 41.

Townsend, J. P., and Hartl, D. L. (2002). Bayesian analysis of gene expression levels: Statistical quantification of relative mRNA level across multiple strains or treatments. *Genome Biol.* **3,** RESEARCH0071.

Vinciotti, V., Khanin, R., D'Alimonte, D., Liu, X., Cattini, N., Hotchkiss, G., Bucca, G., de Jesus, O., Rasaiyaah, J., Smith, C. P., Kellam, P., and Wit, E. (2005). An experimental evaluation of a loop versus a reference design for two-channel microarrays. *Bioinformatics* **21**(4), 492–501.

Wicks, S., Yeh, R., Gish, W., Waterson, R., and Plasterk, H. (2001). Rapid gene mapping in *C. elegans* using a high density polymorphism map. *Nat. Genet.* **28,** 160–164.

Winzeler, E. A., Shoemaker, D. D., Astromoff, A., Liang, H., Anderson, K., Andre, B., Bangham, R., Benito, R., Boeke, J. D., Bussey, H., Chu, A. M., Connelly, C., *et al.* (1999). Functional characterization of the *S. cerevisiae* genome by gene deletion and parallel analysis. *Science* **285** (5429), 901–906.

Wolfinger, R. D., Gibson, G., Wolfinger, E. D., Bennett, L., Hamadeh, H., Bushel, P., Afshari, C., and Paules, R. S. (2001). Assessing gene significance from cDNA microarray expression data via mixed models. *J. Comput. Biol.* **8**(6), 625–637.

Yang, Y. H., and Speed, T. (2002). Design issues for cDNA microarray experiments. *Nat. Rev. Genet.* **3,** 579–588.

Young, J. L., and Marzluf, G. A. (1991). Molecular comparison of the negative-acting nitrogen control gene, *nmr*, in *Neurospora crassa* and other *Neurospora* and fungal species. *Biochem. Genet.* **29**(9–10), 447–459.

Zeitlinger, J., Simon, I., Harbison, C. T., Hannett, N. M., Volkert, T. L., Fink, G. R., and Young, R. A. (2003). Program-specific distribution of a transcription factor dependent on partner transcription factor and MAPK signaling. *Cell* **113**(3), 395–404.

Zhu, H., Nowrousian, M., Kupfer, D., Colot, H., Berrocal-Tito, G., Lai, H., Bell-Pedersen, D., Roe, B., Loros, J. J., and Dunlap, J. C. (2001). Analysis of expressed sequence tags from two starvation, time-of-day-specific libraries of *Neurospora crassa* reveals novel clock-controlled genes. *Genetics* **157,** 1057–1065.

3 Genomics of the Plant Pathogenic Oomycete *Phytophthora*: Insights into Biology and Evolution

Howard S. Judelson
Department of Plant Pathology, Center for Plant Cell Biology
University of California, Riverside, California 92521

Advances in Genetics, Vol. 57
0065-2660/07 $35.00
DOI: 10.1016/S0065-2660(06)57003-8

ABSTRACT

The genus *Phytophthora* includes many destructive pathogens of plants. Although having "fungus-like" appearances, *Phytophthora* species reside in a eukaryotic kingdom separate from that of true fungi. Distinct strategies are therefore required to study and defend against *Phytophthora*. Large sequence databases have recently been developed for several species, and tools for functional genomics have been enhanced. This chapter will review current progress in understanding the genome and transcriptome of *Phytophthora*, and provide examples of how genomics resources are advancing molecular studies of pathogenesis, development, transcription, and evolution. A better understanding of these remarkable pathogens should lead to new approaches for managing their diseases.   © 2007, Elsevier Inc.

I. INTRODUCTION

Members of the genus *Phytophthora* cause devastating diseases on a wide range of plants throughout the world, and are arguably the most important pathogens of dicots (Erwin and Ribeiro, 1996). *Phytophthora* (Greek for "plant destroyer") is classified as an oomycete, along with other plant pathogenic genera such as *Pythium* and the downy mildews, animal pathogens including *Lagenidium* and *Saprolegnia,* and saprophytes such as *Achlya.* Taxonomists initially placed oomycetes in the Kingdom Fungi due to their typically filamentous growth habits. However, contemporary biochemical and molecular data demonstrate that oomycetes have little affinity with "true" fungi such as ascomycetes and basidiomycetes, and instead reside in a distinct branch of the eukaryotic tree (Baldauf *et al.*, 2000). The closest relatives of oomycetes are heterokont (brown) algae and diatoms, which belong to the eukaryotic kingdom *Stramenopila* (Fig. 3.1; Sogin and Silberman, 1998).

     The most notorious and earliest member of the genus to be identified is *P. infestans,* which was responsible for the Irish potato famine of the 1840s. The late blight diseases caused by *P. infestans* on potato and tomato continue to cost billions of dollars per year in terms of lost harvests and the expense of fungicides, which are generally necessary to sustain production (Fry and Goodwin, 1997). Significant losses are also caused by the 65 or so other *Phytophthora* species, such as *P. sojae* which causes soybean root rot, *P. palmivora* and *P. megakarya* which cause black pod of cacao, and *P. parasitica* which affects diverse herbaceous and deciduous hosts (Erwin and Ribeiro, 1996). Others have major deleterious impacts on native plant communities such as *P. cinnamomi* and *P. ramorum.* The latter is responsible for a relatively new disease termed sudden oak death, which is afflicting oaks along the western coast of the United States and may be

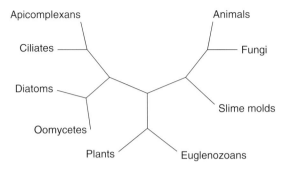

Figure 3.1. Schematic representation of phylogenetic relationships between major eukaryotic groups. Adapted from data of Arisue *et al.* (2002) and Baldauf *et al.* (2000).

capable of attacking other native and ornamental trees and shrubs throughout North America and Europe (Davidson *et al.*, 2003). Reflecting the taxonomic differences between oomycetes and true fungi, most fungicides developed for battling the latter are ineffective against *Phytophthora*. Consequently, there is a strong need to better understand *Phytophthora* biology, since this may lead to new strategies for protecting plants.

Although the importance of *Phytophthora* has been recognized for more than a century, its fundamental biology remained understudied for many years compared to that of many other eukaryotes, particularly true fungi. This was largely due to a historical dearth of tools for the classical genetic analysis of oomycetes. In contrast, true fungi such as *Neurospora* were found to be easily crossed and to yield useful auxotrophic or pigmentation markers, which enhanced the appeal of such species to early geneticists. The biology underlying why oomycetes were less-attractive model systems is now understood: oomycetes are diploid, so most mutations will be masked in the heterozygous state, and the types of genes that produce pigments in true fungi are lacking in oomycetes (Randall *et al.*, 2005; Shaw and Khaki, 1971). Despite such challenges, a nucleus of scientists interested in the fundamental biology of oomycetes persisted which has begun to grow substantially over the past 15 years. Much of their focus has been on species of *Phytophthora* with the highest agronomic impact, such as *P. infestans* and *P. sojae*, but others have also been the subject of notable genetic and molecular analyses. Downy mildews, despite their importance, have been less-attractive oomycete models since they are culturable only on a host plant, unlike *Phytophthora* which can complete its life cycle on plants or artificial media (Fig. 3.2). Noteworthy effort has nevertheless involved *Hyaloperonospora parasitica*, which causes downy mildew on the model plant *Arabidopsis thaliana* (Rehmany *et al.*, 2003).

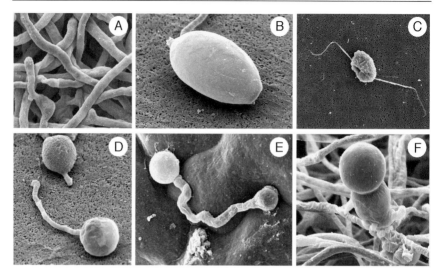

Figure 3.2. Major stages of the *Phytophthora* life cycle, illustrated using *P. infestans*. Shown are vegetative hyphae (A), an asexual sporangium (B), one of the biflagellated zoospores normally released from each asexual sporangium (C), germ tubes emerging from encysted zoospores (D), an appresorium produced on the surface of a host plant following germination of a zoospore cyst (E), and an oospore, including attached antheridium, produced from hyphae (F).

Despite its slow beginnings, *Phytophthora* research has entered an exciting phase in recent years. Tools for functional genomics have been developed for several species such as methods for transformation and gene silencing (Kamoun, 2003). Also, structural genomic studies have generated draft chromosome sequences and expressed sequence tag (EST) databases. This chapter summarizes those advances and illustrates how they are helping to address issues in the biology and evolution of *Phytophthora*. The emphasis will be on covering the progress and prospects in species for which significant genomics data exist rather than comprehensively reviewing all molecular and classical genetic studies within the genus.

II. ADVANCES IN STRUCTURAL GENOMICS

A. Current datasets

The evolution of more economical and high-throughput sequencing capabilities, and an increased interest in *Phytophthora*, facilitated the recent entry of several members of the genus into the genomics age. This initially involved sequencing clones from cDNA libraries to generate EST databases, followed by whole-genome sequencing efforts. To date, major sequencing projects have been executed for *P. infestans*, *P. ramorum*, and *P. sojae*, and a few other species have been examined to a lesser extent. Table 3.1 lists *Phytophthora* genomics resources currently in the public domain, either in GenBank or other accessible databases.

B. Expressed sequence tags

An EST approach was a natural first option for the community since *Phytophthora* genomes are relatively large and high in repetitive DNA. Haploid genome sizes are estimated by Feulgen absorbance photometry and DAPI micro-fluorescence to range from about 60 to 240 Mb (Tooley and Therrien, 1987; Voglmayr and Greilhuber, 1998). Although the majority of species have not yet been examined, *P. infestans* contains the largest known genome. In comparison, filamentous true fungi such as *Neurospora crassa* and *Magnaporthe grisea* have haploid genomes of about 40 Mb (Dean *et al.*, 2005; Galagan *et al.*, 2003).

Table 3.1. *Phytophthora* Genome Data in the Public Domain

Species	Disease	Genome size	Estimate of gene content	ESTs	Whole-genome sequencing
P. infestans	Late blight of potato and tomato	237 Mb	22K	94K	9× draft completed
P. ramorum	Sudden oak death, other diseases on woody plants	65 Mb	16K	–	7× draft completed
P. sojae	Root and stem rot of soybean	95 Mb	18K	41K	9× draft completed
P. parasitica	Foliar blights, fruit rots, stem and root rots	–[a]	–	3568	–
P. nicotianae	Foliar blights, fruit rots, stem and root rots	–[a]	–	755	–
P. capsici	Foliar blights, fruit rots, stem and root rots	–[a]	–	–	"2–30×" sequencing in progress[b]

[a] Probably 60–70 Kb.
[b] Plans are for 2× conventional WGS sequencing plus 30× bead-based picoliter-scale parallel sequencing (Margulies *et al.*, 2005).

Prior to the EST projects, only a handful of genes had been identified by conventional approaches, such housekeeping genes which were cloned by heterologous hybridization, infection and mating-specific genes obtained by subtraction or differential hybridization, and spore components isolated by immuno-screening (see Fabritius et al., 2002; Goernhardt et al., 2000; Pieterse et al., 1993; Unkles et al., 1991). Therefore, the EST projects dramatically increased the number of known genes and enabled many studies relevant to development and pathology. For example, while biochemical approaches showed that Phytophthora secretes an acidic and a basic form of a plant necrosis-inducing protein called elicitin, EST data revealed that they belong to a larger family of diverse secreted and membrane-associated proteins (Kamoun et al., 1999). Examples of other findings from the EST studies will be introduced later in this chapter, integrated with results from genome sequencing.

P. infestans was the focus of the first reported EST project in the genus, which sequenced 1000 cDNA clones from hyphae (Kamoun et al., 1999). This was expanded in later projects to 93,000 high-quality ESTs, which assembled into ~16,000 unigenes (Randall et al., 2005; Tani and Judelson, unpublished data). To maximize coverage of the transcriptome, cDNA from 20 distinct tissues were sampled. These included hyphae from rich, defined, and starvation media and exposed to environmental stress, mating cultures, asexual sporangia, zoospores, and germinated zoospore cysts (Fig. 3.2). To help reveal genes involved in plant colonization, other libraries were from hyphae exposed to plant extracts, infected leaves of potato and tomato, and colonized tubers. While <10% of cDNAs from infected tissues were typically derived from P. infestans, suppressive subtraction hybridization-PCR (SSH-PCR) could raise this to nearly 50% (Tani and Judelson, unpublished data). Bioinformatic approaches could also separate plant from pathogen sequences. As first shown in P. sojae-soybean interaction libraries, plant transcripts are about 46% G + C, while Phytophthora transcripts average 57% G + C (Qutob et al., 2000). Therefore, a 50% G + C threshold distinguished the origin of most transcripts. Once the list of potential pathogen transcripts was narrowed, plant transcripts that were atypically GC-rich were excluded by comparisons to plant databases.

Substantial ESTs also exist for P. sojae. An initial study identified 2189 unique transcripts by sequencing clones from axenically grown mycelia, zoospores, and infected soybean hypocotyls, with the G + C content approach used with the latter to distinguish pathogen from plant transcripts (Qutob et al., 2000). Interestingly, the fraction of pathogen transcripts in infected hypocotyls was exceptionally high, 60–70% of the total. This was exploited in a following study, which expanded the number of ESTs from infected soybean and also obtained more from zoospores and hyphae, including cultures grown on rich and nutrient-limited media. For the new infection libraries, plant sequences were removed by searching against a soybean database, and scoring for TA/CG dinucleotide and codon patterns typical of *Phytophthora*. This raised the number of high-quality *P. sojae* ESTs to 28,913, which assembled into 13,234 contigs.

Other species have also been subjected to the EST approach. For example, 755 ESTs representing up to 386 genes expressed in germinated cysts or zoospores were obtained from *P. nicotianae*. Despite the modest number of ESTs, this was sufficient to help identify stage-specific transcripts (Shan *et al.*, 2004b; Skalamera *et al.*, 2004). Using a closely related species, *P. parasitica*, 3568 ESTs corresponding to 2269 putatively unique genes were obtained from hyphae grown in defined media (Panabieres *et al.*, 2005). Defined media was employed since the prior *P. infestans* and *P. sojae* EST data suggested that many genes involved in plant interactions are expressed in nonrich broth. This is related to observations that nutrient limitation induces many infection-specific genes in both *Phytophthora* and non-oomycetes (Pieterse *et al.*, 1994a). The nutrient-limited cultures also yielded EST libraries with a relatively high complexity.

In the *P. infestans* EST projects, sequence complexity was noted to vary significantly between libraries. For example, nitrogen-starved, mating, zoospore, and germinating cyst libraries yielded a high proportion of new sequences even after many thousands of clones had been sequenced, greater than other stages such as hyphae from rich media or directly germinating sporangia (Randall *et al.*, 2005). Conversely, ESTs from heat-shocked and peroxide-treated hyphae had a low complexity. Such patterns likely reveal important clues about transcriptional programming in *Phytophthora*. Moreover, it was shown that a computational approach could predict stage-specific genes. For example, by comparing EST frequencies in hyphal, sporangial, and zoospore libraries, genes were identified which were later confirmed experimentally to be stage-specific (Ah Fong and Judelson, 2003; Judelson and Roberts, 2002).

C. Genome sequencing

Practical interest in sequencing *Phytophthora* genomes arose at a time when whole-genome shotgun (WGS) approaches for complex genomes were still somewhat controversial. It was unclear how effective WGS would be with *Phytophthora* genomes due to their relatively large size and high repeat content. Also unknown was the extent of heterozygosity of these diploids, which might confound assembly. The significance of such issues remains to be fully assessed, but the success of WGS with other eukaryotes made this the method of choice for *Phytophthora*.

Annotated draft genome sequences are now available for both *P. ramorum* and *P. sojae* (http://genome.jgi-psf.org). For *P. ramorum*, sevenfold coverage was generated and assembly data indicated that the size of the genome was about 65 Mb. As of November 2005, the assembly represented 54 Mb of DNA present on 2576 scaffolds containing 3831 gaps in which gene-prediction programs identified 15,743 genes. For *P. sojae*, ninefold coverage was achieved and the genome estimated at 95 Mb, which agrees well with data from Feulgen image analysis of nuclei (Voglmayr and Greilhuber, 1998). As of November 2005, the assembled *P. sojae* sequence exists as 78 Mb of DNA spread over 1810 scaffolds bearing 2298 gaps, with 19,027 predicted genes.

P. infestans has recently been sequenced to 9-fold coverage, and funding has been obtained for sequencing *P. capsici*. For *P. infestans*, substantial data already exists from several pilot projects, including a "genome skim" which achieved about onefold coverage (Randall *et al.*, 2005; Zody *et al.*, 2005). When the genome data was matched to the unigene set from *P. infestans* ESTs, 2330 new coding regions were identified, increasing the total number of predicted unigenes to 18,256.

The gene contents cited above should be taken as preliminary estimates. Continued refinement of assemblies and annotations will likely alter the values for *P. ramorum* and *P. sojae*, as has been the case in all genome projects. Also, no ESTs yet exist for *P. ramorum*, which may depress its current gene inventory since EST support can be critical for gene prediction. Also, as discussed later many genes appear to be unique to each *Phytophthora* species, so interspecific comparisons may be an incomplete tool for annotating their genomes. More genes will also be identified from *P. infestans* as it is sequenced to a greater depth; one estimate suggests that the total number will rise to 22,000 (Randall *et al.*, 2005).

Regardless of the absolute number of genes, it is clear that the gene content of *Phytophthora* is substantially higher than that of phytopathogenic true fungi. For example, the rice blast pathogen M. *grisea* and the cereal head blight agent *Fusarium graminearum* are reported to encode about 11,109 and 11,994 genes, respectively (Dean *et al.*, 2005; Güldener *et al.*, 2006). As noted later, a large portion of the "extra" genes in *Phytophthora* result both from novel genes and expanded families, and many are believed to participate in plant interactions such as those in the "crinkler" (CRN) group (Torto *et al.*, 2003). Even so, the number of genes in these species is not in proportion to their genome sizes, consistent with the presence of more noncoding DNA in *Phytophthora*.

III. ORGANIZATION OF *PHYTOPHTHORA* GENOMES

A. Genetic maps

Phytophthora species are diploid, with chromosome numbers generally ranging between 8 and 20 (Sansome and Brasier, 1973, 1974). Most chromosomes are large, and consequently pulsed-field gel analysis can only partially separate chromosomes of most species (Tooley and Carras, 1992). For species with the larger genomes such as *P. infestans*, attempts to separate chromosomes electrophoretically have so far been unsuccessful.

Detailed genetic maps exist for both *P. infestans* and *P. sojae*. The latest published maps contain 508 and 386 molecular markers, respectively, plus avirulence genes (May *et al.*, 2002; van der Lee *et al.*, 2004). Most loci reside on about 13 and 20 major linkage groups in the two species, respectively. This is slightly larger than the numbers of chromosomes visualized microscopically, possibly because some regions were devoid of polymorphic markers and could therefore not be joined genetically. Genetic sizes of *P. infestans* and *P. sojae* are estimated at about 1200 and 2590 cM, averaging 197 and 35.3 Kb/cM (May *et al.*, 2002; Van der Lee *et al.*, 1997). The ratio of physical to genetic distance is often reduced in gene-dense intervals. For example, the regions containing *Avr4* in *P. sojae* and the mating-type determinant of *P. infestans* are 3 and 4 cM/kb, respectively (Judelson, 1996b; Whisson *et al.*, 2004). However, such values vary between crosses, possibly due to chromosomal features that impair recombination such as inverted or hemizygous regions (Ah Fong and Judelson, 2004; Judelson, 1996a). Also, some regions may be prone to mitotic gene conversion (Chamnanpunt *et al.*, 2001), leading to differences in the calculated genetic distances.

Polymorphisms identified through the sequencing projects will facilitate the placement of further loci on the maps. For example, 200,000 SNPs were identified between haplotypes of *P. ramorum* or 0.3% of bases (Tyler *et al.*, 2005). An SNP rate of only 0.05% was predicted from *P. infestans*, however this still represents over 100,000 loci (Zody *et al.*, 2005). The mapping of such polymorphisms will help develop a detailed framework for finishing the *Phytophthora* genomes. Moreover, the maps have already been instrumental for cloning avirulence genes, the loci that trigger defense responses in a plant host that contains the matching resistance gene. For example, fine-structure mapping combined with walking in a bacterial artificial chromosome library led to the identification of *Avr1b* from *P. sojae* (Shan *et al.*, 2004a).

Due to limitations in the cytogenetic and genetic analysis of oomycetes, centromere locations are unknown. In other species such regions are gene-poor and repeat-rich, so integration of the genetic maps with other data from the genome projects may suggest the sites of centromeres. The structures of *Phytophthora* telomeres are poorly defined, but they cross-hybridize with the TTTAGGG telomeric DNA repeat of *A. thaliana*. Such telomeric probes detect exceptionally high rates of interisolate polymorphism in *P. infestans* (Pipe and Shaw, 1997).

In addition to high variation near telomeres, other aspects of the genome exhibit considerable plasticity. For example, polyploidy is common in certain isolates of some species such as *P. infestans* and *P. megasperma* (Sansome and Brasier, 1974; Tooley and Therrien, 1987). Segregation abnormalities appear common, resulting in distorted inheritance or strains that are aneuploid or tertiary trisomics (Forster and Coffey, 1990; Judelson et al., 1995; van der Lee et al., 2004). Translocations and deletions have been observed in the course of cytogenetic and genetic linkage studies, often associated with important pheno-types. For example, in the normally heterothallic species P. infestans, trisomy or duplications of the mating-type locus can generate self-fertile (homothallic) isolates (Judelson, 1996b; Mortimer et al., 1977). Such strains may therefore form oospores, thick-walled sexual structures that can carry over viable inoculum between growing seasons, in the absence of a mating partner. Another example of abnormalities concerns the generation of virulent strains. Some P. infestans isolates virulent against potato resistance genes R3, R10, and R11 appear to be hemizygous for the region that would otherwise contain the matching avirulence alleles (Van der Lee et al., 2001). Other researchers reported virulence changes in isolates which could also be possibly caused by deletion (Al-Kherb et al., 1995). Such events would resemble that seen for avirulence loci Avr9 and Pi-ta in the ascomycetes Cladosporium fulvum and M. grisea, respectively, where virulence also results from deletion (Orbach et al., 2000; Van Kan et al., 1991). However, such events are not the universal cause of virulence, since virulence in P. sojae against Rps1b-containing soybean was attributed to amino acid substitutions in Avr1b (Shan et al., 2004a).

B. Gene distribution and structure

Analyses of the draft genomes of P. ramorum and P. sojae indicate a tight packing of genes, with an average of one every 1.8 and 2.3 kb, respectively. A high density was also observed in selected BAC clones from P. infestans (Randall et al., 2003). Interestingly, these values are closer to that observed in a fungus with a compact genome, Saccharomyces cerevisiae (2.1 kb per gene) than those with larger genomes such as M. grisea (3.5 kb per gene; Dean et al., 2005). The adjacency of genes in Phytophthora is reflected in their tightly clustered occurrence along the scaffolds assembled to date. As illustrated in Fig. 3.3A for a typical stretch of DNA from P. sojae, intergenic regions are typically small and sometimes less than 300 bp. Regions with lower densities are frequently interspersed with transposon-like sequences (Fig. 3.3B; Jiang et al., 2005a). Overall, genes in the current scaffolds of the P. ramorum and P. sojae draft genomes span 47% and 43% of DNA. The lower density in P. sojae suggests that an increase in noncoding sequences has contributed more to its greater genome size than its larger complement of genes. Comparisons between regions of P. infestans, P. ramorum, and P. sojae chromosomes indicate substantial synteny, which may prove useful in finishing and annotating their genomes.

There is currently some disparity in the average gene sizes and intron contents determined for the different Phytophthora species. The majority of genes studied in the "pregenomics era" had few introns, leading to suggestions that

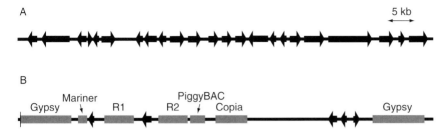

Figure 3.3. Typical distributions of genes along *Phytophthora* chromosomes. (A) High-density cluster of predicted genes (arrows) along a 64-kb region from *P. sojae*. (B) The 64-kb portion of a *P. infestans* chromosome with a lower density of genes; adapted from Jiang et al. (2005a). Gray rectangles indicate sequences representing various Class I and Class II transposable elements.

only 10–20% might bear more than one exon (Judelson, 1997). A study performed after the *P. infestans* EST database and partial genome sequences became available found a total of 19 introns in 68 genes, or 28% of genes (Win et al., 2006). Full-length EST-derived unigenes average about 1.2 kb and intron sizes average 84 bp, so a mean transcription unit size of 1.2–1.3 kb can be predicted for *P. infestans*. In contrast, current annotations of the *P. ramorum* and *P. sojae* genomes register averages of 1.58 and 1.81 introns per gene and mean gene sizes of 1.43 and 1.53 kb. Some of the differences between *P. infestans* and the other species could reflect bias in the nature of genes identified through the EST approach. The gene-finding programs so far applied to the draft genomes of *P. ramorum* and *P. sojae* may also overestimate the number and size of introns, a situation that will improve as better training sets for those programs are established. Regardless, it is interesting that *Phytophthora* genes are typically larger than those predicted from their closest sequenced relative, the diatom *Thalassiosira pseudonana*, which average 992 bp (Armbrust et al., 2004).

  Phytophthora introns contain the same GT/AG termini as do other eukaryotes, although most lack the branch point sequence (CTRAC) involved in lariat formation (Win et al., 2006). Sequences flanking the splice junctions are variable, with a consensus from 55 introns calculated as $A_{48}G_{64}/G_{100}$ $T_{100}A_{66}A_{44}G_{86}T_{60}$ and $C_{58}A_{100}G_{100}/G_{46}$ for the 5′ donor and 3′ acceptor sites, respectively (Cvitanich and Judelson, 2003b). Their utility for predicting introns and thus improving computational gene predictions is therefore uncertain. Interestingly, the introns of *P. infestans* genes lack high similarity to those of their *P. ramorum* and *P. sojae* homologues (Win et al., 2006). In *P. infestans*, the G + C content of introns is about 45%, lower than the 57% value of coding regions and the 54% of 5′ UTRs. The 3′ UTRs are 45% G + C, have a mean size of about 110 nt, and generally contain AAUAAA or AAUGAA polyadenylation signals.

Codon usage tables developed from *P. infestans* ESTs indicate a moderately biased pattern of codon choice, largely due to a preference for guanine or cytosine in the third position (Randall *et al.*, 2005; Win *et al.*, 2006). However, this is not the only contributor to codon bias. For example, GGC was employed eight times more than GGG to encode glycine. This trend is also seen in other G + C-rich organisms, possible due to selection against G-homotetramers which may exert detrimental effects on mRNA stability (De Amicis and Marchetti, 2000). Sequences flanking the start codon showed modest but significant conservation, with the consensus (ACC<u>ATG</u>A) matching the bias observed in other eukaryotes (Win *et al.*, 2006).

C. Overview of gene content

Roughly 90% of *Phytophthora* genes have similarity with proteins in public databases, with the rest (at least 1500) appearing unique to *Phytophthora*. Frequencies of database matches reported previously for the EST projects were often lower due to the short lengths of many of the unigenes in those analyses. As in most other eukaryotes, proteins involved in metabolism and signal transduction are generally most common. Within the 15,743 predicted *P. ramorum* proteins, for example, about 22,000 matches are detected against the InterPro database of protein domains, with the top 10 domains including protein kinases, FYVE zinc fingers, Ankyrin, AAA ATPase, Integrase, CCHC zinc fingers, and the G-protein beta (Gβ) WD-40 repeat. Many of these were also in the top matches detected in *P. infestans* (Randall *et al.*, 2005). Interesting differences with non-oomycetes were noted in several categories (Table 3.2). For example, like plants but unlike the other species, Myb-like transcription factors appear most common in *Phytophthora*. Conversely, homeobox transcription factors appear less prevalent in *Phytophthora* than plants. Compared to *M. grisea*, *Phytophthora* appears to have more ABC transporters. Such proteins are expected to be common in *Phytophthora* due to their roles in eliminating host defense molecules and environmental toxicants, but such tasks are also required by *M. grisea*. Several chemical stresses were shown to upregulate the transcription of ABC transporters in *P. infestans*, although none are yet proved to be needed for host colonization (Judelson and Senthil, 2006).

A notable contrast between *Phytophthora* and the other species in Table 3.2 concerns domains related to retrotransposons. Integrase and reverse transcriptase were in the top 15 domains in the *P. ramorum* proteome, substantially higher than in the other species. As noted below, this is consistent with the presence of numerous *gypsy-* and *copia*-like sequences in *Phytophthora*; many more are also present within the genome, but undetected by the gene-calling algorithm due to

Table 3.2. Genome-Wide Comparisons of Selected InterPro Domains

InterPro domain	P. ramorum No.	P. ramorum Rank	M. grisea No.	M. grisea Rank	A. thaliana No.	A. thaliana Rank	H. sapiens No.	H. sapiens Rank
Retroelement related								
Integrase, catalytic region	218	8	77	28	146	52	41	>200
Retrotransposon gag protein	88	25	29	76	100	91	4	>200
Reverse transcriptase	171	14	108	11	110	80	2	>200
Selected protein kinase domains								
Serine/threonine protein kinase	381	2	99	14	318	20	424	10
Tyrosine protein kinase	360	3	77	27	111	79	332	25
Transporters								
ABC transporter	136	13	46	38	160	48	81	129
Major facilitator superfamily	104	21	175	2	124	67	131	72
Major eukaryotic transcription factor families								
bZIP	23	141	21	93	98	90	19	133
Forkhead	0	>200	2	>200	0	0	58	183
Fungal-specific TF	0	>200	33	61	1	>200	0	>200
Helix-loop-helix	7	>200	11	178	202	36	151	65
Homeobox	9	>200	6	>200	120	76	387	16
HSF/ETS	14	>200	3	>200	7	>200	75	144
Myb, DNA-binding	95	24	35	62	400	12	62	173
Zinc finger, C2H2 type	67	33	95	15	221	30	1089	1

Indicated are the total number of proteins with the indicated domains and the ranking compared to all Interpro domains. *P. ramorum* data are summarized from analyses of predicted proteins in version 1.0 of the *P. ramorum* assembly (http://genome.jgi-psf.org), except for ABC transporters which are based on gene models developed by P. Morris. Data from other species are from the Munich Information Center for Protein Sequences (MIPS; http://mips.gsf.de) or the European Bioinformatics Institute (EMBL; http://www.ebi.ac.uk).

degeneracy. Such domains were even more common in *P. infestans*, ranking in the top five (Randall *et al.*, 2005). Their enhanced prevalence in *P. infestans* compared to *P. ramorum* is commensurate with the fourfold larger genome of *P. infestans* (237 vs 65 Mb), but this is a shaky comparison since their gene catalogs were developed differently. Analyzing *P. ramorum* versus *P. sojae* should be valid, however, as their databases were generated by the same pipeline. Both integrase and reverse transcriptase domains were higher in *P. sojae*: 218 and 171 in *P. ramorum*, respectively, compared to 574 and 430 for *P. sojae*. In contrast, matches to other domains are roughly the same between *P. ramorum* and *P. sojae* (506 and 517 total protein kinases, respectively; 195 and 198 Gβ WD-40; 127 and 134 pleckstrin-like, and so on). This is consistent with a model in which growth of the basal *Phytophthora* genome to 95 Mb in *P. sojae* involved transposon activity or the expansion of retroelement arrays by unequal crossing-over.

Expansions of families have been a common occurence in *Phytophthora*. Members of some families are dispersed in the genome, but many reside in clusters. The mating-induced M96 family of *P. infestans*, for example, contains 21 genes in a single array of near-identical tandem repeats (Fig. 3.4A; Cvitanich *et al.*, 2006). Genes encoding ABC transporters (Fig. 3.4B) and elicitins (Fig. 3.4C) are typically also clustered. Many of the clustered genes, including some elicitins, belong to families of secreted proteins that are associated with plant pathogenesis.

Overall, secreted proteins represent a significant proportion of the proteome, with 1256 and 1570 predicted in *P. ramorum* and *P. sojae*, respectively (Jiang *et al.*, 2005b). These values are similar to that of the phytopathogenic fungus *M. grisea*, 1258, but more than the 642 found in the saprophyte *N. crassa*. This may indicate a shared requirement in oomycete and fungal pathogens for proteins used to colonize plants such as degradative enzymes, proteins for adhering appressoria to plant surfaces, or effectors of plant defense responses.

P. infestans, *P. ramorum*, and *P. sojae* each contain many species-specific genes, that is, genes lacking homologues in the others based on a BLASTN cutoff E value of 10^{-5}. For example, when genes from *P. ramorum* and *P. sojae* were compared, the former appeared to have 625 unique genes and the latter 1755 (Tyler *et al.*, 2005). Also, of 97 mating-induced genes from *P. infestans* 23 lacked homologues in *P. sojae* (Prakob, 2005). Of 24 genes misregulated in an appressorium-minus mutant of *P. infestans*, about half lack homologues in *P. ramorum* and *P. sojae*

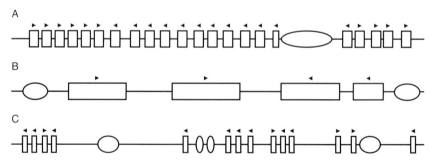

Figure 3.4. Examples of clustered gene families in *Phytophthora*. (A) The M96 mating-specific family of *P. infestans*, which is distributed over a 52-kb region of a single chromosome. M96 ORFs are indicated by rectangles, with orientations indicated by arrowheads. Ovals represent unrelated genes, which in this case are *gypsy*-like elements; adapted from Cvitanich *et al.* (2006). (B) Genes encoding four ABC transporters along a 25-kb region of a *P. ramorum* chromosome (C) Cluster of elicitin genes on a 60-kb region of a *P. ramorum* chromosome; adapted from Jiang *et al.* (2005b).

(Blanco and Judelson, unpublished data). The varied number of genes in the species is therefore not simply the result of the differential expansion of gene families, but instead reflects the presence of novel or rapidly evolving genes. To what extent this involves divergence in the biology of the species is unknown; however, variation in gene content may reflect differences in the repertoire of proteins needed for specific developmental or pathogenic events. For example, while *P. infestans* and *P. ramorum* are heterothallic, *P. sojae* is homothallic, and while *P. infestans* generally infects hosts using appressoria that mode of infection is not believed to be common in *P. sojae* (Hardham, 2001). Some of the nonconserved genes may also encode effectors targeted to specific host plants.

    For proteins with homologues in all three species, the extent of amino acid conservation varies depending on the type of protein. For example, for 10 metabolic enzymes examined, the *P. infestans* proteins averaged 94% and 93% amino acid identity with homologues from *P. ramorum* and *P. sojae*, respectively (Cvitanich *et al.*, 2006). Of 10 *P. infestans* genes induced during asexual sporulation, 81% and 80% conservation was observed, and 68% and 72% was measured for genes induced during sexual development. Another analysis indicated that secreted proteins from *P. infestans* average 67% identity to *P. ramorum* and *P. sojae* homologues (Win *et al.*, 2006).

D. Noncoding and repetitive DNA

Stretches of *Phytophthora* chromosomes lacking significant numbers of protein-coding genes contain a mixture of single-copy and repeated DNA with either no notable structure, simple sequence repeats, or sequences related to transposons. Such regions appear as gene-poor islands in the existing scaffolds of *P. ramorum* and *P. sojae*. The ~15% of their genomes that are not yet assembled into a scaffold are likely to be especially rich in repeated DNA.

    That the genomes would contain many repetitive sequences was revealed by work in the pregenomic era. DNA-reassociation and reverse-Southern hybridization studies indicated that repeats comprised 50% or more of *P. infestans* and *P. sojae* chromosomes (Judelson and Randall, 1998; Mao and Tyler, 1991). In *P. infestans*, repeats representing 51% of the genome were classified into 33 distinct families ranging from 70 to 10,000 copies per haploid genome (Judelson and Randall, 1998). The elements ranged in size from a few hundred bases to 10.2 kb, and were dispersed or tandemly repeated. Other studies identified moderately repeated elements such as RG57 that due to high polymorphism is useful for population studies (Goodwin *et al.*, 1995). Some repeats are specific to *P. infestans* and close relatives, while others are widely distributed (Judelson and Randall, 1998). Species-specific repeats were also identified in *P. cryptogea* (Panabieres and Le Berre, 1999). By measuring copy numbers of the *P. infestans*-like repeats in 26 species, it was concluded that differential expansion or contraction of the families was a major feature of genome evolution in *Phytophthora* (Judelson and Randall, 1998).

Also present are simple sequence repeats (microsatellites), a common feature of eukaryotic genomes. In *P. ramorum* and *P. sojae*, about 1000 and 2128 microsatellites having 2- to 6-nt repeat units were detected (Tyler *et al.*, 2005). A scan of the partial *P. infestans* genome data revealed 333 microsatellite loci (van der Lee *et al.*, 2005). Since microsatellite regions are typically highly mutable, these represent a source of useful markers for linkage and population studies.

E. Transposable elements

1. Class I elements

Much of the genome is comprised of Class I elements, which replicate using RNA intermediates. Members of the Tc3/*gypsy* class of long terminal repeat (LTR) retroelements are most common, and data support ancient and more recent events of their horizontal transfer into *Phytophthora*. *P. infestans* contains at least five distinct families with sizes of about 6–13 kb (Jiang *et al.*, 2005a; Judelson, 2002). Some are found throughout the genus, while others are narrowly distributed. For example, GypsyPi-1, GypsyPi-3, and GypsyPi-5 of *P. infestans* have homologues in *P. ramorum* and *P. sojae*, while GypsyPi-2 is absent in the latter two species (Jiang *et al.*, 2005a). *Gypsy*-like family PiRETA, which has 30 copies per haploid *P. infestans* genome, was detected in 29 of 37 species (Judelson, 2002). Sequence relationships of PiRETA in the 29 species parallel phylogenies based on other loci such as rRNA, suggesting that the retroelement was introduced into a progenitor of *Phytophthora*. However, PiRETA-like sequences are absent from some species with strong affinity to *P. infestans*, suggesting that they were expunged through processes such as unequal crossing-over or diverged rapidly. Within species containing PiRETA, copy numbers range from 10 to 11,000. The most are in *P. iranica* and *P. cactorum*, where the PiRETA family may represent up to a quarter of the genome. Interestingly, most *gypsy*-like elements in *Phytophthora* have plant sequences as their closest relatives, raising the possibility that they were acquired by horizontal transfer from a host.

Other retroelement relatives within *Phytophthora* include the Tc1/*copia* family, LINEs (long interspersed sequences), and SINES (short interspersed sequences). Of four *copia*-like families detected in *P. infestans*, two have homologues in *P. ramorum* and *P. sojae*, one has a homologue in *P. sojae* but not *P. ramorum*, and one is absent from both of the two latter species (Jiang *et al*, 2005a; Tooley and Garfinkel, 1996). A LINE from *P. infestans* was shown to lack relatives in the *P. sojae* and *P. ramorum* databases (Jiang *et al.*, 2005a). Fifteen types of SINES were also found in *P. infestans*, ranging in number from 1 to 2000 per genome (Whisson *et al.*, 2005a). Relatives of about half the SINE families are in *P. ramorum* and *P. sojae*.

There is no evidence for the recent movement of any Class I element in a *Phytophthora* genome. DNA blot data reveals some instances of polymorphism, but there is no proof that this is due to transposition. Most copies are degenerate due to frameshifts or deletions in the polyprotein gene or terminal repeats. However, a few elements appear intact, ESTs resembling *gypsy*- and *copia*-like elements have been detected, and transposition could be an explanation for the phenotypic plasticity reported for many isolates (Caten and Jinks, 1968). Nevertheless, such elements have likely played roles in restructuring *Phytophthora* genomes, and transposon invasions may be linked to evolutionary bursts within the genus.

2. Class II elements

These move through a "cut-and-paste" process involving a transposase that acts on flanking terminal inverted repeats (TIRs). DNA transposons have been classified into nine superfamilies in eukaryotes called TC1/*mariner*, piggyBac, hAT, merlin, mutator, pogo, CACTA, P, PIF/Harbinger, and transib (reviewed in Feschotte, 2004). To date, elements belonging to the first five classes have been identified in *Phytophthora*. Copy numbers are typically lower than that of retroelement-like sequences, with each having about 5–100 copies per genome.

The most diversity has been identified in the TC1/*mariner* family. In *P. infestans*, at least two distinct types have been detected, both ~1500 bp in size and containing a few copies per genome (Jiang *et al.*, 2005a). At least one appears to be transcribed, but no evidence for movement was reported. Many copies appear degenerate, containing frameshifts in the transposase and/or lacking TIRs. No obvious homologues appear in *P. sojae* or *P. ramorum*, which interestingly contain 44 and 34 other *mariner*-like elements, respectively. As noted for the *gypsy*-like elements, the *mariner*-like sequences in *Phytophthora* have plant transposons as their closest relatives, raising the possibility of horizontal transfer.

Diversity between the *Phytophthora* species also exists in hAT and mutator-like sequences. *PiDodo*, a 2722-bp hAT element containing 12-bp TIRs, has five copies in *P. infestans* (Ah Fong and Judelson, 2004). However, the *P. ramorum* and *P. sojae* databases lack relatives. Like the *mariner*-like elements, three *PiDodo* elements bear deletions that would render them inviable. Two appear intact, but extensive tests failed to detect transcription and only minor polymorphism exists between isolates. Two unrelated *mutator*-like elements are also in *P. infestans*, sized at 6199 and 4702 bp with 220- and 26-bp TIRs (Ah Fong and Judelson, 2004). They also lack obvious relatives in *P. ramorum* and *P. sojae*. None of the elements also have significant affinity at the DNA level with sequences in the diatom *T. pseudonana*, indicating that they are a recent acquisition by *Phytophthora*.

Like the Class I elements, there are no data to indicate that any Class II element is currently mobile. However, some data suggest recent activity, since one *PiDodo* element appears to have inserted within a gene (Ah Fong and Judelson, 2004). Moreover, although most members of the PCIS family of *P. cryptogea* appear to be degenerate, many have maintained TIRs and are flanked by duplications of host DNA, which suggests that they were mobilized by a transposase provided *in trans* (Panabieres and Le Berre, 1999). Interestingly, Class I and Class II elements frequently reside near each other in the genome (Ah Fong and Judelson, 2004; Jiang *et al.*, 2005a), in apparent hot spots for transposon insertions.

IV. OTHER GENETIC ELEMENTS

While the focus of *Phytophthora* genomics has largely concerned nuclear genes, other genetic elements exist including mitochondrial DNA and parasitic or viruslike entities. Although the latter remain poorly characterized, they have the potential to influence the biology of *Phytophthora* and provide insight into its evolution. Understanding mitochondrial genomes is also of interest since such molecules have been a traditional source of markers for population studies (Gavino and Fry, 2002; Ristaino *et al.*, 2001), and since some crop protection chemicals target mitochondrial components (Bartlett *et al.*, 2002).

Mitochondrial genomes of *P. infestans*, *P. ramorum*, and *P. sojae* have each been sequenced. All are circular with sizes ranging from 37,922 bp for some strains of *P. infestans* to 42,975 bp for *P. sojae* (Martin, 2005). As in other taxa, the genome is dense in coding sequences, covering more than 90% of the chromosome, and has a low $G + C$ content of about 22%. A total of 37 protein-coding genes and 19 tRNA genes are conserved between the species, but open reading frames (ORFs) having unknown functions are also present. The number of these ORFs varies, for example *P. sojae* has five ORFs not in *P. ramorum*. Gene orders are generally the same between *P. sojae* and *P. ramorum*, but in *P. infestans* orders are inverted at two sites. Substantial intraspecific variation also exists. In *P. infestans*, four mitochondrial haplotypes are described. Three were sequenced, revealing size polymorphisms (37,922, 39,870, and 39,840 bp for type Ia, IIa, and IIb haplotypes, respectively) and variation in the novel ORFs (Avila-Adame *et al.*, 2006). None of the polymorphisms is associated with a phenotype, but they have served as useful markers for tracing the worldwide movement of isolates and connecting contemporary strains with those in the historical record (Gavino and Fry, 2002; Ristaino *et al.*, 2001).

Viruslike particles, double-stranded RNA (dsRNA), and poorly characterized RNAs of nonchromosomal origin have been identified in several species including *P. infestans* and *P. drechsleri* (Corbett and Styer, 1976; Judelson and Fabritius, unpublished data; Newhouse *et al.*, 1992; Shaw, 1983a). One of the dsRNAs, sized at 13.9 kb, is related to plant endornaviruses (Hacker *et al.*, 2005). None are known to affect the biology of *Phytophthora*, as is usually the case with dsRNA in fungi (Buck, 1986). Most may be evidence of ancient, degenerate parasites.

One intriguing entity discovered in *P. infestans* is an unusual RNA element found in a minority of isolates (Judelson and Fabritius, 2000). It exists predominantly as a 625-nt single-stranded RNA lacking obvious ORFs. *P. infestans* DNA does not contain cross-hybridizing sequences, and the RNA appears unencapsidated, so it was described as an autonomously replicating linear RNA plasmid. Complementarity between its 5′ and 3′ termini suggested an ability to circularize, and the RNA resides mostly in the nucleus, making the element reminiscent of plant viroids. The RNA has little effect on growth or pathogenicity, but does silence selected host genes, possibly through an RNAi-like mechanism (Judelson and Fabritius, unpublished data).

Searches for DNA plasmids have so far only identified a mitochondrial plasmid. This ∼3.8-kb circular element was found in a strain of *P. meadii*, where it is unconnected to any discernable phenotype. Attempts to exploit the molecule to develop an autonomously replicating transformation plasmid were unsuccessful (H. Förster and H. S. Judelson, unpublished data).

V. TOOLS FOR FUNCTIONAL GENOMICS

A. Transformation systems for *Phytophthora*

DNA sequencing has identified most genes within several species of *Phytophthora*, but connecting these to biological functions requires experimentation. Critical for such studies is a reliable transformation system. Such a method was first developed for *P. infestans* in which plasmids expressing selectable marker genes under the control of oomycete promoters were introduced into protoplasts using polyethylene glycol–calcium chloride treatment (Judelson *et al.*, 1991). Transformation of *P. infestans* can also be achieved by microprojectile bombardment of young tissue (Cvitanich and Judelson, 2003a), electroporation of zoospores (Latijnhouwers *et al.*, 2004), or transferring DNA using *Agrobacterium* (Vijn and Govers, 2003). A talented individual can generate hundreds of transformants in a week, but rates are typically more modest. Although most experiments involving transformation have involved *P. infestans*, transformation has also been reported for *P. palmivora*, *P. nicotianae*, and *P. sojae* (Gaulin *et al.*, 2002; Judelson *et al.*, 1993a; van West *et al.*, 1999b). Selection for G418 or hygromycin-resistance is usually preferred, depending on the species. In *P. infestans*, selection is possible using both of these as well as genes for resistance to streptomycin, which enables sequential rounds of transformation to be performed. Which method for introducing DNA is best for a given application is not yet known, but the nature of transgene integration does vary between approaches. *Agrobacterium* transformation generates single integrations, while the others usually insert multiple copies of transgenes into chromosomes, often as tandem arrays (Judelson, 1993; Vijn and Govers, 2003). Bombardment typically yields heterokaryons (Cvitanich and Judelson, 2003a).

A method for gene targeting is not yet demonstrated due to a low rate of homologous integration. Neither insertion site nor copy number can therefore be controlled. Combined with the complications of diploidy, gene inactivation by disruption is currently not feasible. This may change, however, if genes responsible for nonhomologous recombination can be identified from the genome databases and silenced. The inability to target genes also means that position effects are strong. When β-glucuronidase (GUS) or green fluorescent protein (GFP) reporters are placed behind a strong promoter, frequently only 25–75% of transformants express those at visible levels (Judelson, 1993). Many transformants also stop expressing transgenes during continued culture due to epigenetic silencing. This was demonstrated in *P. infestans*, where silenced transgenes were cloned and found to be active on retransformation (Judelson *et al.*, 1993b).

Gene silencing based on homology-dependent or RNAi strategies has proved useful, albeit not yet routine and inadequately understood at the molecular level. Silencing has been achieved by introducing sense, antisense, or hairpin constructs. This was demonstrated initially for the GUS reporter (Judelson *et al.*, 1993b), and then for native *Phytophthora* genes. The first case of the latter involved the *inf1* elicitin gene from *P. infestans* (van West *et al.*, 1999a). Silencing was shown to occur at the level of transcription and to involve a signal that could be passed between nuclei. Other studies reported the silencing of genes encoding Gα and Gβ subunits, a *Cdc14* mitotic regulator, and a bZIP transcription factor in *P. infestans*, and the CBEL cellulose-binding elicitor protein in *P. nicotianae* (Ah Fong and Judelson, 2003; Blanco and Judelson, 2005; Gaulin *et al.*, 2002; Latijnhouwers and Govers, 2003; Latijnhouwers *et al.*, 2004). Silencing frequencies in these studies ranged from about 3% to 75%, although failures to silence certain genes are also reported (Judelson, 2003). There is currently no consensus for the best approach for silencing, although one study reported that the protoplast method for transformation induced silencing more efficiently than electroporation (Blanco and Judelson, 2005).

Since many genes exist in families, silencing a single gene may not yield informative phenotypes due to functional redundancy. Therefore, it is useful to ascertain if the expression of a family can be silenced in concert. This was tested against the three-member *PiNifC* family of transcription regulators of *P. infestans* (Tani *et al.*, 2004a). Although the genes diverged in DNA sequence by 5–23%, a *PiNifC1* transgene silenced the other two members of the family. Silencing of each gene was associated with the establishment of more tightly-packed chromatin. (Tani and Judelson, unpublished data).

Recently, a transient silencing system which has the potential to accelerate functional studies was reported. By treating protoplasts of *P. infestans* with dsRNA matching a target gene, downregulation of the target was observed in some of the regenerated hyphae (Whisson *et al.*, 2005b). Silencing was only detectable 12 to 17 days after dsRNA treatment, but this allowed time for small colonies to form in which phenotypes could be assessed. Silencing was slower to develop than in similar experiments involving plants and animals. This was proposed to reflect a slower accumulation in *Phytophthora* of the characteristic molecules of RNAi, siRNAs, although whether silencing actually involves siRNAs was not determined.

Gain-of-function experiments are not yet widely described. In one of the few examples, a gene for a basic elicitin from *P. cryptogea* was transformed into *P. infestans*. The gene was expressed and it altered the interaction of *P. infestans* with tobacco (Panabieres *et al.*, 1998). Another experiment showed that a constitutively active Gα subunit could be expressed in *P. infestans*, although this engendered no phenotypic alterations (Latijnhouwers *et al.*, 2004).

B. Heterologous systems for functional genomics

Functional studies of *Phytophthora* genes in organisms from other kingdoms are possible or even preferable in certain cases. This is particularly true when another species offers useful marked strains or higher transformation rates. For example, the availability of appropriate mutants in the yeast *S. cerevisiae* enabled complementation studies that proved that the *PiCdc14* phosphatase from *P. infestans* is a mitotic regulator (Ah Fong and Judelson, 2003). Similarly, the *P. infestans* ras-like gene *Piypt* was confirmed to participate in vesicle transport (Chen and Roxby, 1997). *S. cerevisiae* was also used to test the function of an ABC transporter from *P. sojae* in toxicant resistance (Connolly *et al.*, 2005).

Agroinfiltration and agroinfection approaches have proved useful for testing *Phytophthora* genes *in planta*, particularly those believed to be inducers or suppressors of host defenses (Huitema *et al.*, 2004). In agroinfiltration, a gene of interest in a plant expression cassette is placed in a T-DNA vector and transformed into *Agrobacterium tumefaciens*, which is infiltrated into leaf tissue. This transiently expresses the *Phytophthora* gene in plant cells, which are monitored for responses (Fig. 3.5A). Agroinfection involves cloning a candidate gene within the genome of *Potato virus* X (PVX), which is expressed from an *A. tumefaciens* T-DNA (Fig. 3.5B). Large numbers of *A. tumefaciens* strains can

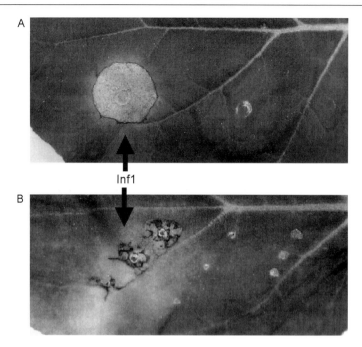

Figure 3.5. Agroinfiltration and agroinfection. (A) Portion of leaf from *N. benthamiana* treated with *A. tumefaciens* expressing proteins from *P. infestans*. The necrotic zone indicates the site of expression of elicitin Inf1. (B) Symptoms on *N. benthamiana* after inoculation with *A. tumefaciens* containing a *Potato virus X* (PVX) vector expressing *P. infestans* genes, also showing necrosis due to Inf1 expression. Images courtesy of C. Young and S. Kamoun.

be toothpick-inoculated into leaf tissue; while only a few leaf cells are initially infected, the PVX chimera can spread into surrounding plant cells where responses may be observed. The latter is particularly suited for high-throughput functional assays of genes from the *Phytophthora* genomics projects.

VI. SELECTED AREAS OF *PHYTOPHTHORA* RESEARCH

A. Plant–*Phytophthora* interactions

Understanding how *Phytophthora* species interact with their hosts is a very active area of investigation as this may lead to novel strategies for protecting crops from disease. Prior to genomics, a number of genes potentially involved in plant colonization were identified through strategies such as subtraction cloning and cDNA-AFLP (Avrova *et al.*, 2004; Beyer *et al.*, 2001; Goernhardt *et al.*, 2000;

Pieterse *et al.*, 1994a). More recently, the sequence databases have been found to be a rich source of factors that may also play such roles. These include proteins that may break down plant cell walls, mobilize nutrients from the plant, induce or suppress host defense responses, or allow *Phytophthora* propagules to adhere to plant surfaces.

Numerous enzymes that are presumably involved in degrading the walls of potential plant hosts have been identified through molecular cloning or database mining. These include extracellular forms of cellulases, cutinases, pectate lyases, and polygalacturonases (Randall *et al.*, 2005), many of which exist as gene families. For example, more than 20 predicted pectate lyases are in the *P. ramorum* database and many were also in the *P. infestans* EST database. At least 17 polygalacturonases are also made (Gotesson *et al.*, 2002; Torto *et al.*, 2002).

A variety of *Phytophthora* proteins that induce defense responses in plants, such as programmed cell death, have been identified through database mining, functional genomics, and biochemistry. Classical names for such proteins have been "elicitors," or "avirulence proteins" in the case of determinants of host or cultivar-specificity. However, a more appropriate term may be "effectors," since to what extent the induction of host defenses is deleterious to the pathogen is controversial (van Dijk *et al.*, 1999). While defense responses induced early in an interaction may hinder colonization, dying plant cells may also release nutrients to *Phytophthora*.

One of the first proteinaceous elicitors known from *Phytophthora* is a 42-kDa extracellular glycoprotein, which was purified biochemically based on the ability of *P. sojae* to cause necrosis in a nonhost, parsley (Sacks *et al.*, 1995). Once the *P. infestans* EST database was developed, it became evident that the elicitor belonged to a multigene family in which different members were expressed during hyphal growth, zoosporogenesis, or oosporogenesis (Fabritius and Judelson, 2003). The proteins exhibit transglutaminase activity (Brunner *et al.*, 2002), so their normal roles may be to strengthen the various types of walls seen in the different developmental stages of *Phytophthora*. The same part of the protein, Pep-13, is critical for both elicitor and transglutaminase activity. Related proteins are not found outside of oomycetes, and its sequence is distinct from transglutaminases in other kingdoms. Plants appear to have evolved to recognize the *Phytophthora* transglutaminase, according to the pathogen-associated molecular pattern (PAMP) paradigm in which plants and animals develop abilities to recognize conserved features of disease agents (Nürnberger and Brunner, 2002).

Another elicitor is the CBEL protein, a lectin-like cellulose-binding factor first isolated from *P. nicotianae*. The 34-kDa cell wall protein was identified based on its elicitation of defense genes and necrosis in tobacco, a host of *P. nicotianae* (Séjalon-Delmas *et al.*, 1997). Gene silencing studies later indicated that CBEL may help the pathogen bind to plant surfaces, although the silenced strains retained their ability to infect tobacco (Gaulin et al., 2002). Surface attachment probably occurs through a PAN domain, which functions in protein-protein or protein-carbohydrate interactions. Other proteins with PAN modules can be found both in the sequence databases of the other Phytophthora species and of the animal-parasitic apicomplexans.

Another protein family that induces programmed cell death in plants encode the elicitins. These affect only a narrow range of plants such as Nicotiana and the radish family (Ponchet et al., 1999). Early biochemical studies showed that Phytophthora and a related phytopathogenic oomycete, Pythium, produce acidic and basic forms of these small (about 10 kDa) cysteine-rich secreted proteins. Molecular cloning and database studies later indicated that each species expresses a diverse family of elicitin (ELI) and elicitin-like (ELL) proteins. In P. ramorum and P. sojae, for example, 48 and 57 family members were identified in their assembled draft genomes (Jiang et al., 2006). Some are secreted and others apparently membrane-associated, the latter by virtue of a predicted C-terminal glycosylphosphatidylinoistol anchor. Many ELI and ELL genes are clustered. For example, five are in a 31-kb syntenous region of the P. ramorum and P. sojae genomes, while others are dispersed to other clusters. This implies that duplications resulting from translocations and unequal crossing-over contributed to the expansion of the family.

Elicitins may participate in defining the host range of Phytophthora. For example, while potato does not respond necrotically to the major P. infestans elicitin Inf1, this protein induces necrosis in a species traditionally considered to be a nonhost, Nicotiana benthamiana. P. infestans silenced for inf1 was shown in one study to gain the ability to colonize N. benthamiana (Kamoun et al., 1998). However, the role of Inf1 may be subtle since another wild-type strain of P. infestans was shown by others to colonize N. benthamiana (Restrepo et al., 2005). Other members of the elicitin family of P. infestans also have necrosis-inducing activities, which may contribute quantitatively to the overall plant response (Huitema et al., 2005; Qutob et al., 2003).

Besides their function as inducers of plant cell death, elicitins may serve roles in binding or processing lipid-like molecules. The major elicitin of P. cryptogea, cryptogein, has sterol-binding activity (Osman et al., 2001a); its function may therefore be to translocate sterols into Phytophthora which cannot manufacture such compounds. However, P. infestans strains silenced for the major elicitin gene (inf1) grow normally. Sterol-binding activity is also not required to trigger plant cell death (Osman et al., 2001b). Another family member, ELI-4 of P. capsici, displays phospholipase activity (Nespoulous et al., 1999). Like the 42-kDa elicitor, elicitins are unique to oomycetes. No relatives are in the genome sequence of the diatom T. pseudonana, so it appears that the protein family is relatively new to the stramenopiles. Lipid transfer in some other species is also mediated by small secreted cysteine-rich proteins, but these are otherwise unrelated to elicitins (Blein et al., 2002).

Some of the *Phytophthora* necrosis-inducing effectors have homologues in other kingdoms. One identified using the *P. infestans* and *P. sojae* EST databases resembles the Nep1 necrosis and ethylene-inducing protein of a fungal pathogen of cacao, *Fusarium oxysporum* (Qutob et al., 2002). Simultaneously, the homologue from *P. parasitica* was identified using a biochemical approach and named NPP1 (Fellbrich et al., 2002). Relatives are also in other fungi such as *M. grisea*, bacteria such as *Bacillus halodurans*, and the oomycete *Pythium aphanidermatum* (Fig. 3.6). These are collectively named Nep1-like proteins or NLPs (Pemberton and Salmond, 2004). Assays using the PVX system in *N. benthamiana* and particle bombardment in soybean indicated that cell death is strongly induced by the NLP from *P. sojae*, a 237-aa cysteine-rich secreted protein called PsojNIP (Qutob et al., 2002). Like many other *Phytophthora*

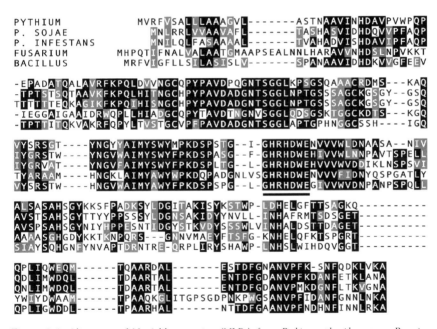

Figure 3.6. Alignment of Nep1-like proteins (NLPs) from *Pythium aphanidermatum*, *P. sojae*, *P. infestans*, *Fusarium oxysporum*, and *Bacillus halodurans* (top to bottom). A few residues from the nonconserved C-terminus are omitted to save space. The heptapeptide motif "GHRHDWE" (underlined) is conserved in nearly all known NLPs. Most species with NLPs are pathogens, although there are exceptions (Gijzen and Nürnberger, 2006). The presence of similar proteins in pathogens from diverse kingdoms is suggested to result from horizontal transfer.

species, *P. sojae* is hemibiotrophic, first infecting living tissues which then become necrotic. PsojNIP is proposed to be used to manipulate the physiology of the plant, since the gene starts to be expressed when plant cells begin to die and theoretically release their contents for consumption by *P. sojae*. Analyses of the *P. ramorum* and *P. sojae* genome sequences detect 50–60 relatives, although more than half may be pseudogenes (Gijzen and Nürnberger, 2006). The family appears to be evolving rapidly, with several subgroups being unique to each species. The large size of the family in *Phytophthora* contrasts with other fungi containing NLPs. For example, *M. grisea* and *N. crassa* appear to encode only four and one NLPs, respectively.

A trend noted in the studies described above was that effector proteins are extracellular, and often small and rich in cysteines that could enhance their stability in the protease-rich plant apoplast by forming disulfide bridges. Such observations were applied to bioinformatic approaches for identifying additional participants in pathogenesis. By searching *P. infestans* ESTs for sequences encoding signal peptides, 142 nonredundant Pex (*Phytophthora* extracellular protein) sequences were identified, of which 55% are novel to *Phytophthora* (Torto *et al.*, 2003). The proteins included several of classes of potential effectors including the CRN group of elicitors, extracellular protease inhibitors, and relatives of known avirulence proteins. An example of the latter was Pex147, a small cysteine-rich protein which appeared similar to Avr1b of *P. sojae*. Association genetics suggested that Pex147 may be Avr3a of *P. infestans*, which was confirmed by agroinfiltration and gene bombardment experiments in potato having or lacking resistance gene R3a (Armstrong *et al.*, 2005). Interestingly, the region containing *Avr3a* exhibited synteny to the chromosomal interval in the downy mildew *H. parasitica* that contained avirulence gene $ATR1^{NdWsB}$. The two avirulence genes share sequence similarity, and surrounding *Avr3a* in *P. infestans* are three relatives. This suggests that *Avr3a* evolved from a gene that duplicated and then diversified to new specificities. The evasion of plant recognition would be expected to confer strong selection for such diversification.

Strong diversifying selection is a feature of the evolution of many effector families, except for elicitins. The presence of such selection may suggest that the genes are functionally important. Such diversification was also observed in the CRN group, which has about 50–100 members in each species (Torto *et al.*, 2003). CRN proteins are roughly 450 aa in size, only found in *Phytophthora*, and induce a crinkling and necrosis phenotype on *N. benthaniana*, tobacco, and tomato. Unlike avirulence proteins, they appear to be noncultivar-specific elicitors. Strong diversifying selection was also observed in SCR74, one of a group of secreted small cysteine-rich (SCR) proteins identified from *P. infestans* (Liu *et al.*, 2005). SCR74 has significant similarity to the 52-aa PcF protein first purified from *P. cactorum,* which is a strong phytotoxin (Orsomando *et al.*, 2001).

Although the precise role of SCR74 in pathogenesis is undefined, it is upregulated during plant colonization and exists as a family, like many other *Phytophthora* effectors. Of 21 copies identified from *P. infestans*, most if not all are clustered. Diversification involved both single site mutations and intergenic recombination, which would be facilitated by the clustered distribution of the family.

A prime example of the value of comparative genomics came from comparing effectors of different oomycetes. Proteins such as $ATR1^{NdWsB}$ of *H. parasitica*, Avr3a of *P. infestans*, and many other predicted effectors were found to contain near the N-terminus of their mature peptides the motif $RXLR-X_{5-21}$-ddEER (Fig. 3.7; Rehmany *et al.*, 2005). This resembles a host target signal found in the malaria parasite *Plasmodium*, which is proposed to mediate the transport of several proteins across the outer membrane of host blood cells. While the signal peptide allows the various effectors to be secreted from *Phytophthora*, the RXLR region would theoretically then enable its uptake into the plant. It is notable that *Plasmodium* is an apicomplexan, which are near stramenopiles in eukaryotic phylogenies (Fig. 3.1).

While effectors such as the RXLRs may act in the cytoplast of the plant, others function in the apoplast. Two families of Kazal-like protease inhibitors, EPI1 and EPI10, were identified from *P. infestans* by mining the Pex database of secreted proteins (Tian *et al.*, 2004; Tian *et al.*, 2005). Like many other effectors, they are members of families. EPI1 and EPI10 bind and inhibit a subtilisin-like serine protease secreted by tomato during its defense response, and EPI10 is upregulated during infection. The inhibitors therefore appear to represent a counter-defense mechanism of *Phytophthora*. Interestingly, this is not the only such case. *P. sojae* is known to secrete glucanase inhibitor proteins which inhibit endo-β-1,3-glucanases of soybean (Rose *et al.*, 2002). The plant enzymes are thought to play a role in defense, either directly by degrading the pathogen cell wall or indirectly by releasing oligosaccharides that elicit additional defenses. The inhibitor proteins also exist as a family in each *Phytophthora*, and appear to be coevolving with the matching glucanases in a range of plant species (Bishop *et al.*, 2005).

```
          Signal peptide                                         RXLR                      ddEER
AA024654  MKLLHHLLVSAACVSVLAATSATPSAAAHVAEGTFASP------ATDVGVSRLLRGPDEDDNDSSESAAGLEEEDDEDS
CAI72329  MRLAIMLSTTAVATYLTTCSA---VDQTKVLMYGSP------AHYIHDSAGRRLLRKNEES-----------EETSEER
AAN31507  MRLAIMLSATAVAINEATCSA---IDQTKVLVYGTP------AHYIHDSAGRRLLRKNEEN-----------EETSEER
AAR05402  MRLSFVLSLVVAIGYVVTCNATEYSDETNIAMVESPDLVRRSLRNGDIAGGRFLRAHE--------------EDDAGER
AAA21422  MRSLLLTVLLNLVVLLATTGAVS-SNLNTAVNYASTSKIRFLSTEYNADEKRSLRGIDYNNEVTKEP-----NTSDEER
BE776768  MRLSCVYLVVATVTTIIASANAAAEASEPMPNIAKYASPEVSVELGAEREKRLLRPDSNDYRD--------DDDEEER
```

Figure 3.7. Representative effector proteins from *Phytophthora* containing the RXLR motif. Shown are alignments of the N-terminal portions of six proteins, identified by their GenBank accession numbers. Indicated in the shaded regions are signal peptides predicted by SignalP 3.0, plus RXLR and ddEER blocks as predicted by Rehmany *et al.* (2005).

In addition to the database mining, functional testing, and map-based approaches for identifying genes involved in pathogenesis described above, expression profiling has also been employed. For example, in one experiment microarrays representing about 1000 *P. sojae* genes were hybridized with probes from soybean tissue harvested 0–48 h after inoculation (Moy *et al.*, 2004). Low concentrations of *P. sojae* early in infection posed a challenge to quantitating all genes, but nevertheless many were shown to accumulate including those encoding CRN, elicitin, and CBEL-like effectors, plus glucanases and proteases that might be employed to break down host tissues. ABC transporters possibly involved in the efflux of host defense molecules were also detected. During a *P. infestans*-potato interaction, an ABC transporter was also shown to be induced, along with transporters of amino acids, phosphate, and sucrose that may be involved in the uptake of such compounds from the host (Beyer *et al.*, 2001).

B. Developmental biology

Knowing how *Phytophthora* transitions between developmental stages is integral to understanding its disease cycles. For example, forming asexual sporangia and zoospores is critical for dissemination. Moreover, genes activated during asexual sporulation or zoosporogenesis likely encode proteins needed for infection such as participants in the chemotaxis of zoospores (van West *et al.*, 2003). The sexual cycle is also relevant to disease, as oospores of both hetero- and homothallic species serve as long-lived reservoirs of inoculum.

Early physiological and cytological studies detailed these pathways, with asexual sporulation and zoosporogenesis being most-studied (Elliott, 1983; Hardham, 2001; Hardham and Hyde, 1997). Genomic and molecular genetic approaches are now making connections to that initial work. For example, matches to many flagella-associated proteins, such as centrin and dyneins, can be found in the genome and EST databases (Narayan, 2005; Randall *et al.*, 2005). Also, genes corresponding to proteins stored in zoospore vacuoles were cloned by screening cDNA libraries with antibodies (Marshall *et al.*, 2001).

Numerous genes differentially expressed in the asexual spore cycle have been identified using cDNA macroarrays or microarrays. In *P. infestans*, arrays of cDNAs representing 4000 genes were used to identify 60 and 71 genes induced >fivefold during sporulation or zoosporogenesis, respectively (Kim and Judelson, 2003; Tani *et al.*, 2004b). One-third genes were expressed exclusively during those stages. Such genes appear to serve diverse metabolic, regulatory, and structural roles. Those in the latter class include those encoding mucin-like proteins, which may protect zoospores or cysts against desiccation, and potential plant adhesion molecules. Such structural proteins typically contain repeated motifs of threonine, serine, and/or proline, which may form linear, rigid, and highly glycosylated molecules. More recently, Affymetrix microarray studies of ~14,000 *P. infestans* genes compared hyphae, sporulating hyphae, sporangia, zoospores, germinating zoospore cysts, and appressoria. This expanded the number of transcripts induced dramatically during the spore cycle to over 1600. A clear lesson from such studies is that major changes occur during both the development and germination of asexual spores. Due to the magnitude of changes, even small EST projects have yielded useful stage-specific genes. For example, a clone sequenced from a *P. nicotianae* zoospore cDNA library encoded a proline biosynthesis enzyme, which fits into a model in which proline helps to osmotically stabilize zoospores (Ambikapathy *et al.*, 2002).

A surprise among the sporulation-specific genes was one encoding Cdc14. In budding yeast and animals, such proteins are known to regulate mitotic exit by suppressing cyclin B and its kinases, and may also regulate the spindle apparatus; their genes are constitutively transcribed and regulated post-translationally. In striking contrast, Cdc14 mRNA in *P. infestans* is absent from vegetative mycelia. It is only present during sporulation and disappears rapidly after zoospore cysts germinate (Ah Fong and Judelson, 2003). Complementation studies in *cdc14*[ts] budding yeast demonstrated that the *P. infestans* protein can regulate mitosis. Therefore, its absence during vegetative growth might explain why mitosis is asynchronous in the coenocytic hyphae of *P. infestans* (Whittaker *et al.*, 1992). Silencing of the gene in *P. infestans* blocks sporulation, suggesting that it may regulate sporulation either due to its control of nuclear behavior or effects of its phosphatase activity on other cellular pathways.

Another gene turned on during asexual sporulation is a member of the Puf (pumilio) family of RNA-binding proteins, which sequester or deadenylate transcripts to block their translation (Cvitanich and Judelson, 2003b). Interestingly, a common target of Puf proteins in other species is cyclin B (Nakahata *et al.*, 2001). The *P. infestans* Cdc14 and Puf proteins may theoretically act in concert to block mitosis and thereby establish the dormancy of sporangia. Interestingly, the Puf gene is also expressed during the formation of *P. infestans* oospores, which like asexual sporangia also need to enter mitotic dormancy for normal development.

Two other *P. infestans* genes upregulated during asexual sporulation, *Pigpa1* and *Pigpb1*, encode α- and β-subunits of the heterotrimeric G-protein. These were identified as part of a strategy of testing the role of conserved signal transduction pathways in *Phytophthora* development (Laxalt *et al.*, 2002). Silencing *Pigpb1* blocks sporulation (Latijnhouwers and Govers, 2003), as also seen when its homologue, was disrupted in the true fungus *Cryphonectria parasitica* (Kasahara and Nuss, 1997). Strains silenced for *Pigpa1* exhibit impaired zoospore swimming and chemotaxis, and have difficulty colonizing plants (Latijnhouwers *et al.*, 2004). Due to the importance of G-protein signaling, efforts are now underway to identify both upstream activators such as G-protein-coupled receptors and downstream targets.

Potential targets were identified by cDNA-AFLP (Dong *et al.*, 2005). Several appear to participate in protein processing, such as a ubiquitin protein lyase homologue, while half had no homology to known sequences. The absence of homologues is not surprising since endpoints of conserved signaling pathways may be quite species-specific.

Several protein kinases induced during asexual sporulation and zoosporogenesis were identified by expression profiling in *P. infestans* (Kim and Judelson, 2003; Tani *et al.*, 2004b). These include apparently constitutively active kinases and those with calcium, phospholipid, or cyclic-nucleotide regulatory domains. One novel kinase has the catalytic domain signature of the eukaryotic calcium-regulated family yet curiously lacks a regulatory domain (Judelson and Roberts, 2002). It is induced in sporangia within minutes of a cold treatment, which causes zoospores to organize from the previously multinucleate sporangial cytoplasm. A bZip transcription factor (Pibzp1) interacting with the kinase was identified by yeast two-hybrid, and when silenced resulted in "rudder-lock" zoospores (zoospores that perpetually turn) and an appressorium-minus phenotype (Fig. 3.8). Using GFP fusions, it was shown that the protein kinase is cytoplasmic while Pibzp1 is both cytoplasmic and nuclear, suggesting

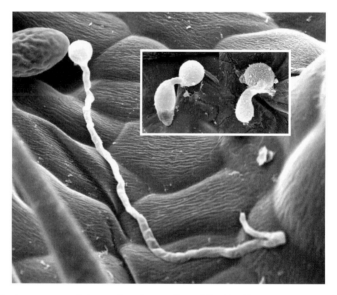

Figure 3.8. Appressorium defect in *P. infestans* silenced for the *Pibzp1* transcription factor. The inset shows two wild-type zoospore cysts which germinated and produced appressoria on the surface of a tomato leaflet. The larger image shows a germinated zoospore cyst from a *Pibzp1*-silenced strain, which fails to produce appressoria on either plant or artificial hydrophobic surfaces (Blanco and Judelson, 2005).

that the kinase modifies Pibzp1 which enters the nucleus to regulate appressorium-essential genes. Candidates for the latter were identified using microarrays (Blanco and Judelson, unpublished data). Most lack homology to known proteins, as was the case for the downstream targets of the Pigpa1 α-subunit (Dong *et al.*, 2005).

Several studies identified genes upregulated during the germination of zoospore cysts and the formation of appressoria. A combination of subtraction cloning and 2D protein gel analysis showed that several genes encoding amino acid biosynthesis enzymes were upregulated during such stages (Grenville-Briggs *et al.*, 2005). This correlated with a decrease in the concentrations of most free amino acids following asexual sporulation and zoospore release, and a rise during appressorium formation. Genes encoding other types of proteins upregulated during cyst germination and appressorium development, such as heat-shock proteins and metabolic enzymes, were identified by cDNA-AFLP (Avrova *et al.*, 2004).

During the development of sexual spores in *P. infestans*, subtraction cloning and microarray studies identified 103 genes induced >tenfold, of which 46 were totally mating-specific (Fabritius *et al.*, 2002; Prakob, 2005). Their BLAST hits suggested a variety of roles in regulation, metabolism, structure, and meiosis. A disproportionate number matched RNA-binding or metabolizing proteins. These included the Puf protein mentioned earlier, another RNA-binding protein in the KH-domain family, an RNAse, a DOM3Z RNAse regulator, and a cap-removing enzyme. This suggests that posttranscriptional regulation may play key roles in oosporogenesis. Such proteins may help commit *P. infestans* to sexual development by degrading specific substrates, which is conceptually related to the observation that several asexual sporulation-induced genes and Pigpa1 targets are involved in protein turnover. Alternatively, the proteins may be involved in storing RNA for germination.

As with many pathogenesis-associated proteins, several stage-specific genes belong to families in which expansions appears to have been accompanied by diversification in function or expression pattern. For example, the NIF family of transcriptional regulators exists as a four-gene cluster with one copy induced during asexual sporulation and three during zoosporogenesis (Tani *et al.*, 2004a). Also, five members of a transglutaminase family were described in which individual genes diverge in both expression pattern (having hyphal-, zoospore-, and oospore-specific forms) and structure (Fabritius and Judelson, 2003). These range from a minimal 426-aa transglutaminase to a 1600-aa protein in which the transglutaminase domain is fused with a large region of repeated proline/threonine-rich motifs.

C. Transcription mechanisms

Knowing what drives gene expression during *Phytophthora* development is integral to understanding its disease and life cycles. A better definition of promoter structure can also aid the prediction of coding regions from genome data. Promoter studies can proceed more rapidly now that methods for transformation, expression profiling data, and genome sequences are available. The prevalence of different DNA-binding motifs in the *Phytophthora* proteome suggest deviation in the spectrum of transcription factors that are utilized compared to other organisms (Table 3.2). The basic transcriptional machinery is also diverged, since non-oomycete promoters lack function in *Phytophthora* (Judelson *et al.*, 1992). There is, however, some functional conservation within oomycetes. In fact, the most-used promoters for expressing transgenes in *Phytophthora* come from a downy mildew, *Bremia lactucae* (Judelson *et al.*, 1991).

The tight spacing of genes noted in the genome projects implies that most sequences regulating transcription are likely within a few hundred bases of the transcriptional start point. Start sites are typically 50- to 100-nt upstream of the start codon, although some are in excess of 200 nt (Cvitanich and Judelson, 2003b; Kamoun, 2003). Canonical promoter elements of other eukaryotes, such as the TATA box, are absent from most *Phytophthora* promoters.

A key feature of the core promoter appears to be a pyrimidine tract near the initiation site, which usually is at an adenosine. One study showed that a CTCCTTCT motif in the *Piypt1* promoter of *P. infestans* resembles the Initiator (Inr) sequence of many TATA-less mammalian genes and that it bound nuclear proteins from *P. infestans* (Chen and Roxby, 1997). Others defined a 16-nt consensus for this region, GCTCATTYYNCAWTTT, which was later expanded to a 19-nt consensus by aligning 18 other promoters (McLeod *et al.*, 2004; Pieterse *et al.*, 1994b). Tests in transient assays of mutagenized versions of the 19-nt region from *Piexo1* of *P. infestans* suggested that it contained two functional elements. Its 5' end had sequence and functional homology to the Inr, while its 3' end was dubbed the flanking promoter region (FPR). Proteins from cell extracts were shown to bind both Inr and FPR sequences. These sequences are not in all *Phytophthora* genes, however, especially those that are stage-specific. This contrasts with many other eukaryotes, where Inr is mostly in regulated genes. To date, too few transcripts have been mapped to enable an effective bioinformatic dissection of the diversity of core promoters in *Phytophthora*. However, the regions surrounding the start sites of genes can be exchanged without blocking expression, even if one lacks the 19-nt consensus (Ah Fong and Judelson, unpublished data).

Progress has been made toward identifying the binding sites of transcription factors required for expressing stage-specific genes in *P. infestans*. Fragments of about 300 bp from four sporulation, one mating, and one zoosporogenesis-specific promoters were each shown to confer wild-type expression to a GUS transgene, and in several cases the sites essential for developmental regulation were identified (Tani, Cvitanich, Kim, Ah Fong, and Judelson, unpublished data). For example, deletions and site-directed mutagenesis showed that an 8-nt site 107-nt upstream of the transcription start of *PiPks1* was required for its induction during sporulation (Kim and Judelson, unpublished data). Similar studies identified a 7-nt motif dubbed the "cold box" at position −139 of the *PiNifC1* zoosporogenesis-induced gene (Tani and Judelson, 2006). The cold box is sufficient for activating genes during the low temperature-induced process of zoosporogenesis, functions in different orientations and distances from the start site, and can convert a formerly mating-induced promoter to a zoosporogenesis-specific pattern.

Phylogenetic footprinting, which involves searching for conservation within promoters of orthologs, appears promising in *Phytophthora*. Such an approach often succeeds when moderately related genomes are available such as human and mouse (Qiu *et al.*, 2003). As shown in Fig. 3.9, blocks of conservation are frequently observed within promoters of *Phytophthora* orthologs. These typically include a region spanning the transcription start and one or more upstream locations, the latter potentially representing binding sites for transcription factors important in development. Regions upstream of coexpressed genes can also be searched for overrepresented sequences. The *PiNifC1* cold box, for example, is within the promoters of some other zoospore-induced genes (Tani and Judelson, 2006). Such sequences can be used to isolate the cognate transcription factors by affinity methods, which can be sequenced by mass spectroscopy and matched to genes within the *Phytophthora* databases.

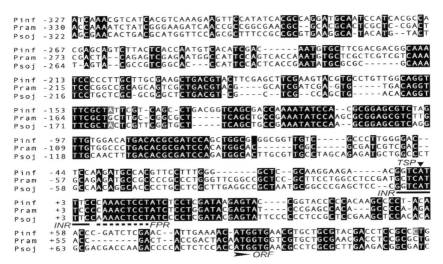

Figure 3.9. Alignment of promoters from endo-1,3-β-glucanase homologues of *P. ramorum* (Pram), and *P. sojae* (Psoj), and *P. infestans* (Pinf), the latter being *Piendo1* (McLeod *et al.*, 2004). Regions of conservation are indicated by a black background. Conserved blocks are found near the transcription start site (TSP, +1), which include the Inr and FPR (indicated by solid and dashed lines, respectively), and other positions between −80 and −210 which represent potential binding sites for transcription factors.

D. Evolution

Since oomycetes reside in an understudied clade of the eukaryotic tree, it is natural to use genome data to address their evolution. Analyses based on rRNA and protein sequences (Baldauf *et al.*, 2000; Gunderson *et al.*, 1987), mitochondrial gene orders (Sankoff *et al.*, 2000), and gene fusions (Stechmann and Cavalier-Smith, 2002) support firmly the premise that *Phytophthora* and true fungi are distinct despite their related filamentous and pathogenic growth habits. Nevertheless, in light of the phenomena of horizontal transfer and multiple endosymbioses during eukaryotic evolution (Yoon *et al.*, 2005), sequences in the *Phytophthora* databases can be used to both support and refine current models, and look for the unexpected.

Support for some of the early biochemical evidence distinguishing oomycetes from true fungi can be obtained from the genome data. For example, while true fungi synthesize lysine using the α-aminoadipate (AAA) pathway, oomycetes use the diaminopimilate (DAP) pathway like bacteria and plants. This can now be explained by the absence of AAA pathway genes in *Phytophthora* (Randall *et al.*, 2005). Another criterion used to distinguish the groups is the absence (or minimal presence) of chitin in oomycete cell walls. However, two chitin synthase-like genes are present in each of the *P. infestans*, *P. ramorum*, and *P. sojae* databases.

Absent from *Phytophthora* databases are the genes employed by true fungi to make the polyketide and nonribosomal peptide toxins that play important roles during their colonization of plants (Randall *et al.*, 2005). On the other hand, the polygalacturonases expressed by *Phytophthora* to presumably break down plant cell walls best resemble enzymes of fungal origin, while pectate lyases match either fungal or bacterial proteins (Gotesson *et al.*, 2002; Randall *et al.*, 2005; Torto *et al.*, 2002). None of the *Phytophthora* enzymes were close to plant proteins, even though affinities between oomycetes and plants have been previously suggested by several data including their primarily cellulosic cell walls and use of the DAP pathway for lysine.

Sequence comparisons, however, do yield some strong affinities with plants. For example, glutamate dehydrogenase of *P. infestans* was shown to reside within a clade of green plant enzymes, far from those of apicomplexans which are otherwise closer to stramenopiles (Andersson and Roger, 2003). The *Phytophthora* species also encode an FtsZ homologue, a key component of the mechanism of division of plant chloroplasts and bacteria (Randall *et al.*, 2005). FtsZ homologues are also encoded by other stramenopiles such as the diatom, suggesting that the group employs an ancient bacterial-like method to divide mitochondria. Interestingly, quite a number of *Phytophthora* genes have strong affinity to sequences from photosynthetic organisms, including cyanobacteria. However, the photosynthetic genes found in plants and diatoms are not detected in *Phytophthora* (Randall *et al.*, 2005). This supports the concept that such genes were introduced into diatoms by a secondary endosymbiosis event, after radiation of heterotrophs such as *Phytophthora*. An alternative hypothesis is that the ancestral stramenopile was photosynthetic but that capacity was lost by *Phytophthora*.

A few surprising similarities to animals also exist in *Phytophthora* such as the presence of two phosphagen kinases (Randall *et al.*, 2005). Such enzymes, which are best known for their roles in the buffering or metabolic channeling of ATP in muscles, were formerly believed to be restricted to animals. However, besides *Phytophthora* they are now known to be encoded by trypanosomes, apicomplexans, and diatoms (Noguchi *et al.*, 2001; Pereira *et al.*, 1999). One of the *P. infestans* genes is induced at the stage when zoospore components are being made (Kim and Judelson, 2003). Since all of these species have flagellated life-stages, it appears that phosphagen kinase systems have coevolved with that high-energy-utilizing apparatus.

VII. CONCLUSIONS AND PROSPECTS

When the author began his studies of *Phytophthora* 15 years ago, the group was regarded by some as "a fungal geneticist's nightmare" (Shaw, 1983b). It is fair to say that the situation has improved dramatically, as genomic resources and improved tools for gene manipulation have enabled relatively rapid progress to be made toward understanding many aspects of their biology and evolution. Enhancements in technologies for functional genomics are nevertheless desirable to further accelerate progress. Much of the current genetic and expression-profiling studies would also benefit from an improved integration with protein data. Due to the focus of this chapter on genomics, several notable studies using proteomics were omitted (Ebstrup *et al.*, 2005; Grenville-Briggs *et al.*, 2005; Shepherd *et al.*, 2003). Such studies will be complementary to the genome projects and necessary since differences in mRNA abundance may not be reflected at the level of protein. Posttranslational modifications may also play important roles.

Pathogenesis by *Phytophthora* appears to involve the standard toolkit of factors needed to colonize plants, such as cell wall-degrading enzymes, plus an extensive repertoire of novel effectors that manipulate plant physiology. Many of the hemibiotrophic *Phytophthora* species, like their biotrophic downy mildew cousins, enter plants by stealth and thus avoid strong defense reactions. It will therefore be important to learn if any of the effectors participate in defense suppression. In the case of effector families that trigger plant cell death, it should prove possible to determine the roles of those that interact with known resistance factors or other plant proteins. However, many could be orphans or evolutionary relics with no current counterparts in plants. Those effectors proved required for pathogenesis and the mechanisms for their secretion are potential targets for measures to control *Phytophthora* diseases, as are the spore pathways needed to disseminate *Phytophthora* to infection sites.

Studies of spore development have identified processes conserved in other eukaryotes such as the role of G-protein signaling. Novel twists on otherwise conserved features have also been revealed such as the diverged pattern of Cdc14 expression in *P. infestans* compared to other eukaryotes. That both similarities and differences exist is logical considering that all eukaryotes evolved from the same core set of protein domains and pathways. Understanding the novel combinations, modifications, and expression patterns of such domains in *Phytophthora* will help illuminate the full complexity of eukaryotic evolution.

A lesson from the genomics data is that a simple linear pathway of *Phytophthora* evolution from other eukaryotes may be too simple. The presence of dsRNAs resembling plant endornaviruses, for example, is a reminder that *Phytophthora* shares an environmental niche with plants, fungi, prokaryotes, and others. Horizontal transfers from such species may have brought in genes with novel activities or replaced genes of similar functions. A final thought on evolution is that most genomic studies have only examined one isolate of each *Phytophthora* species. Intraspecific variation in chromosome structure and gene content, such as with avirulence genes, is already known from such limited sampling. Outcrossing is not essential in *Phytophthora*, so major genetic differences may accumulate within each "species." Within *P. infestans*, for example, host specialization exists among tomato and potato strains (Legard *et al.*, 1995). Intraspecific sequence comparisons may illuminate the bases of such differences.

Acknowledgments

The *Phytophthora* research community was invited to contribute unpublished or in press data to this chapter, and the author thanks the many groups that shared such information. Results lacking formal attributions were extracted by the author from data on the Web sites of the relevant genome projects at the Department of Energy Joint Genome Institute and Virginia Bioinformatics Institute. Work in the author's laboratory has been supported by the Department of Agriculture and National Science Foundation of the United States, the University of California Discovery Grant program, and Syngenta Corporation.

References

Ah Fong, A., and Judelson, H. S. (2003). Cell cycle regulator Cdc14 is expressed during sporulation but not hyphal growth in the fungus-like oomycete *Phytophthora infestans*. *Mol. Microbiol.* **50,** 487–494.

Ah Fong, A., and Judelson, H. S. (2004). The haT-like DNA transposon DodoPi resides in a cluster of retro and DNA transposons in the stramenopile *Phytophthora infestans*. *Mol. Gen. Genomics* **271,** 577–585.

Al-Kherb, S. M., Fininsa, C., Shattock, R. C., and Shaw, D. S. (1995). The inheritance of virulence of *Phytophthora infestans* to potato. *Plant Pathol.* **44,** 552–562.

Ambikapathy, J., Marshall, J. S., Hocart, C. H., and Hardham, A. R. (2002). The role of proline in osmoregulation in *Phytophthora nicotianae*. *Fungal Genet. Biol.* **35,** 287–299.

Andersson, J. O., and Roger, A. J. (2003). Evolution of glutamate dehydrogenase genes: Evidence for lateral gene transfer within and between prokaryotes and eukaryotes. *BMC Evol. Biol.* **3,** 14.

Arisue, N., Hashimoto, T., Yoshikawa, H., Nakamura, Y., Nakamura, G., Nakamura, F., Yano, T. A., and Hasegawa, M. (2002). Phylogenetic position of *Blastocystis hominis* and of stramenopiles inferred from multiple molecular sequence data. *J. Eukaryot. Microbiol.* **49,** 42–53.

Armbrust, E. V., Berges, J. A., Bowler, C., Green, B. R., Martinez, D., Putnam, N. H., Zhou, S., Allen, A. E., Apt, K. E., Bechner, M., Brzezinski, M. A., Chaal, B. K., *et al.* (2004). The genome of the diatom *Thalassiosira pseudonana*: Ecology, evolution, and metabolism. *Science* **306,** 79–86.

Armstrong, M. R., Whisson, S. C., Pritchard, L., Bos, J. I., Venter, E., Avrova, A. O., Rehmany, A. P., Bohme, U., Brooks, K., Cherevach, I., Hamlin, N., White, B., *et al.* (2005). An ancestral oomycete locus contains late blight avirulence gene Avr3a, encoding a protein that is recognized in the host cytoplasm. *Proc. Natl. Acad. Sci. USA* **102,** 7766–7771.

Avila-Adame, C., Gómez-Alpizar, L., Zismann, V., Jones, K. M., Buell, C. R., and Ristaino, J. B. (2006). Mitochondrial genome sequences and molecular evolution of the Irish potato famine pathogen, *Phytophthora infestans*. *Curr. Genet.* **49,** 39–46.

Avrova, A. O., Venter, E., Birch, P. R. J., and Whisson, S. C. (2004). Profiling and quantifying differential gene transcription in *Phytophthora infestans* prior to and during the early stages of potato infection. *Fungal Genet. Biol.* **40,** 4–14.

Baldauf, S. L., Roger, A. J., Wenk-Siefert, I., and Doolittle, W. F. (2000). A kingdom-level phylogeny of eukaryotes based on combined protein data. *Science* **290,** 972–977.

Bartlett, D. W., Clough, J. M., Godwin, J. R., Hall, A. A., Hamer, M., and Parr-Dobrzanski, B. (2002). The strobilurin fungicides. *Pest Manag. Sci.* **58,** 649–662.

Beyer, K., Binder, A., Boller, T., and Collinge, M. (2001). Identification of potato genes induced during colonization by *Phytophthora infestans*. *Mol. Plant Pathol.* **2,** 125–134.

Bishop, J. G., Ripoll, D. R., Bashir, S., Damasceno, C. M., Seeds, J. D., and Rose, J. K. (2005). Selection on *Glycine* beta-1,3-endoglucanase genes differentially inhibited by a *Phytophthora* glucanase inhibitor protein. *Genetics* **169,** 1009–1019.

Blanco, F. A., and Judelson, H. S. (2005). A bZIP transcription factor from *Phytophthora* interacts with a protein kinase and is required for zoospore motility and plant infection. *Mol. Microbiol.* **56,** 638–648.

Blein, J. P., Coutos-Thevenot, P., Marion, D., and Ponchet, M. (2002). From elicitins to lipid-transfer proteins: A new insight in cell signalling involved in plant defence mechanisms. *Trends Plant Sci.* **7,** 293–296.

Brunner, F., Rosahl, S., Lee, J., Rudd, J. J., Geiler, C., Kauppinen, S., Rasmussen, G., Scheel, D., and Nürnberger, T. (2002). Pep-13, a plant defense-inducing pathogen-associated pattern from *Phytophthora* transglutaminases. *EMBO J.* **21,** 6681–6688.

Buck, K. W. (1986). "Fungal Virology." CRC Press, Boca Raton, FL.

Caten, C. E., and Jinks, J. L. (1968). Spontaneous variability in isolates of *Phytophthora infestans*. I. Cultural variation. *Can. J. Bot.* **46,** 329–348.

Chamnanpunt, J., Shan, W.-X., and Tyler, B. M. (2001). High frequency mitotic gene conversion in genetic hybrids of the oomycete *Phytophthora sojae*. *Proc. Natl. Acad. Sci. USA* **98,** 14530–14535.

Chen, Y., and Roxby, R. (1997). Identification of a functional CT-element in the *Phytophthora infestans* piypt1 gene promoter. *Gene* **198,** 159–164.

Connolly, M. S., Sakihama, Y., Phuntumart, V., Jiang, Y., Warren, F., Mourant, L., and Morris, P. F. (2005). Heterologous expression of a pleiotropic drug resistance transporter from *Phytophthora sojae* in yeast transporter mutants. *Curr. Genet.* **48,** 356–365.

Corbett, M. K., and Styer, E. L. (1976). Intranuclear viruslike particles in *Phytophthora infestans*. *Proc. Am. Phytopath. Soc.* **3,** 332.

Cvitanich, C., and Judelson, H. S. (2003a). Stable transformation of the oomycete, *Phytophthora infestans*, using microprojectile bombardment. *Curr. Genet.* **42**, 228–235.

Cvitanich, C., and Judelson, H. S. (2003b). A gene expressed during sexual and asexual sporulation in *Phytophthora infestans* is a member of the Puf family of translational regulators. *Eukaryot. Cell* **2**, 465–473.

Cvitanich, C., Salcido, M., and Judelson, H. S. (2006). Concerted evolution of a tandemly arrayed family of mating-specific genes in *Phytophthora* analyzed through inter- and intraspecific comparisions. *Mol. Genet. Genomics* **275**, 169–184.

Davidson, J. M., Werres, S., Garbelotto, M., Hansen, E. M., and Rizzo, D. M. (2003). Sudden Oak Death and associated diseases caused by *Phytophthora ramorum*. *Plant Health Prog.* **1**, 1–21.

De Amicis, F., and Marchetti, S. (2000). Intercodon dinucleotides affect codon choice in plant genes. *Nucleic Acids Res.* **28**, 3339–3345.

Dean, R. A., Talbot, N. J., Ebbole, D. J., Farman, M. L., Mitchell, T. K., Orbach, M. J., Thon, M., Kulkarni, R., Xu, J. R., Pan, H., Read, N. D., Lee, Y. H., *et al.* (2005). The genome sequence of the rice blast fungus *Magnaporthe grisea*. *Nature* **434**, 980–986.

Dong, W., Latijnhouwers, M., Jiang, R. H. Y., Meijer, H. J. G., and Govers, F. (2005). Downstream targets of the *Phytophthora infestans* G-alpha subunit PiGPA1 revealed by cDNA-AFLP. *Molec. Plant Pathol.* **5**, 483–494.

Ebstrup, T., Saalbach, G., and Egsgaard, H. (2005). A proteomics study of in vitro cyst germination and appressoria formation in *Phytophthora infestans*. *Proteomics* **5**, 2839–2848.

Elliott, C. G. (1983). Physiology of sexual reproduction in *Phytophthora*. In "*Phytophthora*: Its Biology, Taxonomy, Ecology, and Pathology" (D. C. Erwin, S. Bartnicki-Garcia, and P. H. Tsao, eds.), pp. 71–80. APS Press, St. Paul, MN.

Erwin, D. C., and Ribeiro, O. K. (1996). "*Phytophthora* Diseases Worldwide." APS Press, St. Paul, MN.

Fabritius, A.-L., Cvitanich, C., and Judelson, H. S. (2002). Stage-specific gene expression during sexual development in *Phytophthora infestans*. *Mol. Microbiol.* **45**, 1057–1066.

Fabritius, A.-L., and Judelson, H. S. (2003). A mating-induced protein of *Phytophthora infestans* is a member of a family of elicitors with divergent structures and stage-specific patterns of expression. *Mol. Plant Microbe Interact.* **16**, 926–935.

Fellbrich, G., Romanski, A., Varet, A., Blume, B., Brunner, F., Engelhardt, S., Felix, G., Kemmerling, B., Krzymowska, M., and Nürnberger, T. (2002). NPP1, a *Phytophthora*-associated trigger of plant defense in parsley and *Arabidopsis*. *Plant J.* **32**, 375–390.

Feschotte, C. (2004). Merlin, a new superfamily of DNA transposons identified in diverse animal genomes and related to bacterial IS1016 insertion sequences. *Mol. Biol. Evol.* **21**, 1769–1780.

Forster, H., and Coffey, M. D. (1990). Mating behavior of *Phytophthora parasitica*: Evidence for sexual recombination in oospores using DNA restriction fragment length polymorphisms as genetic markers. *Exp. Mycol.* **14**, 351–359.

Fry, W. E., and Goodwin, S. B. (1997). Resurgence of the Irish potato famine fungus. *Bioscience* **47**, 363–371.

Galagan, J. E., Calvo, S. E., Borkovich, K. A., Selker, E. U., Read, N. D., Jaffe, D., FitzHugh, W., Ma, L.-J., Smirnov, S., Purcell, S., Rehman, B., Elkins, T., *et al.* (2003). The genome sequence of the filamentous fungus *Neurospora crassa*. *Nature* **422**, 859–868.

Gaulin, E., Jauneau, A., Villalba, F., Rickauer, M., Esquerre-Tugaye, M. T., and Bottin, A. (2002). The CBEL glycoprotein of *Phytophthora parasitica* var. *nicotianae* is involved in cell wall deposition and adhesion to cellulosic substrates. *J. Cell Sci.* **115**, 4565–4575.

Gavino, P. D., and Fry, W. E. (2002). Diversity in and evidence for selection on the mitochondrial genome of *Phytophthora infestans*. *Mycologia* **94**, 781–793.

Gijzen, M., and Nürnberger, T. (2006). Nep1-like proteins from plant pathogens: Recruitment and diversification of the NPP1 domain across taxa. *Phytochemistry* **67**, 1800–1807.

Goernhardt, B., Rouhara, I., and Schmelzer, E. (2000). Cyst germination proteins of the potato pathogen *Phytophthora infestans* share homology with human mucins. *Mol. Plant Microbe Interact.* **13,** 32–42.

Goodwin, S. B., Sujkowski, L. S., Dyer, A. T., Fry, B. A., and Fry, W. E. (1995). Direct detection of gene flow and probable sexual reproduction of *Phytophthora infestans* in northern North America. *Phytopathology* **85,** 473–479.

Gotesson, A., Marshall, J. S., Jones, D. A., and Hardham, A. R. (2002). Characterization and evolutionary analysis of a large polygalacturonase gene family in the oomycete plant pathogen *Phytophthora cinnamomi. Mol. Plant Microbe Interact.* **15,** 907–921.

Grenville-Briggs, L. J., Avrova, A. O., Bruce, C. R., Williams, A., Whisson, S. C., Birch, P. R., and van West, P. (2005). Elevated amino acid biosynthesis in *Phytophthora infestans* during appressorium formation and potato infection. *Fungal Genet. Biol.* **42,** 244–256.

Güldener, U., Seong, K.-Y., Boddu, J., Cho, S., Trail, F., Xu, J.-R., Adam, G., Mewes, H.-W., Muehlbauer, G. J., and Kistler, H. C. (2006). Development of a *Fusarium graminearum* Affymetrix GeneChip for profiling fungal gene expression *in vitro* and in planta. *Fungal Genet. Biol.* **43,** 316–325.

Gunderson, J. H., Elwood, H., Ingold, A., Kindle, K., and Sogin, M. L. (1987). Phylogenetic relationships between chlorophytes, chrysophytes, and oomycetes. *Proc. Natl. Acad. Sci. USA* **84,** 5823–5827.

Hacker, C. V., Brasier, C. M., and Buck, K. W. (2005). A double-stranded RNA from a *Phytophthora* species is related to the plant endornaviruses and contains a putative UDP glycosyltransferase gene. *J. Gen. Virol.* **86,** 1561–1570.

Hardham, A. R. (2001). The cell biology behind *Phytophthora* pathogenicity. *Austral. Plant Pathol.* **30,** 91–98.

Hardham, A. R., and Hyde, G. J. (1997). Asexual sporulation in the oomycetes. *Adv. Bot. Res.* **24,** 353–398.

Huitema, E., Bos, J. I., Tian, M., Win, J., Waugh, M. E., and Kamoun, S. (2004). Linking sequence to phenotype in *Phytophthora*-plant interactions. *Trends Microbiol.* **12,** 193–200.

Huitema, E., Vleeshouwers, V. G., Cakir, C., Kamoun, S., and Govers, F. (2005). Differences in intensity and specificity of hypersensitive response induction in *Nicotiana* spp. by INF1, INF2A, and INF2B of *Phytophthora infestans. Mol. Plant Microbe Interact.* **18,** 183–193.

Jiang, R. H., Dawe, A. L., Weide, R., van Staveren, M., Peters, S., Nuss, D. L., and Govers, F. (2005a). Elicitin genes in *Phytophthora infestans* are clustered and interspersed with various transposon-like elements. *Mol. Genet. Genomics* **273,** 20–32.

Jiang, R. H. Y., Tyler, B. M., and Govers, F. (2005b). Comparative genomics and synteny studies revealing the reservoir of secreted proteins in *Phytophthora* [Abstract]. *Phytophthora* Molecular Genetics Network Workshop, Asilomar Conference Grounds, Pacific Grove, CA, March 13–15, 2005, p. 14.

Jiang, R. H. Y., Tyler, B. M., Whisson, S. C., Hardham, A. R., and Govers, F. (2006). Ancient origin of elicitin gene clusters in *Phytophthora* genomes. *Mol. Biol. Evol.* **23,** 338–351.

Judelson, H. S. (1993). Intermolecular ligation mediates efficient cotransformation in *Phytophthora infestans. Mol. Gen. Genet.* **239,** 241–250.

Judelson, H. S. (1996a). Chromosomal heteromorphism linked to the mating type locus of the oomycete *Phytophthora infestans. Mol. Gen. Genet.* **252,** 155–161.

Judelson, H. S. (1996b). Genetic and physical variability at the mating type locus of the oomycete, *Phytophthora infestans. Genetics* **144,** 1005–1013.

Judelson, H. S. (1997). The genetics and biology of *Phytophthora infestans*: Modern approaches to a historical challenge. *Fungal Genet. Biol.* **22,** 65–76.

Judelson, H. S. (2002). Sequence variation and genomic amplification of a family of Gypsy-like elements in the oomycete genus *Phytophthora. Mol. Biol. Evol.* **19,** 1313–1322.

Judelson, H. S., and Fabritius, A. L. (2000). A linear RNA replicon from the oomycete *Phytophthora infestans*. *Mol. Gen. Genet.* **263,** 395–403.

Judelson, H. S., and Randall, T. A. (1998). Families of repeated DNA in the oomycete *Phytophthora infestans* and their distribution within the genus. *Genome* **41,** 605–615.

Judelson, H. S., and Roberts, S. (2002). Novel protein kinase induced during sporangial cleavage in the oomycete *Phytophthora infestans*. *Eukaryot. Cell* **1,** 687–695.

Judelson, H. S., and Senthil, G. S. (2006). Investigating the role of ABC transporters in multi-fungicide insensitivity in *Phytophthora infestans*. *Molec. Plant Pathol.* **7,** 17–29.

Judelson, H. S., Tyler, B. M., and Michelmore, R. W. (1991). Transformation of the oomycete pathogen, *Phytophthora infestans*. *Mol. Plant Microbe Interact.* **4,** 602–607.

Judelson, H. S., Tyler, B. M., and Michelmore, R. W. (1992). Regulatory sequences for expressing genes in oomycete fungi. *Mol. Gen. Genet.* **234,** 138–146.

Judelson, H. S., Coffey, M. D., Arredondo, F. R., and Tyler, B. M. (1993a). Transformation of the oomycete pathogen *Phytophthora megasperma* f. sp. *glycinea* occurs by DNA integration into single or multiple chromosomes. *Curr. Genet.* **23,** 211–218.

Judelson, H. S., Dudler, R., Pieterse, C. M. J., Unkles, S. E., and Michelmore, R. W. (1993b). Expression and antisense inhibition of transgenes in *Phytophthora infestans* is modulated by choice of promoter and position effects. *Gene* **133,** 63–69.

Judelson, H. S., Spielman, L. J., and Shattock, R. C. (1995). Genetic mapping and non-Mendelian segregation of mating type loci in the oomycete, *Phytophthora infestans*. *Genetics* **141,** 503–512.

Kamoun, S. (2003). Molecular genetics of pathogenic oomycetes. *Eukaryot. Cell* **2,** 191–199.

Kamoun, S., van West, P., Vleshouwers, V. G. A. A., de Groot, K. E., and Govers, F. (1998). Resistance of *Nicotiana benthamiana* to *Phytophthora infestans* is mediated by the recognition of the elicitor protein INF1. *Plant Cell* **10,** 1413–1425.

Kamoun, S., Hraber, P., Sobral, B., Nuss, D., and Govers, F. (1999). Initial assessment of gene diversity for the oomycete pathogen *Phytophthora infestans* based on expressed sequences. *Fungal Genet. Biol.* **28,** 94–106.

Kasahara, S., and Nuss, D. L. (1997). Targeted disruption of a fungal G-protein beta subunit gene results in increased vegetative growth but reduced virulence. *Mol. Plant Microbe Interact.* **10,** 984–993.

Kim, K. S., and Judelson, H. S. (2003). Sporangia-specific gene expression in the oomyceteous phytopathogen *Phytophthora infestans*. *Eukaryot. Cell* **2,** 1376–1385.

Latijnhouwers, M., and Govers, F. (2003). A *Phytophthora infestans* G-Protein β subunit is involved in sporangium formation. *Eukaryot. Cell* **2,** 971–977.

Latijnhouwers, M., and Ligterink, W. Vleeshouwers, V. G. A. A., van West, P., and Govers, F. (2004). A G-alpha subunit controls zoospore motility and virulence in the potato late blight pathogen *Phytophthora infestans*. *Mol. Microbiol.* **51,** 925–936.

Laxalt, A. M., Latijnhouwers, M., van Hulten, M., and Govers, F. (2002). Differential expression of G protein alpha and beta subunit genes during development of *Phytophthora infestans*. *Fungal Genet. Biol.* **36,** 137–146.

Legard, D. E., Lee, T. Y., and Fry, W. E. (1995). Pathogenic specialization in *Phytophthora infestans*: Aggressiveness on tomato. *Phytopathology* **85,** 1355–1361.

Liu, Z., Bos, J. I., Armstrong, M., Whisson, S. C., da Cunha, L., Torto-Alalibo, T., Win, J., Avrova, A. O., Wright, F., Birch, P. R., and Kamoun, S. (2005). Patterns of diversifying selection in the phytotoxin-like scr74 gene family of *Phytophthora infestans*. *Mol. Biol. Evol.* **22,** 659–672.

Mao, Y., and Tyler, B. M. (1991). Genome organization of *Phytophthora megasperma* f. sp *glycinea*. *Exp. Mycol.* **15,** 283–291.

Margulies, M., Egholm, M., Altman, W. E., Attiya, S., Bader, J. S., Bemben, L. A., Berka, J., Braverman, M. S., Chen, Y. J., Chen, Z., Dewell, S. B., Du, L., *et al.* (2005). Genome sequencing in microfabricated high-density picolitre reactors. *Nature* **437,** 376–380.

Marshall, J. S., Wilkinson, J. M., Moore, T., and Hardham, A. R. (2001). Structure and expression of the genes encoding proteins resident in large peripheral vesicles of *Phytophthora cinnamomi* zoospores. *Protoplasma* **215,** 226–239.

Martin, F. (2005). Mitochondrial genome organization in *Phytophthora*; What *P. infestans, P. ramorum* and *P. sojae* tell us [Abstract]. *Phytophthora* Molecular Genetics Network Workshop, Asilomar Conference Grounds, Pacific Grove, CA, March 13–15, 2005, p. 4.

May, K. J., Whisson, S. C., Zwart, R. S., Searle, I. R., Irwin, J. A., Maclean, D. J., Carroll, B. J., and Drenth, A. (2002). Inheritance and mapping of 11 avirulence genes in *Phytophthora sojae*. *Fungal Genet. Biol.* **37,** 1–12.

McLeod, A., Smart, C. D., and Fry, W. E. (2004). Core promoter structure in the oomycete *Phytophthora infestans*. *Eukaryot. Cell* **3,** 91–99.

Mortimer, A. M., Shaw, D. S., and Sansome, E. R. (1977). Genetical studies of secondary homothallism in *Phytophthora dreschsleri*. *Arch. Microbiol.* **111,** 255–259.

Moy, P., Qutob, D., Chapman, B. P., Atkinson, I., and Gijzen, M. (2004). Patterns of gene expression upon infection of soybean plants by *Phytophthora sojae*. *Mol. Plant Microbe Interact.* **17,** 1051–1062.

Nakahata, S., Katsu, Y., Mita, K., Inoue, K., Nagahama, Y., and Yamashita, M. (2001). Biochemical identification of *Xenopus* Pumilio as a sequence-specific cyclin B1 mRNA-binding protein that physically interacts with a Nanos homolog, Xcat-2, and a cytoplasmic polyadenylation element-binding protein. *J. Biol. Chem.* **276,** 20945–20953.

Narayan, R. (2005). Characterisation of the pre-sporangium stage sporulation genes in the oomycete plant pathogen *Phytophthora cinnamomi*. Ph.D., Thesis Australian National University.

Nespoulous, C., Gaudemer, O., Huet, J. C., and Pernollet, J. C. (1999). Characterization of elicitin-like phospholipases isolated from *Phytophthora capsici* culture filtrate. *FEBS Lett.* **452,** 400–406.

Newhouse, J. R., Tooley, P. W., Smith, O. P., and Fishel, R. A. (1992). Characterization of double-stranded RNA in isolates of *Phytophthora infestans* from Mexico, the Netherlands, and Peru. *Phytopathology* **82,** 164–169.

Noguchi, M., Sawada, T., and Akazawa, T. (2001). ATP-regenerating system in the cilia of *Paramecium caudatum*. *J. Exp. Biol.* **204,** 1063–1071.

Nürnberger, T., and Brunner, F. (2002). Innate immunity in plants and animals: Emerging parallels between the recognition of general elicitors and pathogen-associated molecular patterns. *Curr. Opin. Plant Biol.* **5,** 318–324.

Orbach, M. J., Farrall, L., Sweigard, J. A., Chumley, F. G., and Valent, B. (2000). A telomeric avirulence gene determines efficacy for the rice blast resistance gene Pi-ta. *Plant Cell* **12,** 2019–2032.

Orsomando, G., Lorenzi, M., Raffaelli, N., Dalla Rizza, M., Mezzetti, B., and Ruggieri, S. (2001). Phytotoxic protein PcF, purification, characterization, and cDNA sequencing of a novel hydroxyproline-containing factor secreted by the strawberry pathogen *Phytophthora cactorum*. *J. Biol. Chem.* **276,** 21578–21584.

Osman, H., Mikes, V., Milat, M. L., Ponchet, M., Marion, D., Prange, T., Maume, B. F., Vauthrin, S., and Blein, J. P. (2001a). Fatty acids bind to the fungal elicitor cryptogein and compete with sterols. *FEBS Lett.* **489,** 55–58.

Osman, H., Vauthrin, S., Mikes, V., Milat, M. L., Panabieres, F., Marais, A., Brunie, S., Maume, B., Ponchet, M., and Blein, J. P. (2001b). Mediation of elicitin activity on tobacco is assumed by elicitin-sterol complexes. *Mol. Biol. Cell* **12,** 2825–2834.

Panabieres, F., and Le Berre, J. Y. (1999). A family of repeated DNA in the genome of the oomycete plant pathogen *Phytophthora cryptogea*. *Curr. Genet.* **36,** 105–112.

Panabieres, F., Birch, P. R., Unkles, S. E., Ponchet, M., Lacourt, I., Venard, P., Keller, H., Allasia, V., Ricci, P., and Duncan, J. M. (1998). Heterologous expression of a basic elicitin from *Phytophthora cryptogea* in *Phytophthora infestans* increases its ability to cause leaf necrosis in tobacco. *Microbiology* **144,** 3343–3349.

Panabieres, F., Amselem, J., Galiana, E., and Le Berre, J.-Y. (2005). Gene identification in the oomycete pathogen *Phytophthora parasitica* during *in vitro* vegetative growth through expressed sequence tags. *Fungal Genet. Biol.* **42,** 611–623.

Pemberton, C. L., and Salmond, G. P. C. (2004). The Nep1-like proteins-a growing family of microbial elicitors of plant necrosis. *Mol. Plant Pathol.* **5,** 353–359.

Pereira, C. A., Alonso, G. D., Paveto, M. C., Flawia, M. M., and Torres, H. N. (1999). L-arginine uptake and L-phosphoarginine synthesis in *Trypanosoma cruzi. J. Eukaryot. Microbiol.* **46,** 566–570.

Pieterse, C. M. J., Riach, M. B. R., Bleker, T., Van Den Berg-Velthuis, G. C. M., and Govers, F. (1993). Isolation of putative pathogenicity genes of the potato late blight fungus *Phytophthora infestans* by differential hybridization of a genomic library. *Physiol. Mol. Plant Pathol.* **43,** 69–79.

Pieterse, C. M. J., Derksen, A.-M. C. E., Folders, J., and Govers, F. (1994a). Expression of the *Phytophthora infestans ipiB* and *ipiO* genes in planta and *in vitro. Mol. Gen. Genet.* **244,** 269–277.

Pieterse, C. M. J., van West, P., Verbakel, H. M., Brasse, P. W. H. M., Van Den Berg-Velthuis, G. C. M., and Govers, F. (1994b). Structure and genomic organization of the *ipiB* and *ipiO* gene clusters of *Phytophthora infestans. Gene* **138,** 67–77.

Pipe, N. D., and Shaw, D. S. (1997). Telomere-associated restriction fragment length polymorphisms in *Phytophthora infestans. Mol. Plant Pathol.* Available online http://www.bspp.org.uk/mppol/1997/1124pipe.

Ponchet, M., Panabieres, F., Milat, M. L., Mikes, V., Montillet, J. L., Suty, L., Triantaphylides, C., Tirilly, Y., and Blein, J. P. (1999). Are elicitins cryptograms in plant-Oomycete communications? *CMLS Cell. Mol. Life Sci.* **56,** 1020–1047.

Prakob, W. (2005). Gene expression during oosporogenesis in *Phytophthora* spp. M.S. Thesis, University of California, Riverside.

Qiu, P., Qin, L., Sorrentino, R. P., Greene, J. R., Wang, L., and Partridge, N. C. (2003). Comparative promoter analysis and its application in analysis of PTH-regulated gene expression. *J. Mol. Biol.* **326,** 1327–1336.

Qutob, D., Hraber, P. T., Sobral, B. W. S., and Gijzen, M. (2000). Comparative analysis of expressed sequences in *Phytophthora sojae. Plant Physiol.* **123,** 243–253.

Qutob, D., Kamoun, S., and Gijzen, M. (2002). Expression of a *Phytophthora* sojae necrosis-inducing protein occurs during transition from biotrophy to necrotrophy. *Plant J.* **32,** 361–373.

Qutob, D., Huitema, E., Gijzen, M., and Kamoun, S. (2003). Variation in structure and activity among elicitins from *Phytophthora sojae. Mol. Plant Pathol.* **4,** 119–124.

Randall, T. A., Ah Fong, A., and Judelson, H. (2003). Chromosomal heteromorphism and an apparent translocation detected using a BAC contig spanning the mating type locus of *Phytophthora infestans. Fungal Genet. Biol.* **38,** 75–84.

Randall, T. A., Dwyer, R. A., Huitema, E., Beyer, K., Cvitanich, C., Kelkar, H., Ah Fong, A. M. V., Gates, K., Roberts, S., Yatzkan, E., Gaffney, T., Law, M., *et al.* (2005). Large-scale gene discovery in the oomycete *Phytophthora infestans* reveals likely components of phytopathogenicity shared with true fungi. *Mol. Plant Microbe Interact.* **18,** 229–243.

Rehmany, A. P., Grenville, L. J., Gunn, N. D., Allen, R. L., Paniwnyk, Z., Byrne, J., Whisson, S. C., Birch, P. R. J., and Beynon, J. L. (2003). A genetic interval and physical contig spanning the *Peronospora parasitica* (At) avirulence gene locus *ATR1Nd. Fungal Genet. Biol.* **38,** 33–42.

Rehmany, A. P., Gordon, A., Rose, L. E., Allen, R. L., Armstrong, M. R., Whisson, S. C., Kamoun, S., Tyler, B. M., Birch, P. R., and Beynon, J. L. (2005). Differential recognition of highly divergent downy mildew avirulence gene alleles by *RPP1* resistance genes from two *Arabidopsis* lines. *Plant Cell* **17,** 1839–1850.

Restrepo, S., Myers, K. L., del Pozo, O., Martin, G. B., Hart, A. L., Buell, C. R., Fry, W. E., and Smart, C. D. (2005). Gene Profiling of a compatible interaction between *Phytophthora infestans* and *Solanum tuberosum* suggests a role for carbonic anhydrase. *Mol. Plant Microbe Interact.* **18**, 913–922.

Ristaino, J. B., Groves, C. T., and Parra, G. R. (2001). PCR amplification of the Irish potato famine pathogen from historic specimens. *Nature* **411**, 695–697.

Rose, J. K. C., Ham, K.-S., Darvill, A. G., and Albersheim, P. (2002). Molecular cloning and characterization of glucanase inhibitor proteins: Coevolution of a counterdefense mechanism by plant pathogens. *Plant Cell* **14**, 1329–1345.

Sacks, W., Nürnberger, T., Hahlbrock, K., and Scheel, D. (1995). Molecular characterization of nucleotide sequences encoding the extracellular glycoprotein elicitor from *Phytophthora megasperma*. *Mol. Gen. Genet.* **246**, 45–55.

Sankoff, D., Bryant, D., Deneault, M., Lang, B. F., and Burger, G. (2000). Early eukaryote evolution based on mitochondrial gene order breakpoints. *J. Comput. Biol.* **7**, 521–535.

Sansome, E., and Brasier, C. M. (1973). Diploidy and chromosomal structural hybridity in *Phytophthora infestans*. *Nature* **241**, 344–345.

Sansome, E., and Brasier, C. M. (1974). Polyploidy associated with varietal differentiation in the megasperma complex of *Phytophthora*. *Trans. Br. Mycol. Soc.* **63**, 461–467.

Séjalon-Delmas, N., Villalba, F., Bottin, A., Rickauer, M., Dargent, R., and Esquerré-Tugayé, M. T. (1997). Purification, elicitor activity, and cell wall localisation of a glycoprotein from *Phytophthora parasitica* var. *nicotianae*, a fungal pathogen of tobacco. *Phytopathology* **87**, 899–909.

Shan, W., Cao, M., Leung, D., and Tyler, B. M. (2004a). The *Avr1b* locus of *Phytophthora sojae* encodes an elicitor and a regulator required for avirulence on soybean plants carrying resistance gene. *Rps1b. Mol. Plant. Microbe Interact.* **17**, 394–403.

Shan, W., Marshall, J. S., and Hardham, A. R. (2004b). Gene expression in germinated cysts of *Phytophthora nicotianae*. *Molec. Plant Pathol.* **5**, 317–330.

Shaw, D. S. (1983a). The cytogenetics and genetics of *Phytophthora*. In "*Phytophthora*: Its Biology, Taxonomy, Ecology, and Pathology" (D. C. Erwin, S. Bartnicki-Garcia, and P. H. Tsao, eds.), pp. 81–94. APS Press, St. Paul, MN.

Shaw, D. S. (1983b). The Peronosporales, a fungal geneticist's nightmare. In "Oosporic Plant Pathogens, A Modern Perspective" (S. T. Buczacki, ed.), pp. 85–121. Academic Press, London.

Shaw, D. S., and Khaki, I. A. (1971). Genetical evidence for diploidy in *Phytophthora*. *Genet. Res.* **17**, 165–167.

Shepherd, S. J., van West, P., and Gow, N. A. R. (2003). Proteomic analysis of asexual development of *Phytophthora palmivora*. *Mycol. Res.* **107**, 395–400.

Skalamera, D., Wasson, A. P., and Hardham, A. R. (2004). Genes expressed in zoospores of *Phytophthora nicotianae*. *Mol. Genet. Genomics* **270**, 549–557.

Sogin, M. L., and Silberman, J. D. (1998). Evolution of the protists and protistan parasites from the perspective of molecular systematics. *Int. J. Parasitol* **28**, 11–20.

Stechmann, A., and Cavalier-Smith, T. (2002). Rooting the eukaryote tree by using a derived gene fusion. *Science* **297**, 89–91.

Tani, S., and Judelson, H. S. (2006). Activation of zoosporogenesis-specific genes in *phytophthora infestans* involves a 7-nucleotide promoter motif and cold-induced membrane rigidity. *Eukaryot. Cell* **5**, 745–752.

Tani, S., Kim, K. S., and Judelson, H. S. (2004a). A cluster of NIF transcriptional regulators with divergent patterns of spore-specific expression in *Phytophthora infestans*. *Fungal Genet. Biol.* **42**, 42–50.

Tani, S., Yatzkan, E., and Judelson, H. S. (2004b). Multiple pathways regulate the induction of genes during zoosporogenesis in *Phytophthora infestans*. *Mol. Plant Microbe Interact.* **17**, 330–337.

Tian, M., Benedetti, B., and Kamoun, S. (2005). A Second Kazal-like protease inhibitor from *Phytophthora infestans* inhibits and interacts with the apoplastic pathogenesis-related protease P69B of tomato. *Plant Physiol.* **138,** 1785–1793.

Tian, M., Huitema, E., da Cunha, L., Torto-Alalibo, T., and Kamoun, S. (2004). A Kazal-like extracellular serine protease inhibitor from *Phytophthora infestans* targets the tomato pathogenesis-related protease P69B. *J. Biol. Chem.* **279,** 26370–26377.

Tooley, P. W., and Carras, M. M. (1992). Separation of chromosomes of *Phytophthora* species using CHEF gel electrophoresis. *Exp. Mycol.* **16,** 188–196.

Tooley, P. W., and Garfinkel, D. J. (1996). Presence of Ty1-copia group retrotransposon sequences in the potato late blight pathogen *Phytophthora infestans*. *Mol. Plant Microbe Interact.* **9,** 305–309.

Tooley, P. W., and Therrien, C. D. (1987). Cytophotometric determination of the nuclear DNA content of 23 Mexican and 18 non-Mexican isolates of *Phytophthora infestans*. *Exp. Mycol.* **11,** 19–26.

Torto, T. A., Rauser, L., and Kamoun, S. (2002). The *pipg1* gene of the oomycete *Phytophthora infestans* encodes a fungal-like endopolygalacturonase. *Curr. Genet.* **40,** 385–390.

Torto, T. A., Li, S., Styer, A., Huitema, E., Testa, A., Gow, N. A., van West, P., and Kamoun, S. (2003). EST mining and functional expression assays identify extracellular effector proteins from the plant pathogen *Phytophthora*. *Genome Res.* **13,** 1675–1685.

Tyler, B., Tripathy, S., Gunwald, N., Lamour, K., Ivors, K., Garbelotto, M., Rokhasr, D., Putnam, I., Grigoriev, I., and Boore, J. (2005). Genome sequence of *Phytophthora ramorum*: Implications for management [Abstract]. Sudden Oak Death Science Symposium, Monterey, CA, Jan. 18–21 p. 42.

Unkles, S. E., Moon, R. P., Hawkins, A. R., Duncan, J. M., and Kinghorn, J. R. (1991). Actin in the oomycetous fungus *Phytophthora infestans* is the product of several genes. *Gene* **100,** 105–112.

van der Lee, T., De Witte, I., Drenth, A., Alfonso, C., and Govers, F. (1997). AFLP linkage map of the oomycete *Phytophthora infestans*. *Fungal Genet. Biol.* **21,** 278–291.

van der Lee, T., Testa, A., van 't Klooster, J., van den Berg-Velthuis, G., and Govers, F. (2001). Chromosomal deletion in isolates of *Phytophthora infestans* correlates with virulence on R3, R10, and R11 potato lines. *Mol. Plant Microbe Interact.* **14,** 1444–1452.

van der Lee, T., Testa, A., Robold, A., van 't Klooster, J. W., and Govers, F. (2004). High density genetic linkage maps of *Phytophthora infestans* reveal trisomic progeny and chromosomal rearrangements. *Genetics* **157,** 949–956.

van der Lee, T., Mendes, O., van der Schoot, H., Ruyter-Spira, C., Hekkert, B. T. L., Govers, F., and Kema, G. H. J. (2005). Generation of a large set of microsatellite-markers for *Phytophthora infestans* by mining sequence data. 23rd Fungal Genetics Conference, Abstract 128.

van Dijk, K., Fouts, D. E., Rehm, A. H., Hill, A. R., Collmer, A., and Alfano, J. R. (1999). The Avr (effector) proteins HrmA (HopPsyA) and AvrPto are secreted in culture from *Pseudomonas syringae* pathovars via the Hrp (type III) protein secretion system in a temperature- and pH-sensitive manner. *J. Bacteriol.* **181,** 4790–4797.

Van Kan, J. A. L., Van Den Ackerveken, G. F. J. M., and De Wit, P. J. G. M. (1991). Cloning and characterization of complementary DNA of avirulence gene *avr9* of the fungal pathogen *Cladosporium fulvum* causal agent of tomato leaf mold. *Mol. Plant Microbe Interact.* **4,** 52–59.

van West, P., Kamoun, S., Van 't Klooster, J. W., and Govers, F. (1999a). Internuclear gene silencing in *Phytophthora infestans*. *Mol. Cell* **3,** 339–348.

van West, P., Reid, B., Campbell, T. A., Sandrock, R. W., Fry, W. E., Kamoun, S., and Gow, N. A. R. (1999b). Green fluorescent protein (GFP) as a reporter gene for the plant pathogenic oomycete *Phytophthora palmivora*. *FEMS Microbiol. Lett.* **178,** 71–80.

van West, P., Appiah, A. A., and Gow, N. A. R. (2003). Advances in root pathogenic oomycete research. *Phys. Mol. Plant Pathol.* **62,** 99–113.

Vijn, I., and Govers, F. (2003). *Agrobacterium tumefaciens* mediated transformation of the oomycete plant pathogen *Phytophthora infestans*. *Mol. Plant Pathol.* **4,** 456–467.

Voglmayr, H., and Greilhuber, J. (1998). Genome size determination in peronosporales (Oomycota) by Feulgen image analysis. *Fungal Genet. Biol.* **25**, 181–195.

Whisson, S. C., Basnayake, S., Maclean, D. J., Irwin, J. A., and Drenth, A. (2004). *Phytophthora sojae* avirulence genes *Avr4* and *Avr6* are located in a 24kb, recombination-rich region of genomic DNA. *Fungal Genet. Biol.* **41**, 62–74.

Whisson, S. C., Avrova, A. O., Lavrova, O., and Pritchard, L. (2005a). Families of short interspersed elements in the genome of the oomycete plant pathogen, *Phytophthora infestans*. *Fungal Genet. Biol.* **42**, 351–365.

Whisson, S. C., Avrova, A. O., van West, P., and Jones, J. T. (2005b). A method for double-stranded RNA-mediated transient gene silencing in *Phytophthora infestans*. *Mol. Plant Pathol.* **6**, 153–163.

Whittaker, S. L., Shattock, R. C., and Shaw, D. S. (1992). The duplication cycle and DAPI-DNA contents in nuclei of germinating zoospore cysts of *Phytophthora infestans*. *Mycol. Res.* **96**, 355–358.

Win, J., Kanneganti, T.-D., Torto-Alalibo, T., and Kamoun, S. (2006). Computational and comparative analyses of 150 near full-length cDNA sequences from the oomycete plant pathogen *Phytophthora infestans*. *Fungal Genet. Biol.* **43**, 20–33.

Yoon, H. S., Hackett, J. D., Van Dolah, F. M., Nosenko, T., Lidie, K. L., and Bhattacharya, D. (2005). Tertiary endosymbiosis driven genome evolution in dinoflagellate algae. *Mol. Biol. Evol.* **22**, 1299–1308.

Zody, M., O'Neill, K., Handsaker, B., Karlsson, E., Govers, F., van de Vondervoort, P., Weide, R., Whisson, S., Birch, P., Ma, L.-J., Ristaino, J., Fry, W., *et al.* (2005). Sequencing the *Phytophthora infestans* genome: Preliminary studies [Abstract] *Phytophthora* Molecular Genetic Network Workshop, p. 21.

4

Sex and Virulence of Human Pathogenic Fungi

Kirsten Nielsen* and Joseph Heitman*,†,‡
*Department of Molecular Genetics and Microbiology
Duke University Medical Center, Durham, North Carolina 27710
†Department of Pharmacology and Cancer Biology
Duke University Medical Center, Durham, North Carolina 27710
‡Department of Medicine, Duke University Medical Center
Durham, North Carolina 27710

ABSTRACT

Over the past decade, opportunistic fungal infectious diseases have increased in prevalence as the population of immunocompromised individuals escalated due to HIV/AIDS and immunosuppression associated with organ transplantation and

Advances in Genetics, Vol. 57
Copyright 2007, Elsevier Inc. All rights reserved.

0065-2660/07 $35.00
DOI: 10.1016/S0065-2660(06)57004-X

cancer therapies. In the three predominant human pathogenic fungi (*Candida albicans*, *Cryptococcus neoformans*, and *Aspergillus fumigatus*), a unifying feature is that all three retained the machinery needed for sex, and yet all limit their access to sexual reproduction. While less well characterized, many of the other human pathogenic fungi also appear to have the ability to undergo sexual reproduction. Recent studies with engineered pairs of diploid strains of the model yeast *Saccharomyces cerevisiae*, one that is sexual and the other an obligate asexual, provide direct experimental validation of the benefits of both sexual and asexual reproduction. The obligate asexual strain had an advantage in response to constant environmental conditions whereas the sexual strain had a competitive edge under stressful conditions (Goddard *et al.*, 2005; Grimberg and Zeyl, 2005). The human pathogenic fungi have gone to great lengths to maintain all of the machinery required for sex, including the mating-type locus and the pheromone response and cell fusion pathways. Yet these pathogens limit their access to sexual or parasexual reproduction in unique and specialized ways. Our hypothesis is that this has enabled the pathogenic fungi to proliferate in their environmental niche, but to also undergo genetic exchange via sexual reproduction in response to stressful conditions such as new environments, different host organisms, or changes in the human host such as antimicrobial therapy. Further study of the sexual nature of the human pathogenic fungi will illuminate how these unique microbes have evolved into successful pathogens in humans.   © 2007, Elsevier Inc.

I. THE PREDOMINANT HUMAN PATHOGENIC FUNGI

Of the greater than 100,000 species of fungi identified to date, only a handful routinely cause disease in humans (Perfect, 2005). This chapter will focus on the most common systemic human fungal pathogens: *Candida albicans*, *Cryptococcus neoformans*, and *Aspergillus fumigatus*. These three organisms are among the best characterized of the human pathogenic fungi. In addition, we also consider what is known about mating or mating type in several other fungal pathogens, including *Pneumocystis*, *Histoplasma capsulatum*, *Coccidioides immitis*, *Coccidioides posadasii*, *Candida glabrata*, and *Aspergillus flavus*.

Only two of the human pathogenic fungi are closely associated with humans. *C. albicans* is a normal part of the human gastrointestinal flora. *C. albicans* is probably best known for causing mucosal infections such as thrush or vaginitis. However, when the immune system is compromised, *C. albicans* can become invasive and colonize virtually any human tissue and organ to cause life-threatening systemic infections (Johnson, 2003).

The other fungal pathogen closely associated with humans is *Pneumocystis jiroveci*. *Pneumocystis* species were originally mistaken as protozoan until routine sequencing in the 1980s revealed that they belong to the fungal kingdom and are more closely related to the nonpathogenic fission yeast *Schizosaccharomyces pombe*

(Cushion, 2004; Edman *et al.*, 1988; Stringer *et al.*, 1989a,b). The *Pneumocystis* species complex encompasses a family of organisms that are host specific; thus, the species that infects humans is *Pneumocystis jiroveci* and the species that infects rats is *Pneumocystis carinii* (Frenkel, 1999). In immunocompromised individuals, *Pneumocystis* infection produces a lethal pneumonia if left untreated (Cushion, 2004). While the infectious form of the organism is unclear, *Pneumocystis* has been shown to be transmitted by an airborne route and asymptomatic carriage has been shown for both healthy and immunocompromised individuals (Helweg-Larsen *et al.*, 2002; Hughes, 1982; Icenhour *et al.*, 2001; Wakefield, 1994; Wakefield *et al.*, 2003).

The rest of the predominant human pathogenic fungi are found in the environment and the infectious propagules are thought to be spores. Cryptococcosis is caused by three varieties or sibling species that have been diverging over the past 10–40 million years and which exhibit dramatic differences in their virulence properties and environmental distribution. *C. neoformans* var. *grubii* (serotype A) isolates cause the vast majority (95%) of infections worldwide and greater than 99% of infections in AIDS patients (Casadevall and Perfect, 1998). *C. neoformans* var. *neoformans* (serotype D) strains account for less than 5% of infections worldwide but can cause up to 20% of clinical infections in Europe, many in the context of a hybrid serotype AD genetic background which may be more pathogenic than serotype D alone (Barchiesi *et al.*, 2005; Tintelnot *et al.*, 2004). Unlike varieties *grubii* and *neoformans*, which predominantly infect immunocompromised individuals, the sibling species *Cryptococcus gattii* (serotypes B and C) strains can infect immunocompetent individuals and are found largely in tropical regions of the world. Nevertheless, an outbreak of *C. gattii* is currently underway in British Columbia, Canada.

C. neoformans and *C. gattii* have been cultured from the environment from various tree species, soil, and bird guano. Spores or desiccated yeast cells are thought to be inhaled and lodged in the alveoli of the lungs. This initial pulmonary infection is limited or asymptomatic, and is either cleared or the organism enters a latent form from which it can later be reactivated in response to immunosuppression and spread to the central nervous system (CNS) to cause fatal meningoencephalitis (Hull and Heitman, 2002).

Invasive pulmonary aspergillosis is the most common manifestation of *Aspergillus* infection in humans and is observed in 90% of patients with aspergillosis (Fisher *et al.*, 1981). Development of invasive aspergillosis is an interplay between the inoculating dose of spores, the ability of the host to resist infection on both the local and systemic levels, and the virulence of the organism (Fraser, 1993). Infection of the CNS may follow hematogenous seeding from primary sites of infection such as the lung (Boes *et al.*, 1994; Boon *et al.*, 1990; Hagensee *et al.*, 1994; Walsh *et al.*, 1985).

Coccidioidomycosis is caused by two sibling species, *C. immitis* and *C. posadasii*. *Coccidioides* propagates in soil in the semiarid regions of the southwestern United States, Mexico, and Central and South America. Humans acquire coccidioidomycosis by inhalation of the arthroconidia, which differentiate into large endosporulating spherules in the host. *Coccidioides* is capable of causing progressive pulmonary and disseminated disease in previously healthy individuals (Chiller *et al.*, 2003). *H. capsulatum* is found in soil worldwide, particularly in soil contaminated with bird or bat guano. The infectious propagules are microconidia produced by hyphae which can be inhaled into the lung to cause histoplasmosis.

The human pathogenic fungi do not form an evolutionarily distinct monophyletic class of organisms. Instead, they are polyphyletic and closely related to nonpathogenic fungi (Bowman *et al.*, 1996). Thus, it is unclear what makes a fungus pathogenic in humans and how this pathogenicity evolves. One aspect now at the forefront of the study of human pathogenic fungi is the role mating plays in the evolution and pathogenicity of these fungi. Sexual cycles have been known for some species of human pathogenic fungi for over 30 years; however, other species were thought to be asexual. In this chapter, we discuss recent findings suggesting that these "asexual" fungi have extant sexual cycles. We examine results from classical mycology experiments defining the known sexual cycles. We then discuss the use of genomics to identify mating loci and mating machinery in organisms previously thought to be "asexual." Results of population genetic studies are analyzed to determine the extent to which sexual reproduction may be playing a role in the population structure and survival of the human pathogenic fungi. Finally, we examine the role mating and mating type play in the virulence of human pathogenic fungi.

II. SEX IN FUNGAL PATHOGENS: COST VERSUS BENEFIT

Sexual species dominate our world. Most plants and animals are sexual. In organisms that can normally multiply asexually, such as microbes and fungi, sexual processes are rarely completely absent. While exclusively asexual microorganisms, such as the bdelloid rotifers, are extremely rare, microorganisms with facultative parthenogenesis—the ability to produce offspring sexually or asexually—are quite common.

Most fungi can produce both asexual and sexual progeny. Asexual reproduction involves the generation of progeny by mitotic processes to produce daughter cells in yeasts or asexual spores such as conidia in filamentous fungi. In contrast, sexual reproduction involves nuclear fusion and subsequent spore production by meiotic processes (Williams, 1975). In species that exhibit both asexual and sexual life cycles, these two distinct phases of reproduction often occur under different environmental/nutritional conditions. Many fungi reproduce asexually when conditions are favorable for growth but reproduce sexually in nutrient-limiting conditions (Dyer *et al.*, 1992).

There are several benefits to generating both sexual and asexual progeny. Sexual reproduction can lead to increased genetic variation, which promotes adaptation by allowing beneficial mutations to spread and enables removal of deleterious genes (Burt, 2000; Butlin, 2002; Lenski, 2001). However, asexual reproduction preserves well-adapted combinations of genes that might be lost during sexual recombination (Bell, 1982). Furthermore, the metabolic costs of sexual reproduction are often higher than those for asexual reproduction resulting in an added advantage to asexual reproduction (Grimberg and Zeyl, 2005).

Studies in *S. cerevisiae* elegantly show the benefits of both sexual and asexual reproduction. In two independent studies by Goddard *et al.* (2005) and Grimberg and Zeyl (2005), *S. cerevisiae* strains were developed that differed only in their capacity for sexual reproduction. The asexual strain was still able to generate spores but differed from the sexual strain in its capacity for recombination and random assortment of chromosomes. Thus, both strains would suffer from the metabolic costs of sporulation but only the sexual strain would be able to reap the benefits. The growth of these strains was examined in different environments—both stressful and benign. The asexual strains were able to outcompete the sexual strains in the benign environments but the sexual strains proved to be more fit under stressful growth conditions. It is of interest to note that similar studies in the human pathogenic fungus *C. neoformans* revealed that extended propagation under conditions favorable for asexual growth resulted in a decrease in sexual reproduction (Xu, 2002, 2005).

Despite the advantages of sexual reproduction, a complete sexual cycle has not as yet been defined for many human pathogenic fungi, including *A. fumigatus, C. albicans, C. glabrata, C. immitis,* and *C. posadasii.* This observation has led to the hypothesis that sexual reproduction would disrupt the complex combinations of genes required for virulence by these fungi, resulting in a predominance of "asexual" species as human pathogenic fungi (Whelan, 1987). Yet recent evidence of sexual reproduction in "asexual" species is requiring a revision of our ideas on the role of sex in human pathogenic fungi. Thus, an alternate hypothesis is that the human pathogenic fungi have retained the ability to generate either clonal or recombining population structures in response to either constant or changing environments by preserving their ability to undergo sexual (or parasexual) reproduction but limiting the conditions under which sexual reproduction occurs in unique ways (Fig. 4.1).

In *C. neoformans* and *C. gattii,* sexual reproduction is limited by a nearly unisexual population to rare encounters between **a** and α cells and by a recently discovered α–α homothallic sexual cycle that would promote inbreeding. In the **a**/α obligate diploid *C. albicans,* sex is limited by the need to first undergo homozygosis at the mating-type locus to generate **a**/**a** or α/α strains and second to switch from white to the mating-specialized opaque cell type. Even then, only a parasexual and no true sexual cycle has been discovered in *C. albicans.* Finally,

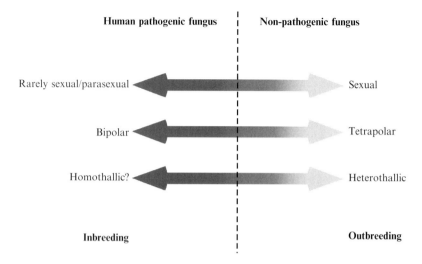

Figure 4.1. A continuum for the role of sexual reproduction in virulence of human pathogenic fungi. Sexual reproduction would disrupt the complex combinations of genes required for virulence by human pathogenic fungi. As a consequence, these fungi may have evolved sexual systems with limited mating that promotes inbreeding or with reduced recombination (parasexual reproduction) to reduce the impact of sexual reproduction on virulence. Thus, one difference between nonpathogenic fungi and human pathogenic fungi would be the pathogenic species' ability to retain sexual machinery to enable responses to drastic environmental changes but still limit sexual reproduction and promote inbreeding and clonality. The human pathogenic fungi appear to tightly regulate their sexual cycles and thus are rarely sexual. All human pathogenic fungi characterized to date are bipolar and, in the case of *Cryptococcus*, homothallic reproduction between strains of the same mating type has also been observed.

in the haploid yeast *C. glabrata*, which like the model fungus *S. cerevisiae* contains both mating-type alleles—one in a silent locus, inefficient silencing in **a** and α haploid cells renders them functionally **a**/α, precluding mating under conditions examined so far.

III. MATING-TYPE LOCI ARE THE SEX-DETERMINING REGIONS IN FUNGI

During sexual reproduction, specialized regions of the genome determine self–nonself interactions. In plants and animals, this region is present as the sex-determining chromosomes such as the X and Y chromosomes in humans. In fungi, the sex-determining region of the genome is the mating-type locus. Similar to mammalian sex chromosomes, recombination within the mating-type locus is suppressed to avoid generating self-fertile, sterile, or inviable offspring (Lengeler *et al.*, 2002). Rearrangements and extensive sequence divergence between alleles of the mating-type locus reduce recombination, akin to balancer chromosomes in *Drosophila* genetics (Steinemann and Steinemann, 2005).

Fungi can have either bipolar or tetrapolar mating systems. In the bipolar mating systems, a single mating-type locus with two alleles controls cell identity by encoding transcription factors. Bipolar mating systems are common in the ascomycetous fungi that include many of the human fungal pathogens, including *C. albicans*, *A. fumigatus*, *C. immitis*, and *H. capsulatum*, as well as the model yeast *S. cerevisiae*. In contrast, tetrapolar mating systems have two independent, unlinked mating loci that can be multiallelic, giving rise to thousands of different mating types. Tetrapolar mating systems are common in the basidiomycetous fungi. In the tetrapolar system, one locus encodes transcription factors and the second encodes pheromones and pheromone receptors. Thus, mating is controlled in a two-step manner. If either the pheromone cannot be recognized by the receptor or the homeodomain proteins cannot come together to form a heterodimer then mating does not proceed (Casselton, 2002; Casselton and Olesnicky, 1998; Feldbrugge *et al.*, 2004). It is of interest to note that pheromone/receptor recognition is not a prerequisite for cell fusion in all basidiomycetes. For example, in the model mushroom *Coprinus cinereus* cell fusion occurs promiscuously and is not controlled by pheromones; instead pheromone signaling regulates nuclear migration and clamp-cell fusion in the heterokaryon (Casselton and Olesnicky, 1998). The homeodomain transcription factors control various aspects of mating such as clamp-cell formation, septation, and pairing of nuclei with coordinated nuclear division (Casselton, 2002; Casselton and Olesnicky, 1998). It is the homeodomain protein-encoding locus that most closely resembles the single mating-type locus that is common in the ascomycetous fungi.

Tetrapolar mating systems promote outbreeding because progeny will only be able to mate with a quarter of their sibling offspring, whereas bipolar mating systems promote inbreeding by allowing the progeny to mate with half of the offspring. Interestingly, available genome sequence evidence suggests that all of the human pathogenic fungi appear to have bipolar mating-type systems, including the basidiomycete *C. neoformans*. Thus, we speculate that the human pathogenic fungi might require a higher level of inbreeding to maintain the complex set of genes required for virulence (Fig. 4.1).

IV. SEX IN *CRYPTOCOCCUS*

A sexual cycle was first reported over 30 years ago for *C. neoformans* var. *neoformans* (serotype D) and the sibling species *C. gattii* (serotypes B and C), and involves fusion of cells of opposite mating type, **a** and α, to produce heterokaryotic filaments (Kwon-Chung, 1975, 1976a,b). In response to nutrient limitation, **a** and α cells produce peptide pheromones that trigger conjugation tube formation in α cells and uniform expansion of **a** cells, leading to cell fusion. Nuclear fusion is delayed, and the resulting heterokaryon adopts a florid filamentous state. The filaments ultimately produce basidia, wherein nuclear fusion and meiosis occur and long chains of basidiospores are produced by repeated rounds of mitosis and budding (Fraser *et al.*, 2003; Kwon-Chung, 1975, 1976a,b). While mating in C. *neoformans* var. *neoformans* enabled classical genetic studies to identify basic virulence factors in *Cryptococcus*, such as the polysaccharide capsule and the pigment melanin, clinical infections with var. *neoformans* serotype D strains account for a limited number of infections compared to var. *grubii* serotype A strains.

Early studies had only identified α mating-type strains in var. *grubii*, leading to the hypothesis that var. *grubii* had evolved to be asexual concomitant with its emergence as the predominant pathogenic variety. In this hypothesis, the complex combinations of genes required for virulence in *Cryptococcus* would no longer be disrupted by meiosis and **a** mating-type strains would become extinct (Lengeler *et al.*, 2000). Instead, recent studies reveal that the **a**–α sexual cycle in var. *grubii* is limited geographically.

C. *neoformans* var. *grubii* **a** mating-type strains have recently been identified. The first strain to be identified (125.91) was found in a collection of four clinical isolates from the cerebrospinal fluid of patients with crytococcal meningitis in Tanzania (Lengeler *et al.*, 2000). The second var. *grubii* **a** mating-type strain identified was an environmental isolate from a collection of 66 strains from Italy (Viviani *et al.*, 2001). While the original and second var. *grubii* **a** mating-type strains would appear to be geographically distinct, Montagna (2002) provided a detailed description of the isolation of the Italian environmental isolate indicating that the strain came from a farm 3 km from a game park housing a variety of plants and animals from Africa—suggesting a possible African link for both var. *grubii* **a** mating-type strains. Subsequently, an additional 14 var. *grubii* **a** strains were identified in a collection of 139 clinical isolates from Botswanan AIDS patients (Litvintseva *et al.*, 2003). Furthermore, analysis of the Botswanan clinical isolates suggests that sexual recombination is recent or ongoing in this population containing both **a** and α mating-type strains (Litvintseva *et al.*, 2003, 2005). Only one additional var. *grubii* **a** strain has been identified to date, a clinical isolate from a Hungarian patient with no foreign travel history. Thus, of the 17 var. *grubii* **a** strains identified to date, 15 of the strains are from African AIDS patients and at least 1 of the other 2 strains might be of African descent (Fig. 4.2). This observation leads us to propose that var. *grubii* **a** strains may be predominantly geographically restricted to sub-Saharan Africa, and thus the extant **a**–α sexual cycle might also be geographically restricted to this region.

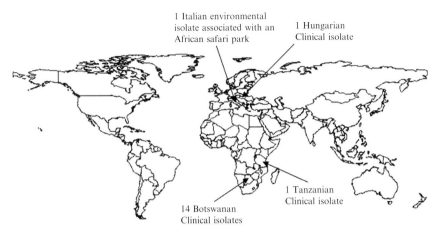

1 Italian environmental
isolate associated with an
African safari park

1 Hungarian
Clinical isolate

1 Tanzanian
Clinical isolate

14 Botswanan
Clinical isolates

Figure 4.2. Geographic distribution of serotype A MATa strains of *C. neoformans* var. *grubii*. Of the 17 serotype A MATa strains identified to date, 15 strains are clinical isolates from sub-Saharan Africa and the other 2 strains are from Europe. The only identified environmental isolate was found in Italy associated with an African game park.

These **a** mating-type strains have been used to identify and characterize mating in C. *neoformans* var. *grubii* (Keller *et al.*, 2003; Nielsen *et al.*, 2003). This mating is highly similar to the sexual cycle described for var. *neoformans* (Kwon-Chung, 1975, 1976b; McClelland *et al.*, 2004); however, some differences have been observed. For example, the var. *grubii* strains have mating specificity that goes beyond mating type. The original var. *grubii* **a** mating-type strain was only able to mate with 3 out of 150 α strains tested and the mating partner preference of progeny strains from a cross differed from the parental strains (Nielsen *et al.*, 2003). These data support the hypothesis that the initial recognition stages in mating may differ between the highly pathogenic var. *grubii* and the less-pathogenic var. *neoformans*. Additional studies examining the initial stages of mating in multiple strains of both *grubii* and *neoformans* will be necessary to further characterize the similarities and differences between these two *Cryptococcus* varieties.

One interesting aspect of *Cryptococcus* ecology that would affect the prevalence of mating in this organism is the observation that α mating-type strains predominate in both the environmental and clinical isolates. In var. *neoformans*, α mating type strains outnumber **a** strains by a ratio of almost 45:1 in environmental isolates and by almost 30:1 in clinical isolates (Kwon-Chung and Bennett, 1978). Thus, **a** strains only account for less than 2% of the var. *neoformans* population. This predominance of α strains is even more dramatic in var. *grubii* where only 17 **a** mating-type strains have been identified from >2000 isolates worldwide. Interestingly, 14 of the var. *grubii* **a** strains are Botswanan clinical isolates where they account for about 10% of clinical isolates (Litvintseva *et al.*, 2003). The other three **a** strains account for only 0.1% of the worldwide var. *grubii* population. At least two theories could explain the predominance of α strains. First, *Cryptococcus* could essentially be a clonal bloom of an α isolate and the low level of recombination observed could be due to a small amount of mating between **a** and α strains which, in the case of var. *grubii*, is likely only occurring in geographically distinct regions such as sub-Saharan Africa.

A second possibility is that the low level of recombination observed in *Cryptococcus* is actually due to sexual reproduction that does not require mating between strains of the opposite mating type, but rather same sex mating between two α strains. Recent evidence supports this hypothesis. Along with classical sexual reproduction, *C. neoformans* var. *neoformans* strains can also undergo a developmental transition known as monokaryotic fruiting (Tscharke *et al.*, 2003; Wickes *et al.*, 1996). This process resembles mating and was thought to be similar to asexual conidia formation in ascomycetes. However, studies by Lin *et al.* (2005) reveal that monokaryotic fruiting in var. *neoformans* α mating-type strains is a sexual process and results in the production of recombinant progeny. While monokaryotic fruiting has not been observed yet in the laboratory for either var. *grubii* or *C. gattii*, evidence suggests that monokaryotic fruiting between two α strains may have produced a novel strain that resulted in the ongoing outbreak of *C. gattii* in British Columbia, Canada (Fraser *et al.*, 2005). Thus, it appears that *Cryptococcus* is both a heterothallic and a facultatively homothallic fungus. However, unlike classical homothallic fungi, such as *S. cerevisiae* and *Aspergillus nidulans*, where one organism will contain both mating types, it appears that two organisms of the same mating type undergo sexual reproduction to produce recombinant progeny. This homothallic mating, by either selfing or outcrossing, likely allows for greater inbreeding in *Cryptococcus* and enables sexual reproduction in geographic regions where only one mating type is present (Fig. 4.1).

Another interesting aspect of *Cryptococcus* biology is the presence of naturally occurring AD hybrid strains between var. *grubii* (serotype A) and var. *neoformans* (serotype D) that arose via intervarietal sexual crosses (Cogliati *et al.*, 2001; Lengeler *et al.*, 2001). These AD hybrid strains often contain both mating types, either αADa or aADα, and thus provided some of the early support for the existence of var. *grubii* serotype A **a** mating-type strains. Interestingly, the vast majority of spores generated from self-fertile AD hybrid strains germinate poorly to produce progeny that are still diploid or aneuploid, suggesting that a complete sexual cycle is not possible between var. *grubii* and var. *neoformans* and that these varieties in fact represent separate species (Lengeler *et al.*, 2001). Population genetic and genomic studies (Kavanaugh *et al.*, 2006) reveal little or no gene flow between var. *grubii* and var. *neoformans* populations, further supporting the presence of cryptic speciation.

V. THE UNUSUAL *CRYPTOCOCCUS* MATING-TYPE LOCUS

Perhaps the most unusual mating-type locus in the human pathogenic fungi belongs to C. *neoformans*. *Cryptococcus* is in the phylum Basidiomycota, and basidiomycetous fungi often have tetrapolar mating systems with two unlinked loci in which strains must differ at both loci for mating to succeed. In *Cryptococcus*, the two mating-type loci normally present in the basidiomycetous tetrapolar mating system have fused to produce one very large mating-type locus that contains the homeodomain transcription factors as well as the pheromones and pheromone receptors (Fig. 4.3) (Lengeler *et al.*, 2002; Fraser *et al.*, 2004; Loftus *et al.*, 2005). Furthermore, the mating-type locus has rearranged dramatically during strain divergence (var. *neoformans*, var. *grubii*, and C. *gattii*), perhaps concomitantly with speciation events (Fraser *et al.*, 2004). Along with the basic elements of mating-type loci, the *Cryptococcus* mating-type locus also contains additional pheromone response pathway elements and genes from many other functional categories, including several essential genes that punctuate the mating-type locus and likely constrain its evolution by preventing large-scale deletions (Fraser and Heitman, 2004; Lengeler *et al.*, 2002). Thus, mating in *Cryptococcus* appears to have evolved from a tetrapolar system to a bipolar system. Interestingly, all of the predominant human pathogenic fungi have bipolar mating systems, even *Cryptococcus* which could conceivably have had a tetrapolar system. What role, if any, the presence of a bipolar mating system may play in the pathogenicity of fungi remains to be elucidated, but may involve shifts in the balance between inbreeding and outbreeding, similar to the shift in balance between clonal and sexual reproduction (Fig. 4.1).

VI. GENOME SEQUENCING IDENTIFIED MATING-TYPE LOCUS IN THE "ASEXUAL" *C. ALBICANS* AND LED TO THE DISCOVERY OF MATING

C. *albicans* is a diploid organism that for more than a century was thought to be strictly asexual. Yet with the advent of genomics, a window was opened on its sexual/parasexual cycle. Using early sequence traces, Hull and Johnson (1999) cloned and sequenced the two alleles of the C. *albicans* mating-type locus based on homology to the model ascomycete S. *cerevisiae* mating-type (Fig. 4.3) locus.

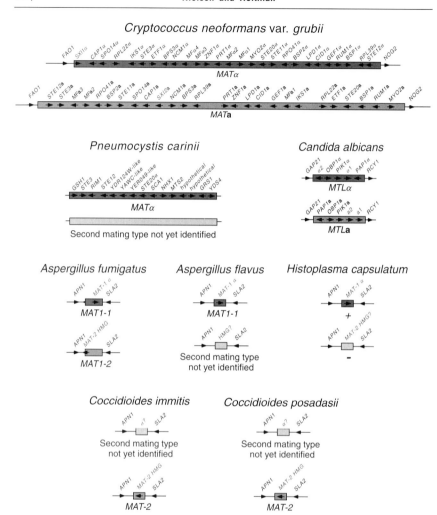

Figure 4.3. Mating-type loci in the human pathogenic fungi. Genome sequencing has identified mating-type loci in all of the predominant human pathogenic fungi. Both alleles of the mating-type locus have been identified in *Cryptococcus neoformans*, *Candida albicans*, and *Aspergillus fumigatus*, and are designated by yellow and blue boxes. Only one allele has been identified in several human pathogenic fungi based on homology and synteny to known mating-type loci in other organisms. A hypothetical second allele for these mating-type loci has been proposed (green boxes) based on homology and synteny with loci from closely related organisms. In cases where only one mating-type allele has been identified, the boundaries of the locus are speculative. Homeodomain, α-box, or HMG box-encoding genes are designated in red. Other genes present in the mating-type locus are designated in black. Not to scale. (See Color Insert.)

Similar to *Cryptococcus*, the *C. albicans* mating-type locus is larger than expected. Along with the transcription factors, the *C. albicans* mating-type locus also contains genes whose protein products have no known function in mating, and which may be essential (Johnson, 2003). Identification of the *C. albicans* mating-type locus also explained why mating had not been previously observed. Due to the diploid nature of *C. albicans*, most strains are **a**/α heterozygous at the mating-type locus and thus contain both mating-type alleles (Hull and Johnson, 1999; Odds *et al.*, 2000). Characterization of several hundred clinical *C. albicans* isolates revealed that only 3% of strains are **a**/**a** or α/α homozygous at the mating-type locus and would therefore potentially be capable of mating (Lockhart *et al.*, 2002).

The characterization of a mating-type locus in *C. albicans* led to the discovery of mating and the characterization of a unique cell type specialized for mating (Johnson, 2003). Because the vast majority of *C. albicans* strains are **a**/α heterozygous, initial mating experiments were performed using hemizygous **a**/o and α/o strains or homozygous **a**/**a** or α/α strains, generated in the laboratory by gene disruption or induced chromosome loss and regain following sorbose selection (Hull *et al.*, 2000; Magee and Magee, 2000). These **a**/**a** and α/α strains were able to mate, both in the laboratory and *in vivo*, but only at an extremely low frequency. However, *C. albicans* has two distinct growth morphologies, designated as white and opaque. White cells are generally round and form dome-shaped colonies on agar plates whereas opaque cells are elongated and form dark, flat colonies (Fig. 4.4). Control of white–opaque switching is regulated in part by the *C. albicans* mating-type locus, suggesting that this switch may be involved in mating. In fact, opaque cells are at least 1 million times more efficient at mating than white cells, and only opaque cells respond to pheromone by initiating conjugation tube formation and cell cycle arrest (Miller and Johnson, 2002; Zhao *et al.*, 2005).

The cell biological events during mating of *C. albicans* have been well documented. Briefly, the diploid **a** and α cells undergo polarized growth (shmooing) and ultimately fuse to produce tetraploids. It was initially unclear whether the two diploid nuclei fuse to exchange genetic information. Studies using laboratory-derived auxotrophic hemizygous or homozygous strains showed mixing of genetic markers after mating (Hull *et al.*, 2000; Magee and Magee, 2000; Miller and Johnson, 2002); however, in initial microscopic studies with naturally occurring strains homozygous at the mating-type locus, nuclear fusion was not observed and it was not possible to detect segregation of mating type or other genetic markers (Lockhart *et al.*, 2003). Studies using fluorescently marked strains and karyogamy mutants provide definitive proof for nuclear fusion during *C. albicans* mating (Bennett and Johnson, 2005). The apparent difference between the early studies was explained when it was observed that

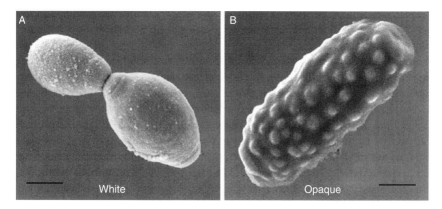

Figure 4.4. *C. albicans* white and opaque cells. Scanning electron micrographs of white (A) and opaque (B) cells. White cells are round and form dome-shaped colonies whereas opaque cells are elongated, "pimply" cells that form flat colonies. Opaque cells are mating competent with levels of mating 10^6 times higher than with white cells. Scale bar = 1 μm. Images courtesy of Karla Daniels and David Soll.

the efficiency of nuclear fusion varies between genetic backgrounds (Bennett and Johnson, 2005). Thus, it has been hypothesized that loci unlinked to the mating-type locus may affect the efficiency of nuclear fusion and ultimately the production of recombinant progeny in *C. albicans*, and these loci could serve to further promote inbreeding (Johnson, 2003).

The tetraploid cells generated during mating have not as yet been observed to undergo meiosis. However, it has been shown that *C. albicans* **a**/**a**/ α/α tetraploid mating products can efficiently return to the diploid state by random chromosome loss via a parasexual process (Bennett and Johnson, 2003). Thus, mating and chromosome loss in *C. albicans* may allow for enough shuffling of genetic material to produce genetic variation without meiosis. In this case, *C. albicans* still retains a parasexual cycle, but a classical sexual cycle involving recombination due to meiosis may have been lost. This parasexual cycle may support a lower level of recombination compared to meiosis (parameiosis), and thereby promote inbreeding. Because spores are often highly antigenic, the commensal *C. albicans* may have evolved to eschew the sporulation phase of the typical budding yeast sexual cycle to avoid provoking host recognition.

The *C. albicans* genome contains homologues of some of the genes involved in meiosis and sporulation in other true sexual yeasts, but the genome appears to lack a few genes thought to be crucial for meiosis and sporulation such as *IME1* and *SPO13* (Tzung *et al.*, 2001). However, an exclusive genomic approach may not identify these genes as the gene sets required for meiosis in budding and fission yeasts are quite distinct (Young *et al.*, 2004). Thus, it is possible that the C. *albicans* genome does contain crucial meiotic genes but they have little homology to those present in model yeasts such as S. *cerevisiae*.

VII. MATING-TYPE LOCUS IN *A. FUMIGATUS*

Fungal infections in humans due to *Aspergillus* species are predominantly caused by A. *fumigatus*. A representative A. *fumigatus* genome has been sequenced and a single mating-type locus with one gene encoding an HMG box transcription factor has been identified (Galagan *et al.*, 2005; Nierman *et al.*, 2005; Ronning *et al.*, 2005) (Fig. 4.3). Subsequent analysis of additional strains has identified the other mating type as encoding an α-box transcription factor (Paoletti *et al.*, 2005) (Fig. 4.3).

Not only is the mating-type locus present in this "asexual" *Aspergillus* species, but genes for pheromones, receptors, and components of the pheromone-signaling pathway have been identified. In A. *fumigatus*, these mating-associated genes are expressed and both mating types are present in the environment in a 1:1 ratio suggesting that the cells may be competent for mating (Dyer and Paoletti, 2005; Paoletti *et al.*, 2005). Furthermore, there is no difference in the ratios of the two mating types between environmental and clinical isolates, and there is no evidence for geographical isolation of one mating type from the other (Paoletti *et al.*, 2005).

The A. *fumigatus* genome also contains a repeat sequence, AFUT1, characteristic of a retrotransposable element. Although this element is defective due to multiple stop codons, the pattern of nucleotide transition is similar to the repeat-induced point mutation (RIP) process used in other fungi to inactivate foreign genetic elements during sexual reproduction and meiosis (Galagan and Selker, 2004; Latge *et al.*, 1991). Thus, A. *fumigatus* has the genetic components involved in mating and the genome shows evidence consistent with ongoing or recent mating but no one has as of yet observed mating to occur under laboratory-defined conditions.

VIII. MATING-TYPE LOCI IN OTHER HUMAN PATHOGENIC FUNGI

Representative genomes from many of the other human pathogenic fungi have also been targeted for genome sequencing because of their rising impact on human health. These genome sequences reveal that all of the common human pathogenic fungi have mating-type loci (Fig. 4.3). However, additional sequencing needs to be carried out to identify the different mating-type alleles in these human pathogenic fungi.

The *H. capsulatum* sexual cycle was defined by June Kwon-Chung 35 years ago (Kwon-Chung, 1972). Mating produces unusual coiled hyphae, but laboratory passage has been reported to result in a loss of fecundity and there have been no contemporary studies on mating in *H. capsulatum*. The two *H. capsulatum* mating types are designated as either + or − and recent genome sequencing of a + strain revealed the presence of a mating-type locus encoding an α-box transcription factor (Fig. 4.3). Similar to *A. fumigatus*, environmental samples of *H. capsulatum* have equal proportions of both mating types (Kwon-Chung, 1973). On the basis of similarity to other ascomycetes, we predict the − mating type will encode an HMG transcription factor.

Like *C. albicans*, the mating type of the *Pneumocystis* and *Coccidioides* strains was unknown prior to genome sequencing. While the mating-type locus of *Pneumocystis* has not been definitively identified, a cluster of genes potentially involved in pheromone sensing and signaling has been sequenced (Fig. 4.3). This region has similarity to the unusual mating-type locus of *Cryptococcus* and thus could represent a mating-type locus allele of *Pneumocystis* (Smulian *et al.*, 2001). Furthermore, a partial expressed sequence tag (EST) database revealed the transcription of many of the genes present in this locus as well as meiosis-specific genes such as *MEI2* (Cushion, 2004). Because of the complexity of the putative *Pneumocystis* mating-type locus, the components or structure of other mating-type alleles cannot be readily predicted.

Analysis of the *C. immitis* and *C. posadasii* genomes reveals a single mating-type locus encoding an HMG box protein but the second mating type has yet to be discovered. On the basis of their structural similarity to the mating-type loci of other ascomycete human pathogenic fungi, one prediction would be that some isolates will contain an opposite mating-type locus encoding an α-box transcription factor (Fig. 4.3). These pathogenic *Coccidioides* species have no as yet defined sexual cycle.

A. flavus is a close relative of *A. fumigatus* which not only causes a significant amount of disease in humans but is also a robust plant pathogen (Steinbach, 2005). Similar to *A. fumigatus*, sexual reproduction has not been characterized in *A. flavus*. The *A. flavus* genome sequence reveals a mating-type locus encoding an α-box transcription factor. Due to the similarity of the structure of the mating-type loci of these two *Aspergillus* species, one can hypothesize that the opposite mating type in *A. flavus* may encode an HMG box transcription factor (Fig. 4.3).

Laboratory-defined mating and/or sexual reproduction still remain elusive in many of the human pathogenic fungi. Both *A. fumigatus* and *A. flavus* exist within groups of closely related species able to undergo sexual reproduction, providing support for the presence of an as yet unidentified sexual cycle in both (Geiser *et al.*, 1998; Peterson, 2000; Poggeler, 1999; Varga *et al.*, 2000). The presence of this sexual state could have profound implications for the development of novel genotypes with enhanced virulence. Furthermore, identification of the sexual cycle in these *Aspergillus* species would allow for classical genetic studies to identify genes involved in virulence and their observed resistance to antifungal drugs. For mating to be characterized in *Pneumocystis*, *Coccidioides*, and *A. flavus*, strains of the opposite mating type first need to be identified. It remains to be seen whether the mating-type alleles in these species are equally prevalent, as in *A. fumigatus* and *C. albicans*, or whether one mating type will predominate, as in *C. neoformans*.

IX. POPULATION GENETIC STUDIES IN "ASEXUAL" FUNGI REVEAL EVIDENCE OF SEX

Studies examining the population structure of the human pathogenic fungi have provided insight into whether sexual recombination may occur in these fungi, and by extension what role this recombination may play in these organisms. Earlier studies in the human pathogenic fungi suggested little, if any, recombination, but as further insight has been gained about the population structure for these organisms, it is becoming apparent that active recombination is likely taking place.

For example, early population genetic studies examining strains from throughout the world suggested only a very low level of recombination in *C. neoformans* (Brandt *et al.*, 1995; Xu and Mitchell, 2003; Xu *et al.*, 2000). However, molecular analysis later identified different molecular subtypes in both *C. neoformans* (VNI, VNII, VNB, and VNIV) and *C. gattii* (VGI, VGII, VGIII, and VGIV) (Ellis *et al.*, 2000; Meyer *et al.*, 1999). Newer population genetic studies examining strains from individual molecular types strongly support the presence of ongoing or very recent recombination in both *C. neoformans* and *C. gattii* (Campbell *et al.*, 2005a; Litvintseva *et al.*, 2003, 2005). Furthermore, Litvintseva *et al.* (2003) showed evidence of recombination in a distinct population of var. *grubii* strains from Botswana (VNB) that contained both **a** and α mating types, suggesting that this recombination could be due to sexual reproduction (Litvintseva *et al.*, 2005). Similar studies also found a link between mating ability and the presence of recombination in the VGII molecular subtype of *C. gattii* that exhibits fertility in the laboratory (Campbell *et al.*, 2005a,b). While mating has now been observed in the laboratory for all of the clinically relevant *Cryptococcus* species, mating has not been directly observed in the environment. However, the isolation of αAD**a** and **a**ADα hybrid strains does provide evidence of at least past mating events. Thus, it seems likely that *Cryptococcus* is capable of sexual reproduction but that mating occurs infrequently in the environment. In the case of *C. neoformans* var. *grubii*, this mating also appears to be geographically restricted to sub-Saharan Africa. Given that var. *grubii* commonly infects AIDS patients, the discovery that the geographic site of ongoing sexual recombination overlaps the origin and current epicenter of the AIDS pandemic has obvious consequences for emergence of more virulent recombinant isolates in a large immunosuppressed patient population. Up to 50% of African AIDS patients are present with cryptococcal meningitis, for whom the prognosis is poor. Thus, further understanding of the sexual evolution of this pathogen may have direct therapeutic or prognostic implications.

A role of mating in population dynamics is also emerging in the *Candida* species. Several *Candida* species have completely defined sexual cycles involving meiosis and sporulation, including *Candida guillermondii*, *Candida krusei*, and *Candida lusitaniae* (Francois *et al.*, 2001; Young *et al.*, 2000). While these *Candida* species are emerging pathogens in humans, they are uncommon and the vast majority of human infections due to *Candida* are caused by the commensal fungi *C. albicans* and, to a lesser degree, *C. glabrata*. Both *C. albicans* and *C. glabrata* have no known complete sexual cycle, although a parasexual cycle has been characterized in *C. albicans* (as described above). Population genetic studies in *C. albicans* revealed that while the vast majority of the *C. albicans* population worldwide is clonal, some degree of genetic exchange can be detected (Graser *et al.*, 1996; Tibayrenc, 1997; Vilgalys *et al.*, 1997; Xu and Mitchell, 2002). However, these studies were carried out on diverse populations of *C. albicans*. Similar to *C. neoformans*, different molecular clades have been identified in *C. albicans* (Tavanti *et al.*, 2005). It remains to be seen whether higher levels of recombination can be detected within individual molecular clades. Alternatively, analysis of naturally occurring **a**/**a** and α/α homozygous strains, which could be the product of mating and parasexual reproduction rather than the precursor, may provide a more robust test of ongoing recombination. Finally, given that the identified cycle is parasexual, the level of recombination may be lower than in organisms with true sexual/meiotic cycles. Given that the human pathogenic fungi appear to limit sexual reproduction, evolution to a parasexual cycle could limit recombination and preclude sporulation to promote inbreeding and prevent production of antigenic spore coat proteins.

The evidence for recombination in *C. glabrata* is tenuous. There have been two studies examining *C. glabrata* populations—one that shows evidence of recombination and the other suggesting the population is clonal (de Meeus *et al.*, 2002; Dodgson *et al.*, 2005). Yet significant karyotype differences have been observed in *C. glabrata* populations that appear to be clonal (Klempp-Selb *et al.*, 2000). Similar to the closely related sexual species *S. cerevisiae*, *C. glabrata* retains both active and silent mating-type locus alleles (Srikantha *et al.*, 2003; Wong *et al.*, 2003). Evidence of mating-type switching has been presented and all of the machinery for sex is conserved (Brockert *et al.*, 2003; Butler *et al.*, 2004). Recent studies reveal that the silent mating-type cassette is incompletely silenced and hence **a** or α strains are functionally **a**/α (Cecile Fairhead, personal communication). This incomplete silencing may be the mechanism by which *C. glabrata* has evolved to limit entry into the sexual cycle; and conditions, mutations, or genome rearrangements that silence or inactivate mating-type loci may be necessary for mating. Alternatively, only some members of the *C. glabrata* population may be sexually active, and studies suggest a few rare isolates may secrete **a** pheromone (Jure Piškur, personal communication). Thus, further studies are necessary to define the sexual reproduction potential of *C. glabrata*.

Population genetic studies in the most commonly pathogenic *Aspergillus* species are consistent with ongoing sexual reproduction. Genetic variation within populations of both *A. fumigatus* and *A. flavus* shows high degrees of genetic diversity among both environmental and clinical isolates with up to 85% of isolates being unique (Bertout *et al.*, 2001; Chazalet *et al.*, 1998; Debeaupuis *et al.*, 1997; Geiser *et al.*, 2000; Latge *et al.*, 1991; Paoletti *et al.*, 2005; Pildain *et al.*, 2004; Tran-Dinh *et al.*, 1999; Varga and Toth, 2003). Using polymorphisms in three intergenic regions in *A. fumigatus*, Paoletti *et al.* (2005) showed that polymorphisms were shared between both mating types, consistent with recent or ongoing recombination by sexual or parasexual means. Interestingly, studies suggest that the single morphological species *A. fumigatus* may in fact contain two or more phylogenetic, or cryptic species, classified as *A. fumigatus* and *Aspergillus lentulus* (Balajee *et al.*, 2005; Pringle *et al.*, 2005). These two cryptic species show a global distribution with no correlation between genotype and geographic location, but may in fact explain clinical examples of difficult-to-treat or drug-resistant isolates.

Population genetic evidence also suggests the existence of mating in the "asexual" *Coccidioides* species. Unlike many of the other human pathogenic fungi, *Coccidioides* appears to be restricted to the Western hemisphere. Molecular phylogenic analysis of isolates from a single hospital reveals evidence suggesting the presence of sexual reproduction within this group (Burt *et al.*, 1996). Similar to *A. fumigatus*, analysis of representative strains from the entire *Coccidioides* species range indicates that sex may be occurring in two reproductively isolated taxa that have been classified as two distinct species, designated *C. immitis* and *C. posadasii* (Koufopanou *et al.*, 1997). No genetic flow has been observed between these two *Coccidioides* species even though their geographic regions overlap, and morphological differences between the species are not apparent (Fisher *et al.*, 2001; Koufopanou *et al.*, 2001).

Taken together, these population genetic studies suggest that the "asexual" human pathogenic fungi may have some form of genetic exchange between individuals. For many of these fungi, it remains to be seen whether this genetic exchange is due to a classical sexual cycle or by other means such as same sex mating (via selfing or outcrossing) or parasexual reproduction.

X. THE ROLE OF SEX IN PATHOGENESIS

While sexual reproduction does not appear to play a direct role in the infections caused by many of the human pathogenic fungi; in both C. *neoformans* and H. *capsulatum*, there is a difference in the pathogenicity of the two mating types. In the diploid C. *albicans*, there is also a difference in virulence between strains that are heterozygous or homozygous at the mating-type locus.

The infectious propagule for C. *neoformans* is thought to be the spore, which can be 100× more infectious than the yeast cell (Sukroongreung *et al.*, 1998). Spores are only produced by *Cryptococcus* during the sexual processes of mating or monokaryotic fruiting. Thus, while sexual reproduction has not been observed in mammalian cryptococcal infections, sexual reproduction may contribute to increase the prevalence of cryptococcal infections. Epidemiological studies show that the vast majority of clinical isolates are α mating type (Casadevall and Perfect, 1998; Kwon-Chung and Bennett, 1978). However, it is unclear whether this bias toward α mating type is due to the predominance of α cells in the environment or because α cells are more virulent than **a** cells. However, due to the large size and complexity of the *MAT* locus, simple gene exchange experiments may not be sufficient to elucidate the role of this large genomic region in virulence. Instead, **a** and α congenic strains have been generated by a series of backcrosses to yield strains that differ only at the mating-type locus.

Early experimental studies on the role of mating type in virulence were conducted with var. *neoformans* because of the availability of congenic strains (Heitman *et al.*, 1999; Kwon-Chung *et al.*, 1992). In var. *neoformans*, a comparison of the levels of virulence in heterogeneous populations revealed that α strains are more virulent at lower inoculum levels than **a** strains (Kwon-Chung *et al.*, 1992; Wickes, 2002) and that the α strain was more virulent than the congenic **a** strain in a murine model of cryptococcosis as well as in the heterologous host *Caenorhabditis elegans* system (Kwon-Chung *et al.*, 1992; Mylonakis *et al.*, 2004). However, clinical infections with var. *neoformans* strains account for a limited number of cases compared to infections caused by var. *grubii* strains.

As discussed previously, in var. *grubii*, mating-type **a** strains are exceptionally rare and have only recently been identified. As with var. *neoformans*, Barchiesi *et al.* (2005) observed that nonisogenic var. *grubii* **a** strains were less virulent than α strains. However, when congenic var. *grubii* strains were compared in standard models of cryptococcosis, the **a** and α congenic strains had equivalent virulence (Nielsen *et al.*, 2003).

These observations led to a number of questions. Are there innate differences in the pathogenicity of var. *grubii* and *neoformans*? Does the genetic background of the strain affect the role of mating type in virulence? To address these questions, two additional congenic strain pairs in var. *neoformans* were generated and examined for their virulence (Nielsen *et al.*, 2005b). These studies revealed that the mating-type virulence potential of congenic strains can differ with genetic background and thus the differences in the pathogenicity of the var. *grubii* and *neoformans* congenic **a** and α strains could be due to the genetic background of the strains. Interestingly, in all cases where a difference in virulence between **a** and α mating-types strains has been observed, the α mating type is more virulent (Barchiesi *et al.*, 2005; Keller *et al.*, 2003; Kwon-Chung *et al.*, 1992; Lengeler *et al.*, 2000; Nielsen *et al.*, 2005b).

The role of mating type in pathogenicity of *C. neoformans* has been further characterized by examining the virulence of the var. *grubii* **a** and α congenic strains at various stages of the infective cycle—from their survival in potential environmental predators to their colonization of various organs in animal models (Nielsen *et al.*, 2005a). No difference in virulence was observed when the strains were infected individually but coinfection of the strains revealed that the **a** and α congenic strains reached equivalent levels in the lungs and spleen, but a significantly higher proportion of α cells infected the brain. Following intracerebral coinfections, the strains exhibited equivalent persistence in the CNS. Thus, these data show that α cells outcompete **a** cells in entry into the CNS which is the predominant site of disease in humans. Taken together, these studies suggest that the prevalence of α strains as clinical isolates is in part due to distinct differences between the **a** and α mating types and that characterization of these differences will likely provide insight into how *Cryptococcus* causes disease in humans.

A difference between the virulence of the two mating types is also seen in *H. capsulatum*. Clinical infections in humans are caused by the − mating type. However, when representative + and − isolates were tested in mouse infection experiments, both mating types had equivalent virulence (Kwon-Chung, 1981). The reason for this difference in virulence between animal models and human infections remains to be elucidated and will likely require an analysis of the growth of strains not only in models of disease but also in the environment. For example, there may be intrinsic differences between human and mouse infections with *H. capsulatum* so that other animal models might prove informative. Alternatively, there may be differences in the ability of the − and + strains to generate the infectious spores. Whether mating type is linked to sporulation or virulence has not been investigated with congenic strains in *H. capulatum*. The generation of congenic strains, such as in *Cryptococcus*, may also help to identify differences between the + and − mating types that could ultimately result in the reduced prevalence of + strains as clinical isolates.

Perhaps most intriguing is the role mating and mating type plays in the pathogenicity of the commensal fungus *C. albicans*. Population genetic studies show that *C. albicans* is primarily a clonal organism with only hints of recombination (Graser *et al.*, 1996; Pujol *et al.*, 1993, 2004; Xu *et al.*, 1999). However, strains heterozygous at the mating-type locus can spontaneously generate mating-type homozygous strains either by loss of the mating-type chromosome followed by duplication of the retained homologue or through mitotic crossing-over (Magee and Magee, 2000; Wu *et al.*, 2005).

If heterozygous strains can spontaneously generate homozygous strains, then why do homozygous strains only account for 3% of the population? Furthermore, if mating generates tetraploid strains, then why are there no reports of tetraploids among clinical isolates? The answers to these questions may lie in the virulence potential of the different types of strains. Ibrahim *et al.* (2005) demonstrated that tetraploid strains are less virulent than isogenic diploid strains. These results, combined with high rates of chromosome loss both *in vitro* and *in vivo*, provide an explanation for the absence of tetraploids as clinical isolates (Bennett and Johnson, 2003; Ibrahim *et al.*, 2005).

The role of mating type in *C. albicans* virulence is less clear. Lockhart *et al.* (2005) showed that naturally occurring *C. albicans* heterozygous a/α strains are substantially more virulent than the homozygous a/a or α/α strains they spontaneously generate. Furthermore, heterozygous strains containing the a mating-type allele and the $\alpha2$ gene from the α mating-type locus were more virulent than strains containing only one of the mating-type alleles (Lockhart *et al.*, 2005). However, when Ibrahim *et al.* (2005) examined the virulence of laboratory-generated close-to-isogenic homozygous and heterozygous strains, they found that the homozygous strains were almost as virulent as the heterozygous strains. If strains homozygous at the mating-type locus are less virulent than strains heterozygous at the mating-type locus, and perhaps this is dependent on strain background, then these results could explain why heterozygous strains dominate the population and how *C. albicans* conserves its mating system by ensuring that both mating-type loci are maintained even when mating is a rare event. While mating has been observed in the laboratory and on the skin of mice (Lockhart *et al.*, 2003), it is still unclear where or when mating might take place in humans (or in the environment) and why *C. albicans* would go to great lengths to retain the ability to mate yet rarely engage in parasexual reproduction.

Sex may also be an integral part of *Pneumocystis* infections. Unfortunately, it is not yet possible to cultivate any of the *Pneumocystis* species *in vitro*, thus studies examining the life cycle of *Pneumocystis* result from observational studies of the growth of cells in infected lung tissue. The life cycle of *Pneumocystis* is thought to include both an asexual phase and a sexual phase (Cushion, 2004). There are three morphological forms of *Pneumocystis*: the haploid trophic form which is the primary proliferative cell type, the diploid precyst which is an intermediate stage before development into the final form, and the mature cyst (Thomas and Limper, 2004). Asexual reproduction by binary fusion is thought to occur in the trophic form. In response to unknown stimuli, it is hypothesized that two cells of opposite mating type fuse and undergo karyogamy to produce the diploid precysts. Synaptonemal complexes have been identified in the precyst by electron microscopy, suggesting meiosis may occur in these cells (Matsumoto and Yoshida, 1984). The precysts appear to undergo meiosis to produce a total of four nuclei. These nuclei then undergo mitosis to generate eight haploid nuclei that are then packaged into individual spores. It is assumed that these spores are released from the mature cyst (ascus) and then enter the vegetative state again in the trophic form. The infectious propagule of *Pneumocystis* is unknown; however, person-to-person transmission has been documented and spontaneous transmission between animals in an infected rat colony occurs (Demanche *et al.*, 2005; Icenhour *et al.*, 2001; Lundgren *et al.*, 1997; Manoloff *et al.*, 2003; Miller *et al.*, 2001; Vargas *et al.*, 2000). In many of the other human pathogenic fungi, spores are highly infectious, thus one hypothesis is that the sexual spores could be the infectious propagule of *Pneumocystis*.

XI. CONCLUDING REMARKS

The genomics age has expanded our knowledge of human pathogenic fungi. Genome sequencing has revealed that all of the human pathogenic fungi have mating-type loci. Some of these loci, such as in *Cryptococcus*, *C. albicans*, and *Pneumocystis*, are quite complex and additional work will be required to define the roles of individual genes in these loci. Other mating-type loci, like those in *A. fumigatus*, *C. immitis*, and *H. capsulatum*, are simpler and will allow the opposite mating-type alleles, if they exist, to be more readily defined. While all of the human pathogenic fungi have genomes that contain mating-type loci, it is still unresolved what role, if any, classical sexual cycles play in the life cycle of these important pathogens. Molecular population genetic studies have suggested that many of the human pathogenic fungi may be actively undergoing recombination, suggesting that these previously "asexual" organisms may in fact be sexual. Future research may discover that unusual cycles such as the same sex homothallic cycle observed in *C. neoformans* or the parasexual cycle characterized in *C. albicans* may provide enough genetic diversity for these organisms while still enabling them to retain their virulence potential.

Of the greater than 100,000 species of fungi identified to date, only a few routinely cause disease in humans. Why is this? With the exception of *Candida* and *Pneumocystis*, most fungal infections of humans are caused by organisms present in the environment. Thus, pathogenicity in humans has been proposed to be accidental on the part of the fungal pathogen. The virulence factors required for human pathogenesis, including mating type, are likely required for environmental survival. What differences distinguish the common human pathogenic fungi and all other fungi present in our environment to enable just a few to be pathogenic? To understand the differences between human pathogenic fungi and other fungi, we need to understand the growth and reproduction of the human pathogenic fungi not only in humans but also in the environment. C. *neoformans* uses many of the same virulence factors to infect nematodes, amoebae, insects, and humans (Mylonakis *et al.*, 2002; Steenbergen *et al.*, 2001). This observation has led to the theory that the virulence factors used by C. *neoformans* to infect humans were originally developed by the fungus to survive predation in the environment (Casadevall *et al.*, 2003). Perhaps this holds true for the other human pathogenic fungi as well. By comparing the life cycles and environmental propagation of the human pathogenic fungi with non-pathogenic fungi, we may be able to identify traits that define human pathogenic fungi and explain how they evolved from otherwise harmless environmental saprophytes into deadly pathogens.

References

Balajee, S. A., Gribskov, J. L., Hanley, E., Nickle, D., and Marr, K. A. (2005). *Aspergillus lentulus* sp. nov., a new sibling species of *A. fumigatus*. *Eukaryot. Cell* **4,** 625–632.

Barchiesi, F., Cogliati, M., Esposto, M. C., Spreghini, E., Schimizzi, A. M., Wickes, B. L., Scalise, G., and Viviani, M. A. (2005). Comparative analysis of pathogenicity of *Cryptococcus neoformans* serotypes A, D and AD in murine cryptococcosis. *J. Infect.* **51,** 10–16.

Bell, G. (1982). "The Masterpiece of Nature: The Evolution and Genetics of Sexuality." University of California Press, Berkeley, California.

Bennett, R. J., and Johnson, A. D. (2003). Completion of a parasexual cycle in *Candida albicans* by induced chromosome loss in tetraploid strains. *EMBO J.* **22,** 2505–2515.

Bennett, R. J., and Johnson, A. D. (2005). Mating in *Candida albicans* and the search for a sexual cycle. *Annu. Rev. Microbiol.* **59,** 233–255.

Bertout, S., Renaud, F., Barton, R., Symoens, F., Burnod, J., Piens, M. A., Lebeau, B., Viviani, M. A., Chapuis, F., Bastide, J. M., Grillot, R., and Mallie, M. (2001). Genetic polymorphism of *Aspergillus fumigatus* in clinical samples from patients with invasive aspergillosis: Investigation using multiple typing methods. *J. Clin. Microbiol.* **39,** 1731–1737.

Boes, B., Bashir, R., Boes, C., Hahn, F., McConnell, J. R., and McComb, R. (1994). Central nervous system aspergillosis. Analysis of 26 patients. *J. Neuroimaging* **4,** 123–129.

Boon, A. P., Adams, D. H., Buckels, J., and McMaster, P. (1990). Cerebral aspergillosis in liver transplantation. *J. Clin. Pathol.* **43,** 114–118.

Bowman, B. H., White, T. J., and Taylor, J. W. (1996). Human pathogenic fungi and their close nonpathogenic relatives. *Mol. Phylogenet. Evol.* **6,** 89–96.

Brandt, M. E., Hutwagner, L. C., Kuykendall, R. J., and Pinner, R. W. (1995). Comparison of multilocus enzyme electrophoresis and random amplified polymorphic DNA analysis for molecular subtyping of *Cryptococcus neoformans*. The Cryptococcal Disease Active Surveillance Group. *J. Clin. Microbiol.* **33,** 1890–1895.

Brockert, P. J., Lachke, S. A., Srikantha, T., Pujol, C., Galask, R., and Soll, D. R. (2003). Phenotypic switching and mating type switching of *Candida glabrata* at sites of colonization. *Infect. Immun.* **71,** 7109–7118.

Burt, A. (2000). Perspective: Sex, recombination, and the efficacy of selection—was Weismann right? *Evolution Int. J. Org. Evolution* **54,** 337–351.

Burt, A., Carter, D. A., Koenig, G. L., White, T. J., and Taylor, J. W. (1996). Molecular markers reveal cryptic sex in the human pathogen *Coccidioides immitis*. *Proc. Natl. Acad. Sci. USA* **93,** 770–773.

Butler, G., Kenny, C., Fagan, A., Kurischko, C., Gaillardin, C., and Wolfe, K. H. (2004). Evolution of the MAT locus and its Ho endonuclease in yeast species. *Proc. Natl. Acad. Sci. USA* **101**, 1632–1637.

Butlin, R. (2002). Opinion–evolution of sex: The costs and benefits of sex: New insights from old asexual lineages. *Nat. Rev. Genet.* **3**, 311–317.

Campbell, L. T., Currie, B. J., Krockenberger, M., Malik, R., Meyer, W., Heitman, J., and Carter, D. A. (2005a). Clonality and recombination in genetically differentiated subgroups of *Cryptococcus gattii*. *Eukaryot. Cell* **4**, 1403–1409.

Campbell, L. T., Fraser, J. A., Nichols, C. B., Dietrich, F. S., Carter, D. A., and Heitman, J. (2005b). Clinical and environmental isolates of *Cryptococcus gattii* from Australia that retain sexual fecundity. *Eukaryot. Cell* **4**, 1410–1419.

Casadevall, A., and Perfect, J. R. (1998). "*Cryptococcus neoformans*." ASM Press, Washington, DC.

Casadevall, A., Steenbergen, J. N., and Nosanchuk, J. D. (2003). "Ready made" virulence and "dual use" virulence factors in pathogenic environmental fungi–the *Cryptococcus neoformans* paradigm. *Curr. Opin. Microbiol.* **6**, 332–337.

Casselton, L. A. (2002). Mate recognition in fungi. *Heredity* **88**, 142–147.

Casselton, L. A., and Olesnicky, N. S. (1998). Molecular genetics of mating recognition in basidiomycete fungi. *Microbiol. Mol. Biol. Rev.* **62**, 55–70.

Chazalet, V., Debeaupuis, J. P., Sarfati, J., Lortholary, J., Ribaud, P., Shah, P., Cornet, M., Vu Thien, H., Gluckman, E., Brucker, G., and Latge, J. P. (1998). Molecular typing of environmental and patient isolates of *Aspergillus fumigatus* from various hospital settings. *J. Clin. Microbiol.* **36**, 1494–1500.

Chiller, T. M., Galgiani, J. N., and Stevens, D. A. (2003). Coccidioidomycosis. *Infect. Dis. Clin. North Am.* **17**, 41–57, viii..

Cogliati, M., Esposto, M. C., Clarke, D. L., Wickes, B. L., and Viviani, M. A. (2001). Origin of *Cryptococcus neoformans* var. *neoformans* diploid strains. *J. Clin. Microbiol.* **39**, 3889–3894.

Cushion, M. T. (2004). Pneumocystis: Unraveling the cloak of obscurity. *Trends Microbiol.* **12**, 243–249.

de Meeus, T., Renaud, F., Mouveroux, E., Reynes, J., Galeazzi, G., Mallie, M., and Bastide, J. M. (2002). Genetic structure of *Candida glabrata* populations in AIDS and non-AIDS patients. *J. Clin. Microbiol.* **40**, 2199–2206.

Debeaupuis, J. P., Sarfati, J., Chazalet, V., and Latge, J. P. (1997). Genetic diversity among clinical and environmental isolates of *Aspergillus fumigatus*. *Infect. Immun.* **65**, 3080–3085.

Demanche, C., Wanert, F., Barthelemy, M., Mathieu, J., Durand-Joly, I., Dei-Cas, E., Chermette, R., and Guillot, J. (2005). Molecular and serological evidence of *Pneumocystis* circulation in a social organization of healthy macaques (*Macaca fascicularis*). *Microbiology* **151**, 3117–3125.

Dodgson, A. R., Pujol, C., Pfaller, M. A., Denning, D. W., and Soll, D. R. (2005). Evidence for recombination in *Candida glabrata*. *Fungal Genet. Biol.* **42**, 233–243.

Dyer, P. S., and Paoletti, M. (2005). Reproduction in *Aspergillus fumigatus*: Sexuality in a supposedly asexual species? *Med. Mycol.* **43**(Suppl. 1), S7–S14.

Dyer, P. S., Ingram, D. S., and Johnstone, K. (1992). The control of sexual morphogenesis in the Ascomycotina. *Biol. Rev.* **67**, 421–458.

Edman, J. C., Kovacs, J. A., Masur, H., Santi, D. V., Elwood, H. J., and Sogin, M. L. (1988). Ribosomal RNA sequence shows *Pneumocystis carinii* to be a member of the fungi. *Nature* **334**, 519–522.

Ellis, D., Marriott, D., Hajjeh, R. A., Warnock, D., Meyer, W., and Barton, R. (2000). Epidemiology: Surveillance of fungal infections. *Med. Mycol.* **38**(Suppl. 1), 173–182.

Feldbrugge, M., Kamper, J., Steinberg, G., and Kahmann, R. (2004). Regulation of mating and pathogenic development in *Ustilago maydis*. *Curr. Opin. Microbiol.* **7**, 666–672.

Fisher, B. D., Armstrong, D., Yu, B., and Gold, J. W. (1981). Invasive aspergillosis. Progress in early diagnosis and treatment. *Am. J. Med.* **71**, 571–577.

Fisher, M. C., Koenig, G. L., White, T. J., and Taylor, J. W. (2001). Molecular and phenotypic description of *Coccidioides posadasii* sp. nov., previously recognized as the non-Californian population of *Coccidioides immitis*. *Mycologia* **94**, 73–84.

Francois, F., Noel, T., Pepin, R., Brulfert, A., Chastin, C., Favel, A., and Villard, J. (2001). Alternative identification test relying upon sexual reproductive abilities of *Candida lusitaniae* strains isolated from hospitalized patients. *J. Clin. Microbiol.* **39**, 3906–3914.

Fraser, J. A., and Heitman, J. (2004). Evolution of fungal sex chromosomes. *Mol. Microbiol.* **51**, 299–306.

Fraser, J. A., Subaran, R. L., Nichols, C. B., and Heitman, J. (2003). Recapitulation of the sexual cycle of the primary fungal pathogen *Cryptococcus neoformans* var. *gattii*: Implications for an outbreak on Vancouver Island, Canada. *Eukaryot. Cell* **2**, 1036–1045.

Fraser, J. A., Diezmann, S., Subaran, R. L., Allen, A., Lengeler, K. B., Dietrich, F. S., and Heitman, J. (2004). Convergent evolution of chromosomal sex-determining regions in the animal and fungal kingdoms. *PLoS Biol.* **2**, 2243–2255.

Fraser, J. A., Giles, S. S., Wenink, E. C., Geunes-Boyer, S. G., Wright, J. R., Diezmann, S., Allen, A., Stajich, J. E., Dietrich, F. S., Perfect, J. R., and Heitman, J. (2005). Same-sex mating and the origin of the Vancouver Island *Cryptococcus gattii* outbreak. *Nature* **437**, 1360–1364.

Fraser, R. S. (1993). Pulmonary aspergillosis: Pathologic and pathogenetic features. *Pathol. Annu.* **28** (Pt. 1), 231–277.

Frenkel, J. K. (1999). *Pneumocystis* pneumonia, an immunodeficiency-dependent disease (IDD): A critical historical overview. *J. Eukaryot. Microbiol.* **46**, 89S–92S.

Galagan, J. E., and Selker, E. U. (2004). RIP: The evolutionary cost of genome defense. *Trends Genet.* **20**, 417–423.

Galagan, J. E., Calvo, S. E., Cuomo, C., Ma, L. J., Wortman, J. R., Batzoglou, S., Lee, S. I., Basturkmen, M., Spevak, C. C., Clutterbuck, J., Kapitonov, V., Jurka, J., *et al.* (2005). Sequencing of *Aspergillus nidulans* and comparative analysis with A. *fumigatus* and A. *oryzae*. *Nature* **438**, 1105–1115.

Geiser, D. M., Frisvad, J. C., and Taylor, J. W. (1998). Evolutionary relationships in *Aspergillus* section *Fumigati* inferred from partial beta-tubulin and hydrophobin sequences. *Mycologia* **90**, 831–845.

Geiser, D. M., Dorner, J. W., Horn, B. W., and Taylor, J. W. (2000). The phylogenetics of mycotoxin and sclerotium production in *Aspergillus flavus* and *Aspergillus oryzae*. *Fungal Genet. Biol.* **31**, 169–179.

Goddard, M. R., Godfray, H. C., and Burt, A. (2005). Sex increases the efficacy of natural selection in experimental yeast populations. *Nature* **434**, 636–640.

Graser, Y., Volovsek, M., Arrington, J., Schonian, G., Presber, W., Mitchell, T. G., and Vilgalys, R. (1996). Molecular markers reveal that population structure of the human pathogen *Candida albicans* exhibits both clonality and recombination. *Proc. Natl. Acad. Sci. USA* **93**, 12473–12477.

Grimberg, B., and Zeyl, C. (2005). The effects of sex and mutation rate on adaptation in test tubes and to mouse hosts by *Saccharomyces cerevisiae*. *Evolution Int. J. Org. Evolution* **59**, 431–438.

Hagensee, M. E., Bauwens, J. E., Kjos, B., and Bowden, R. A. (1994). Brain abscess following marrow transplantation: Experience at the Fred Hutchinson Cancer Research Center, 1984–1992. *Clin. Infect. Dis.* **19**, 402–408.

Heitman, J., Allen, B., Alspaugh, J. A., and Kwon-Chung, K. J. (1999). On the origins of congenic MATα and MATa strains of the pathogenic yeast *Cryptococcus neoformans*. *Fungal Genet. Biol.* **28**, 1–5.

Helweg-Larsen, J., Jensen, J. S., Dohn, B., Benfield, T. L., and Lundgren, B. (2002). Detection of *Pneumocystis* DNA in samples from patients suspected of bacterial pneumonia—a case-control study. *BMC Infect. Dis.* **2**, 28.

Hughes, W. T. (1982). Natural mode of acquisition for *de novo* infection with *Pneumocystis carinii*. *J. Infect. Dis.* **145**, 842–848.

Hull, C. M., and Heitman, J. (2002). Genetics of *Cryptococcus neoformans*. *Annu. Rev. Genet.* **36**, 557–615.

Hull, C. M., and Johnson, A. D. (1999). Identification of a mating type-like locus in the asexual pathogenic yeast *Candida albicans*. *Science* **285**, 1271–1275.

Hull, C. M., Raisner, R. M., and Johnson, A. D. (2000). Evidence for mating of the "asexual" yeast *Candida albicans* in a mammalian host. *Science* **289**, 307–310.

Ibrahim, A. S., Magee, B. B., Sheppard, D. C., Yang, M., Kauffman, S., Becker, J., Edwards, J. E., Jr., and Magee, P. T. (2005). Effects of ploidy and mating type on virulence of *Candida albicans*. *Infect. Immun.* **73**, 7366–7374.

Icenhour, C. R., Rebholz, S. L., Collins, M. S., and Cushion, M. T. (2001). Widespread occurrence of *Pneumocystis carinii* in commercial rat colonies detected using targeted PCR and oral swabs. *J. Clin. Microbiol.* **39**, 3437–3441.

Johnson, A. (2003). The biology of mating in *Candida albicans*. *Nat. Rev. Microbiol.* **1**, 106–116.

Kavanaugh, L. A., Fraser, J. A., and Dietrich, F. S. (2006). Recent evolution of the human pathogen *Cryptococcus neoformans* by intervarietal transfer of a 14-gene fragment. *Mol. Biol. Evol.* **23**, 1879–1890.

Keller, S. M., Viviani, M. A., Esposto, M. C., Cogliati, M., and Wickes, B. L. (2003). Molecular and genetic characterization of a serotype A MATa *Cryptococcus neoformans* isolate. *Microbiology* **149**, 131–142.

Klempp-Selb, B., Rimek, D., and Kappe, R. (2000). Karyotyping of *Candida albicans* and *Candida glabrata* from patients with *Candida* sepsis. *Mycoses* **43**, 159–163.

Koufopanou, V., Burt, A., and Taylor, J. W. (1997). Concordance of gene genealogies reveals reproductive isolation in the pathogenic fungus *Coccidioides immitis*. *Proc. Natl. Acad. Sci. USA* **94**, 5478–5482.

Koufopanou, V., Burt, A., Szaro, T., and Taylor, J. W. (2001). Gene genealogies, cryptic species, and molecular evolution in the human pathogen *Coccidioides immitis* and relatives (Ascomycota, Onygenales). *Mol. Biol. Evol.* **18**, 1246–1258.

Kwon-Chung, K. J. (1972). Sexual stage of *Histoplasma capsulatum*. *Science* **175**, 326.

Kwon-Chung, K. J. (1973). Studies on *Emmonsiella capsulata*. I. Heterothallism and development of the ascocarp. *Mycologia* **65**, 109–121.

Kwon-Chung, K. J. (1975). A new genus, *Filobasidiella*, the perfect state of *Cryptococcus neoformans*. *Mycologia* **67**, 1197–1200.

Kwon-Chung, K. J. (1976a). Morphogenesis of *Filobasidiella neoformans*, the sexual state of *Cryptococcus neoformans*. *Mycologia* **68**, 821–833.

Kwon-Chung, K. J. (1976b). A new species of *Filobasidiella*, the sexual state of *Cryptococcus neoformans* B and C serotypes. *Mycologia* **68**, 943–946.

Kwon-Chung, K. J. (1981). Virulence of the two mating types of *Emmonsiella capsulata* and the mating experiments with *Emmonsiella capsulata* var. *duboisii*. In "Sexuality and Pathogenicity of Fungi" (C. de Vroey and R. Vanbreuseghem, eds.), p. 250. Masson, Paris, New York.

Kwon-Chung, K. J., and Bennett, J. E. (1978). Distribution of α and **a** mating types of *Cryptococcus neoformans* among natural and clinical isolates. *Am. J. Epidemiol.* **108**, 337–340.

Kwon-Chung, K. J., Edman, J. C., and Wickes, B. L. (1992). Genetic association of mating types and virulence in *Cryptococcus neoformans*. *Infect. Immun.* **60**, 602–605.

Latge, J. P., Moutaouakil, M., Debeaupuis, J. P., Bouchara, J. P., Haynes, K., and Prevost, M. C. (1991). The 18-kilodalton antigen secreted by *Aspergillus fumigatus*. *Infect. Immun.* **59**, 2586–2594.

Lengeler, K. B., Wang, P., Cox, G. M., Perfect, J. R., and Heitman, J. (2000). Identification of the MATa mating-type locus of *Cryptococcus neoformans* reveals a serotype A MATa strain thought to have been extinct. *Proc. Natl. Acad. Sci. USA* **97**, 14455–14460.

Lengeler, K. B., Cox, G. M., and Heitman, J. (2001). Serotype AD strains of *Cryptococcus neoformans* are diploid or aneuploid and are heterozygous at the mating-type locus. *Infect. Immun.* **69**, 115–122.

Lengeler, K. B., Fox, D. S., Fraser, J. A., Allen, A., Forrester, K., Dietrich, F. S., and Heitman, J. (2002). Mating-type locus of *Cryptococcus neoformans*: A step in the evolution of sex chromosomes. *Eukaryot. Cell* **1,** 704–718.

Lenski, R. E. (2001). Genetics and evolution. Come fly, and leave the baggage behind. *Science* **294,** 533–534.

Lin, X., Hull, C. M., and Heitman, J. (2005). Sexual reproduction between partners of the same mating type in *Cryptococcus neoformans*. *Nature* **434,** 1017–1021.

Litvintseva, A. P., Marra, R. E., Nielsen, K., Heitman, J., Vilgalys, R., and Mitchell, T. G. (2003). Evidence of sexual recombination among *Cryptococcus neoformans* serotype A isolates in sub-Saharan Africa. *Eukaryot. Cell* **2,** 1162–1168.

Litvintseva, A. P., Thakur, R., Vilgalys, R., and Mitchell, T. G. (2005). Multilocus sequence typing reveals three genetically distinct subpopulations of *Cryptococcus neoformans* var. *grubii* (serotype A), including a unique population in Botswana. *Genetics* **172,** 2223–2238.

Lockhart, S. R., Daniels, K. J., Zhao, R., Wessels, D., and Soll, D. R. (2003). Cell biology of mating in *Candida albicans*. *Eukaryot. Cell* **2,** 49–61.

Lockhart, S. R., Pujol, C., Daniels, K. J., Miller, M. G., Johnson, A. D., Pfaller, M. A., and Soll, D. R. (2002). In *Candida albicans*, white-opaque switchers are homozygous for mating type. *Genetics* **162,** 737–745.

Lockhart, S. R., Wu, W., Radke, J. B., Zhao, R., and Soll, D. R. (2005). Increased virulence and competitive advantage of a/α over a/a or α/α offspring conserves the mating system of *Candida albicans*. *Genetics* **169,** 1883–1890.

Loftus, B. J., Fung, E., Roncaglia, P., Rowley, D., Amedeo, P., Bruno, D., Vamathevan, J., Miranda, M., Anderson, I. J., Fraser, J. A., Allen, J. E., Bosdet, I. E., *et al.* (2005). The genome of the basidiomycetous yeast and human pathogen *Cryptococcus neoformans*. *Science* **307,** 1321–1324.

Lundgren, B., Elvin, K., Rothman, L. P., Ljungstrom, I., Lidman, C., and Lundgren, J. D. (1997). Transmission of *Pneumocystis carinii* from patients to hospital staff. *Thorax* **52,** 422–424.

Magee, B. B., and Magee, P. T. (2000). Induction of mating in *Candida albicans* by construction of MTLa and MTLα strains. *Science* **289,** 310–313.

Manoloff, E. S., Francioli, P., Taffe, P., Van Melle, G., Bille, J., and Hauser, P. M. (2003). Risk for *Pneumocystis carinii* transmission among patients with pneumonia: A molecular epidemiology study. *Emerg. Infect. Dis.* **9,** 132–134.

Matsumoto, Y., and Yoshida, Y. (1984). Sporogony in *Pneumocystis carinii*: Synaptonemal complexes and meiotic nuclear divisions observed in precysts. *J. Protozool.* **31,** 420–428.

McClelland, C. M., Chang, Y. C., Varma, A., and Kwon-Chung, K. J. (2004). Uniqueness of the mating system in *Cryptococcus neoformans*. *Trends Microbiol.* **12,** 208–212.

Meyer, W., Marszewska, K., Amirmostofian, M., Igreja, R. P., Hardtke, C., Methling, K., Viviani, M. A., Chindamporn, A., Sukroongreung, S., John, M. A., Ellis, D. H., and Sorrell, T. C. (1999). Molecular typing of global isolates of *Cryptococcus neoformans* var. *neoformans* by polymerase chain reaction fingerprinting and randomly amplified polymorphic DNA-a pilot study to standardize techniques on which to base a detailed epidemiological survey. *Electrophoresis* **20,** 1790–1799.

Miller, M. G., and Johnson, A. D. (2002). White–opaque switching in *Candida albicans* is controlled by mating-type locus homeodomain proteins and allows efficient mating. *Cell* **110,** 293–302.

Miller, R. F., Ambrose, H. E., and Wakefield, A. E. (2001). *Pneumocystis carinii* f. sp. hominis DNA in immunocompetent health care workers in contact with patients with *P. carinii* pneumonia. *J. Clin. Microbiol.* **39,** 3877–3882.

Montagna, M. T. (2002). A note on the isolation of *Cryptococcus neoformans* serotype A MATa strain from the Italian environment. *Med. Mycol.* **40,** 593–595.

Mylonakis, E., Ausubel, F. M., Perfect, J. R., Heitman, J., and Calderwood, S. B. (2002). Killing of *Caenorhabditis elegans* by *Cryptococcus neoformans* as a model of yeast pathogenesis. *Proc. Natl. Acad. Sci. USA* **99,** 15675–15680.

Mylonakis, E., Idnurm, A., Moreno, R., El Khoury, J., Rottman, J. B., Ausubel, F. M., Heitman, J., and Calderwood, S. B. (2004). *Cryptococcus neoformans* Kin1 protein kinase homologue, identified through a *Caenorhabditis elegans* screen, promotes virulence in mammals. *Mol. Microbiol.* **54,** 407–419.

Nielsen, K., Cox, G. M., Wang, P., Toffaletti, D. L., Perfect, J. R., and Heitman, J. (2003). Sexual cycle of *Cryptococcus neoformans* var. *grubii* and virulence of congenic **a** and α isolates. *Infect. Immun.* **71,** 4831–4841.

Nielsen, K., Cox, G. M., Litvintseva, A. P., Mylonakis, E., Malliaris, S. D., Benjamin, D. K., Jr., Giles, S. S., Mitchell, T. G., Casadevall, A., Perfect, J. R., and Heitman, J. (2005a). *Cryptococcus neoformans* α strains preferentially disseminate to the central nervous system during coinfection. *Infect. Immun.* **73,** 4922–4933.

Nielsen, K., Marra, R. E., Hagen, F., Boekhout, T., Mitchell, T. G., Cox, G. M., and Heitman, J. (2005b). Interaction between genetic background and the mating-type locus in *Cryptococcus neoformans* virulence potential. *Genetics* **171,** 975–983.

Nierman, W. C., Pain, A., Anderson, M. J., Wortman, J. R., Kim, H. S., Arroyo, J., Berriman, M., Abe, K., Archer, D. B., Bermejo, C., Bennett, J., Bowyer, P., *et al.* (2005). Genomic sequence of the pathogenic and allergenic filamentous fungus *Aspergillus fumigatus*. *Nature* **438,** 1151–1156.

Odds, F. C., Brown, A. J., and Gow, N. A. (2000). Might *Candida albicans* be made to mate after all? *Trends Microbiol.* **8,** 4–6.

Paoletti, M., Rydholm, C., Schwier, E. U., Anderson, M. J., Szakacs, G., Lutzoni, F., Debeaupuis, J. P., Latge, J. P., Denning, D. W., and Dyer, P. S. (2005). Evidence for sexuality in the opportunistic fungal pathogen *Aspergillus fumigatus*. *Curr. Biol.* **15,** 1242–1248.

Perfect, J. R. (2005). Weird Fungi. *ASM News* **71,** 407–411.

Peterson, S. W. (2000). "Phylogenetic Relationships in *Aspergillus* Based on rDNA Sequence Analysis." Hardwood Academic Publishers, Singapore.

Pildain, M. B., Vaamonde, G., and Cabral, D. (2004). Analysis of population structure of *Aspergillus flavus* from peanut based on vegetative compatibility, geographic origin, mycotoxin, and sclerotia production. *Int. J. Food Microbiol.* **93,** 31–40.

Pöggeler, S. (1999). Phylogenetic relationships between mating-type sequences from homothallic and heterothallic ascomycetes. *Curr. Genet.* **36,** 222–231.

Pringle, A., Baker, D. M., Platt, J. L., Wares, J. P., Latge, J. P., and Taylor, J. W. (2005). Cryptic speciation in the cosmopolitan and clonal human pathogenic fungus *Aspergillus fumigatus*. *Evolution* **59,** 1886–1899.

Pujol, C., Reynes, J., Renaud, F., Raymond, M., Tibayrenc, M., Ayala, F. J., Janbon, F., Mallie, M., and Bastide, J. M. (1993). The yeast *Candida albicans* has a clonal mode of reproduction in a population of infected human immunodeficiency virus-positive patients. *Proc. Natl. Acad. Sci. USA* **90,** 9456–9459.

Pujol, C., Pfaller, M. A., and Soll, D. R. (2004). Flucytosine resistance is restricted to a single genetic clade of *Candida albicans*. *Antimicrob. Agents Chemother.* **48,** 262–266.

Ronning, C. M., Fedorova, N. D., Bowyer, P., Coulson, R., Goldman, G., Kim, H. S., Turner, G., Wortman, J. R., Yu, J., Anderson, M. J., Denning, D. W., and Nierman, W. C. (2005). Genomics of *Aspergillus fumigatus*. *Rev. Iberoam. Micol.* **22,** 223–228.

Smulian, A. G., Sesterhenn, T., Tanaka, R., and Cushion, M. T. (2001). The *ste3* pheromone receptor gene of *Pneumocystis carinii* is surrounded by a cluster of signal transduction genes. *Genetics* **157,** 991–1002.

Srikantha, T., Lachke, S. A., and Soll, D. R. (2003). Three mating type-like loci in *Candida glabrata*. *Eukaryot. Cell* **2,** 328–340.

Steenbergen, J. N., Shuman, H. A., and Casadevall, A. (2001). *Cryptococcus neoformans* interactions with amoebae suggest an explanation for its virulence and intracellular pathogenic strategy in macrophages. *Proc. Natl. Acad. Sci. USA* **98,** 15245–15250.

Steinbach, W. J. (2005). Pediatric aspergillosis: Disease and treatment differences in children. *Pediatr. Infect. Dis. J.* **24,** 358–364.

Steinemann, S., and Steinemann, M. (2005). Y chromosomes: Born to be destroyed. *Bioessays* **27,** 1076–1083.

Stringer, S. L., Hudson, K., Blase, M. A., Walzer, P. D., Cushion, M. T., and Stringer, J. R. (1989a). Sequence from ribosomal RNA of *Pneumocystis carinii* compared to those of four fungi suggests an ascomycetous affinity. *J. Protozool.* **36,** 14S–16S.

Stringer, S. L., Stringer, J. R., Blase, M. A., Walzer, P. D., and Cushion, M. T. (1989b). *Pneumocystis carinii:* Sequence from ribosomal RNA implies a close relationship with fungi. *Exp. Parasitol.* **68,** 450–461.

Sukroongreung, S., Kitiniyom, K., Nilakul, C., and Tantimavanich, S. (1998). Pathogenicity of basidiospores of *Filobasidiella neoformans* var. *neoformans. Med. Mycol.* **36,** 419–424.

Tavanti, A., Davidson, A. D., Fordyce, M. J., Gow, N. A., Maiden, M. C., and Odds, F. C. (2005). Population structure and properties of *Candida albicans*, as determined by multilocus sequence typing. *J. Clin. Microbiol.* **43,** 5601–5613.

Thomas, C. F., Jr., and Limper, A. H. (2004). *Pneumocystis* pneumonia. *N. Engl. J. Med.* **350,** 2487–2498.

Tibayrenc, M. (1997). Are *Candida albicans* natural populations subdivided? *Trends Microbiol.* **5,** 253–254; discussion 254–257.

Tintelnot, K., Lemmer, K., Losert, H., Schar, G., and Polak, A. (2004). Follow-up of epidemiological data of cryptococcosis in Austria, Germany and Switzerland with special focus on the characterization of clinical isolates. *Mycoses* **47,** 455–464.

Tran-Dinh, N., Pitt, J. I., and Carter, D. A. (1999). Molecular genotype analysis of natural toxigenic and non-toxigenic isolates of *Aspergillus flavus* and *Aspergillus parasiticus. Mycological Research* **103,** 1485–1490.

Tscharke, R. L., Lazera, M., Chang, Y. C., Wickes, B. L., and Kwon-Chung, K. J. (2003). Haploid fruiting in *Cryptococcus neoformans* is not mating type α-specific. *Fungal Genet. Biol.* **39,** 230–237.

Tzung, K. W., Williams, R. M., Scherer, S., Federspiel, N., Jones, T., Hansen, N., Bivolarevic, V., Huizar, L., Komp, C., Surzycki, R., Tamse, R., Davis, R. W., *et al.* (2001). Genomic evidence for a complete sexual cycle in *Candida albicans. Proc. Natl. Acad. Sci. USA* **98,** 3249–3253.

Varga, J., and Toth, B. (2003). Genetic variability and reproductive mode of *Aspergillus fumigatus. Infect. Genet. Evol.* **3,** 3–17.

Varga, J., Vida, Z., Toth, B., Debets, F., and Horie, Y. (2000). Phylogenetic analysis of newly described *Neosartorya* species. *Antonie Van Leeuwenhoek* **77,** 235–239.

Vargas, S. L., Ponce, C. A., Gigliotti, F., Ulloa, A. V., Prieto, S., Munoz, M. P., and Hughes, W. T. (2000). Transmission of *Pneumocystis carinii* DNA from a patient with *P. carinii* pneumonia to immunocompetent contact health care workers. *J. Clin. Microbiol.* **38,** 1536–1538.

Vilgalys, R., Gräser, Y., Schönian, G., Presber, W., and Mitchell, T. G. (1997). Are *Candida albicans* populations clonal? *Trends Microbiol.* **5,** 254–257.

Viviani, M. A., Esposto, M. C., Cogliati, M., Montagna, M. T., and Wickes, B. L. (2001). Isolation of a *Cryptococcus neoformans* serotype A MATa strain from the Italian environment. *Med. Mycol.* **39,** 383–386.

Wakefield, A. E. (1994). Detection of DNA sequences identical to *Pneumocystis carinii* in samples of ambient air. *J. Eukaryot. Microbiol.* **41,** 116S.

Wakefield, A. E., Lindley, A. R., Ambrose, H. E., Denis, C. M., and Miller, R. F. (2003). Limited asymptomatic carriage of *Pneumocystis jiroveci* in human immunodeficiency virus-infected patients. *J. Infect. Dis.* **187,** 901–908.

Walsh, T. J., Hier, D. B., and Caplan, L. R. (1985). Aspergillosis of the central nervous system: Clinicopathological analysis of 17 patients. *Ann. Neurol.* **18,** 574–582.

Whelan, W. L. (1987). The genetics of medically important fungi. *Crit. Rev. Microbiol.* **14,** 99–170.

Wickes, B. L. (2002). The role of mating type and morphology in *Cryptococcus neoformans* pathogenesis. *Int. J. Med. Microbiol.* **292,** 313–329.

Wickes, B. L., Mayorga, M. E., Edman, U., and Edman, J. C. (1996). Dimorphism and haploid fruiting in *Cryptococcus neoformans*: Association with the α-mating type. *Proc. Natl. Acad. Sci. USA* **93,** 7327–7331.

Williams, G. C. (1975). "Sex and Evolution." Princeton University Press, Princeton, New Jersey.

Wong, S., Fares, M. A., Zimmermann, W., Butler, G., and Wolfe, K. H. (2003). Evidence from comparative genomics for a complete sexual cycle in the "asexual" pathogenic yeast *Candida glabrata. Genome Biol.* **4,** R10.

Wu, W., Pujol, C., Lockhart, S. R., and Soll, D. R. (2005). Chromosome loss followed by duplication is the major mechanism of spontaneous mating-type locus homozygosis in *Candida albicans. Genetics* **169,** 1311–1327.

Xu, J. (2002). Estimating the spontaneous mutation rate of loss of sex in the human pathogenic fungus *Cryptococcus neoformans. Genetics* **162,** 1157–1167.

Xu, J. (2005). Cost of interacting with sexual partners in a facultative sexual microbe. *Genetics* **171,** 1597–1604.

Xu, J., and Mitchell, T. G. (2002). Strain variation and clonality in *Candida* spp. and *Cryptococcus neoformans. In* "Fungal Pathogenesis: Principles and Clinical Applications" (R. A. Calderone and R. L. Cihlar, eds.), pp. 739–749. Marcel Dekker, New York, New York.

Xu, J., and Mitchell, T. G. (2003). Comparative gene genealogical analyses of strains of serotype AD identify recombination in populations of serotypes A and D in the human pathogenic yeast *Cryptococcus neoformans. Microbiology* **149,** 2147–2154.

Xu, J., Mitchell, T. G., and Vilgalys, R. (1999). PCR-restriction fragment length polymorphism (RFLP) analyses reveal both extensive clonality and local genetic differences in *Candida albicans. Mol. Ecol.* **8,** 59–73.

Xu, J., Vilgalys, R., and Mitchell, T. G. (2000). Multiple gene genealogies reveal recent dispersion and hybridization in the human pathogenic fungus *Cryptococcus neoformans. Mol. Ecol.* **9,** 1471–1481.

Young, J. A., Hyppa, R. W., and Smith, G. R. (2004). Conserved and nonconserved proteins for meiotic DNA breakage and repair in yeasts. *Genetics* **167,** 593–605.

Young, L. Y., Lorenz, M. C., and Heitman, J. (2000). A STE12 homolog is required for mating but dispensable for filamentation in *Candida lusitaniae. Genetics* **155,** 17–29.

Zhao, R., Daniels, K. J., Lockhart, S. R., Yeater, K. M., Hoyer, L. L., and Soll, D. R. (2005). Unique aspects of gene expression during *Candida albicans* mating and possible G(1) dependency. *Eukaryot. Cell* **4,** 1175–1190.

5 | From Genes to Genomes: A New Paradigm for Studying Fungal Pathogenesis in *Magnaporthe oryzae*

Jin-Rong Xu,* Xinhua Zhao,* and Ralph A. Dean[†]

*Department of Botany and Plant Pathology, Purdue University
West Lafayette, Indiana 47907
[†]Department of Plant Pathology, North Carolina State University
Raleigh, North Carolina 27606

Advances in Genetics, Vol. 57
Copyright 2007, Elsevier Inc. All rights reserved.
0065-2660/07 $35.00
DOI: 10.1016/S0065-2660(06)57005-1

ABSTRACT

Magnaporthe oryzae is the most destructive fungal pathogen of rice worldwide and because of its amenability to classical and molecular genetic manipulation, availability of a genome sequence, and other resources it has emerged as a leading model system to study host–pathogen interactions. This chapter reviews recent progress toward elucidation of the molecular basis of infection-related morphogenesis, host penetration, invasive growth, and host–pathogen interactions. Related information on genome analysis and genomic studies of plant infection processes is summarized under specific topics where appropriate. Particular emphasis is placed on the role of MAP kinase and cAMP signal transduction pathways and unique features in the genome such as repetitive sequences and expanded gene families. Emerging developments in functional genome analysis through large-scale insertional mutagenesis and gene expression profiling are detailed. The chapter concludes with new prospects in the area of systems biology, such as protein expression profiling, and highlighting remaining crucial information needed to fully appreciate host–pathogen interactions. © 2007, Elsevier Inc.

I. INTRODUCTION

Rice is the staple food for almost half of the world's population. Rice blast, caused by *Magnaporthe grisea* (Hebert) Barr, is one of the most severe diseases of rice throughout the world, accounting for more than 10 million tons of yield loss each year. Fungi belonging to the M. *grisea* species complex also cause disease on many other grass species, including economically important crops such as barley, wheat, and millet (Valent and Chumley, 1991). Phylogenetic analysis of the M. *grisea* species complex has resulted in the renaming of the rice blast pathogen to M. *oryzae* (Couch and Kohn, 2002). The rice blast fungus invades rice plants in a manner typical of many foliar pathogens by producing specialized infection structures called appressoria. In nature, the rice blast fungus attacks all aboveground parts of rice plants, and seedlings can be killed during epidemics. Under laboratory conditions, root infection of wheat and rice seedlings by M. *oryzae* has been observed (Dufresne and Osbourn, 2001; Sesma and Osbourn, 2004).

Over the past decade, M. *oryzae* has emerged as a model system to study fungal–plant interactions. The vast amount of genetic and genomic resources that are available for M. *oryzae* is unparalleled in any other fungal pathogen. M. *oryzae* is the first fungal plant pathogen for which the genome sequence has been published (Dean *et al.*, 2005). Analysis of the M. *oryzae* genome has revealed valuable information about the basis of fungal pathogenesis. This chapter covers recent progress in molecular and genetic studies of infection-related morphogenesis and fungal–plant interactions. Since genome analysis and genomic studies will have a significant impact on all aspects of M. *oryzae* studies, related information is summarized under specific topics where appropriate. We also attempt to highlight genome features of M. *oryzae* and briefly review ongoing functional genomics studies.

II. ATTACHMENT AND APPRESSORIUM MORPHOGENESIS

A. Attachment and germination

The attachment of M. *oryzae* conidia to rice leaves is mediated by the release of preformed spore tip mucilage. After attachment, conidia germinate when free water is present and produce germ tubes. Endocytosis can be detected within 2–3 min of conidium hydration (Atkinson *et al.*, 2002). Genes involved in endocytosis are conserved between M. *oryzae* and *Neurospora crassa* (Borkovich *et al.*, 2004; Dean *et al.*, 2005). To date, genes involved in conidium and germ tube attachment have not been characterized in M. *oryzae*. Pharmacological studies on several plant pathogens have indicated that integrin may be involved in fungal attachment and appressorium formation (Correa *et al.*, 1996; Tucker and Talbot, 2001). The M. *oryzae* genome contains one predicted gene that is homologous to *Candida albicans INT1*, which is the only known putative fungal integrin (Gale *et al.*, 1996). However, *INT1* and its homologue in M. *oryzae* share only limited sequence homology in the pleckstrin homology (PH) domain. These integrin-like proteins may be more similar to Bud4 in *Saccharomyces cerevisiae* and Mid2 in *Schizosaccharomyces pombe* that are involved in septin formation instead of attachment (Gale *et al.*, 2001).

In M. *oryzae*, distinct pathogenic and saprophytic germination modes such as described in *Colletotrichum gloeosporioides* (Barhoom and Sharon, 2004) have not been reported. When germinated in water drops on plant surfaces or artificial hydrophobic surfaces, germ tubes usually emerge from the apical and/or basal cells of conidia and are unbranched. In contrast, all three conidial cells can germinate in rich media and produce branching germ tubes that differentiate into vegetative hyphae. It is likely that nutrient sensing plays an important role in regulating germination and germ tube growth in M. *oryzae*, but the exact regulatory mechanisms are not defined. Unlike *Aspergillus nidulans* and C. *lagenarium* in which a MAP (mitogen-activated protein) kinase pathway and cAMP signaling have been implicated in conidium germination (Osherov and May, 2001; Yamauchi *et al.*, 2004), M. *oryzae* mutants blocked in various signal transduction pathways have no obvious defect in conidium attachment and germination (Xu, 2000). Although a few mutants are known to be delayed in conidium germination, such as the *icl1* and *gde1* mutants (Balhadere and Talbot, 2001; Wang *et al.*, 2003), mutants blocked in conidium germination have not been reported in M. *oryzae*.

The rice leaf surface is extremely hydrophobic. The M. *oryzae* germ tubes (with hydrophilic surfaces) must be able to modify the leaf surface hydrophobicity before they can adhere. One gene that may play an important role in this step is MPG1, which encodes a secreted protein with characteristics of a class I fungal hydrophobin. Targeted deletion of *MPG1* reduces the efficiency of appressorium formation and virulence. Recombinant Mpg1 proteins can self-assemble *in vitro* and on the conidial surface (Kershaw *et al.*, 1998), indicating that Mpg1 proteins may assemble into an amphipathic layer over the rice leaf surface to assist germ tube adhesion and act as a signal for appressorium formation. Interestingly, when the cysteine-alanine substitution alleles of *MPG1* are expressed in the *mpg1* mutant, the secretion but not the self-assembly of the mutant Mpg1 protein is blocked (Kershaw *et al.*, 2005), indicating that the disulfide bridges are essential only for Mpg1 secretion. A class II hydrophobin Mhp1 was characterized in M. *oryzae* (Kim *et al.*, 2005). Similar to *MPG1*, *MHP1* is highly expressed during plant colonization and conidiation. The *mhp1* mutant is reduced in conidiation, conidial germination, appressorium development, and plant infection. However, the exact role of *MHP1* during rice infection has not been determined, and it is not clear whether Mhp1 proteins are involved in surface attachment or recognition. In the M. *oryzae* genome, at least six other genes encode putative hydrophobins (D. Ebbole, personal communication).

B. Surface recognition and cAMP signaling

On rice leaves or hydrophobic artificial surfaces, germ tubes arrest tip growth and begin the germ tube tip deformation process known as "hooking." In normal situations, one appressorium is formed at the tip of each germ tube (Fig. 5.1). On hydrophilic surfaces, appressorium formation can be induced with exogenous

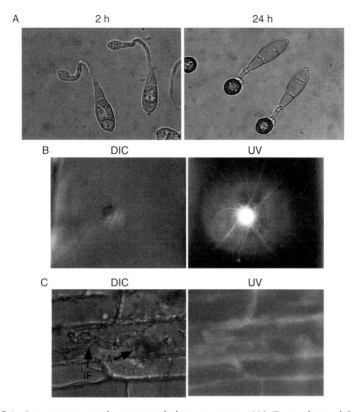

Figure 5.1. Appressorium morphogenesis and plant penetration. (A) Germ tube tip deformation
(2 h) and mature appressoria (24 h) on hydrophobic surfaces. (B) Actin cytoskeleton
rearrangements of onion epidermal cells in response to appressorial penetration.
(C) Infectious hyphae formed inside rice leaf sheath epidermal cells at 48 h after
inoculation. A, appressorium; IF, infectious hypha. Images B and C were copied from Park
et al. (2004) with the courtesy of the authors.

cyclic AMP (cAMP) or IBMX. Targeted deletion analyses of the *MAC1* adenylate cyclase further proved that cAMP signaling is involved in surface recognition
in M. *oryzae* (Choi and Dean, 1997). Defects in *mac1* mutants can be complemented by exogenous cAMP or suppressed by a spontaneous mutation in
SUM1, which encodes the regulatory subunit of PKA (Adachi and Hamer,
1998). Different from *MAC1*, *SUM1* is an essential gene in M. *oryzae*. Mutants
with *SUM1* silenced by RNA interference (RNAi) are significantly reduced in
the production of conidia and aerial hyphae and are nonpathogenic (Li and Xu,
unpublished data). In contrast to *mac1* mutants, deletion mutants of *CPKA*, a
gene encoding a catalytic subunit of PKA, have normal growth and still produce

melanized appressoria (Mitchell and Dean, 1995; Xu *et al.*, 1997). The *cpkA* mutants are nonpathogenic on healthy plants but can infect through wounds. On hydrophilic surfaces, *cpkA* mutants still respond to exogenous cAMP (Xu *et al.*, 1997). Analysis of the M. *oryzae* genome revealed the presence of a second catalytic subunit of PKA named *CPK2* (MG02832), which may have overlapping functions with *CpkA* (Fig. 5.2). However, mutants deleted of *CPK2* have no obvious defects in vegetative growth, appressorium formation, and plant infection. All attempts to generate a *cpkA cpk2* double mutant have failed (Kim and Xu, unpublished data). Unlike S. *cerevisiae*, all the filamentous ascomycetes sequenced have only two genes encoding catalytic subunits of PKA. In *Ustilago maydis* and A. *nidulans*, *CPK2* orthologues play only minor roles in cAMP

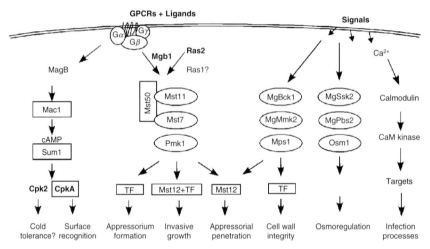

Figure 5.2. Signaling pathways involved in infection-related morphogenesis. Surface recognition is mediated by the cAMP pathway (Mac1-cAMP-PKA). Two catalytic units of PKA, CpkA, and Cpk2, may have overlapping functions during appressorium formation and plant infection. Various GPCRs (e.g., Pth11) may bind to different ligands during plant infection and signaling through the trimeric G-proteins. The Mst11-Mst7-Pmk1 cascade is required for appressorium formation, penetration, and growth of infectious hyphae (invasive growth). Mst50 may function as an adaptor by binding to Mst7, Mst11, Ras2, and Mgb1. Ras1 also directly interacts with Mst50 and Mst11 but its function is not clear. Multiple transcription factors (TFs) may be regulated by Pmk1 (e.g., Mst12) to control different infection processes. Ras2 may be also involved in the cAMP signaling for surface recognition. Pmk1 may also cross-talk with the Mps1 MAP kinase pathway in regulating cell wall integrity and appressorial penetration. The Osm1 pathway is involved in osmoregulation and stress responses but does not play an essential role in plant infection. Ca^{2+} signaling may be also involved in appressorium formation by regulating PKC activities. However, the relationship or interaction between cAMP- and Ca^{2+}-signaling pathways with the Pmk1 MAP kinase cascade is not clear.

signaling (Durrenberger *et al.*, 1998; Ni *et al.*, 2005). Deletion of both catalytic subunits of PKA is lethal in *A. nidulans* but viable in *U. maydis* (Ni *et al.*, 2005). Several Mac1- and CpkA-interacting clones have been isolated using the yeast two-hybrid system (Kulkarni and Dean, 2004), including a putative transcriptional regulator and two different glycosyl hydrolases. However, none of these genes have been further characterized for their interactions with Mac1 or CpkA and their roles in cAMP signaling.

Although membrane proteins with limited homology to *Dictyostelium discoideum* cAMP receptors have been identified in the *M. oryzae* and *N. crassa* genomes (Galagan *et al.*, 2003), these putative G-protein–coupled receptors are not well conserved, and their function in cAMP signaling is not clear. One putative receptor gene *PTH11*, encoding a transmembrane protein localized to the cell membrane and vacuoles, has been implicated in surface recognition in *M. oryzae* (DeZwaan *et al.*, 1999). Germ tubes of the *pth11* mutant undergo hooking and apical swelling, but only 10–15% of them eventually develop into normal appressoria. The *pth11* mutant is nonpathogenic on healthy leaves but it can colonize wounded plant tissues. Exogenous cAMP restores appressorium formation and pathogenicity in the *pth11* mutant (DeZwaan *et al.*, 1999), suggesting that *PTH11* may be involved in signal sensing and regulating Mac1 activities. However, to date direct evidence is lacking to support the sensory role of Pth11 and its regulation of the cAMP signaling. Interestingly, there is one putative intracellular cAMP-binding protein (other than Sum1) in the *M. oryzae* genome. It has no homologue in *S. cerevisiae* and *S. pombe* but is well conserved in the sequenced filamentous ascomycetes, including *N. crassa* and *Fusarium graminearum*.

Activation of adenylate cyclase is mediated by heterotrimeric GTP-binding proteins in several fungi (Kays and Borkovich, 2004). Similar to other fungi, *M. oryzae* has three Gα (MagA, MagB, and MagC), one Gβ (Mgb1), and one Gγ (MG10193) subunits. Among all three Gα subunits characterized, only the *magB* deletion mutant is significantly reduced in appressorium formation and virulence (Fang and Dean, 2000; Liu and Dean, 1997). Transformants expressing the dominant active MAGB^{G42R} allele are reduced in virulence but form appressoria on both hydrophobic and hydrophilic surfaces (Fang and Dean, 2000). They also exhibit a colony autolysis phenotype in aged cultures and misscheduled melanization of hyphal tips. The Gβ subunit encoded by *MGB1* in *M. oryzae* is essential for appressorium formation and plant infection (Nishimura *et al.*, 2003). Exogenous cAMP stimulates appressorium formation in *mgb1* mutants, but these appressoria are morphologically abnormal and nonfunctional for plant penetration. In *Cryphonectria parasitica*, the *BDM1* (beta disruption mimic factor-1) gene is involved in G-protein signaling (Kasahara *et al.*, 2000). Orthologues of *BDM1* are found in *M. oryzae* (MG05200) and other sequenced filamentous ascomycetes. The *M. oryzae*

genome also contains eight RGS proteins but none of them has been functionally characterized.

C. Appressorium formation and maturation

After surface recognition and the initial stages of appressorium formation, a septum
forms to delimit the developing appressorium from the rest of the germ tube.
Appressoria then undergo a maturation process, including deposition of additional
cell wall layers, melanization, and turgor generation (Fig. 5.1). Melanin in appressoria is synthesized from the polyketide precursor 1,8-dihydroxynaphthalene
(DHN) and is essential for turgor generation (Howard and Valent, 1996). *ALB1*,
RSY1, and *BUF1* encoding a polyketide synthase (PKS), a scytalone dehydratase,
and a polyhydroxynaphthalene reductase, respectively, are three major genes
involved in the DHN melanin synthesis. Unlike their homologues for melanin
synthesis in *Alternaria alternata* (Kawamura *et al.*, 1997), these genes are not physically linked. Interestingly, the *pig1* transcription factor deletion mutant produces
melanized appressoria but is defective in hyphal melanization (Tsuji *et al.*, 2000),
indicating that melanin synthesis is regulated by different mechanisms in mycelia and appressoria of M. *oryzae*. Cmr1 of C. *lagenarium*, the orthologue of Pig1,
also is only required for melanin synthesis in hyphae.

When appressoria mature, the conidial and germ tube cells usually
collapse and are no longer viable after 24 h (Veneault-Fourrey *et al.*, 2006a).
Therefore, the septum separating appressorium from the germ tube may be
essential for appressorium development. The M. *oryzae* genome has homologues
of all the genes known to be involved in septation in A. *nidulans* (Table 5.1),
including SepA and SepH (Harris, 2001; Walther and Wendland, 2003). Several genes known to be involved in cytokinesis in S. *cerevisiae*, including Bni5,
Dse2, Skm1, and Bud4, are absent in M. *oryzae* and other sequenced Sordar-
iomycetes (Table 5.1). Deletion of the *SepH* homologue (MG04100) eliminated
septation, conidiation, and plant infection in M. *oryzae* (Zhao and Xu, unpublished data). In contrast, a disruption mutant of the *HEX1* gene still produces misshaped appressoria and is only reduced in penetration and virulence
(Asiegbu *et al.*, 2004; Soundararajan *et al.*, 2004). Hex1 is the major component
of the Woronin body (Tenney *et al.*, 2000) and the septal pores of *hex1* mutants
are not plugged when hyphae are damaged and lead to extensive loss of cytoplasm. These observations indicate that M. *oryzae* must have other mechanisms
to completely plug the septal pore in the septum delimitating appressoria
and Woronin bodies are not essential for infectious hyphae to grow inside plant
cells.

Appressorium formation is known to be regulated by the *PMK1* (path-
ogenicity-MAP kinase 1) gene, which is the only homologue of yeast Fus3 and
Kss1 in M. *oryzae* (Xu and Hamer, 1996). Germ tubes of the *pmk1* mutant still

Table 5.1. *Magnaporthe* Genes Related to Septation in A. *nidulans* and Cytokinesis in S. *cerevisiae*[a]

Function	Name	M. *oryzae* homologue (e-value)	N. *crassa* homologue (e-value)	S. *cerevisiae* or A. *nidulans*
A. nidulans				
Septin (morphogenetic	AspA	MG03087 (9e-158)	NCU02464 (4e-138)	Cdc11 (1e-83)
scaffolds at sites of	AspB	MG01521 (0)	NCU08297 (0)	Cdc3 (2e-118)
cell division and	AspC	MG07466 (0)	NCU03795 (0)	Cdc12 (5e-119)
polarized growth)	AspD	MG06726 (2e-164)	NCU03515 (2e-171)	Cdc10 (4e-105)
	AspE	MG02626 (8e-110)	NCU01998 (6e-119)	Cdc11 (3e-11)
Formin (organizing	SepA	MG04061 (0)	NCU01431 (0)	Bni1 (1e-81)
actin cytoskeleton	SepB	MG06969 (0)	NCU08484 (0)	Chl15 (1e-44)
at cell wall	SepH	MG04100 (0)	NCU01335 (0)	Cdc15 (2e-55)
division site)				
Regulation of	NimX	MG01362 (5e-138)	NCU09778 (2e-139)	Cdc28 (5e-116)
septum formation	RnrA	MG06408 (3e-153)	NCU07887 (3e-164)	Rnr2 (8e-131)
	SNAD	MG04321 (8e-21)	NCU04826 (2e-14)	Uso1 (2e-11)
	ArtA	MG01588 (2e-114)	NCU02806 (3e-114)	Bmh2 (2e-98)
S. cerevisiae				
Septin ring-associated	Axl2	MG01350 (3e-35)	NCU04601 (7e-32)	AN1359 (3e-39)
genes	Bni4	MG08826 (3e-15)	NCU00064 (3e-16)	AN0979 (2e-12)
	Hsl1	MG02810 (4e-74)	NCU09064 (3e-83)	AN8693 (3e-91)
	Shs1	MG03087 (2e-63)	NCU02464 (3e-63)	AN4667 (2e-59)
	Siz1	MG08837 (1e-38)	NCU06213 (3e-32)	AN6498 (9e-39)
	Smt3	MG05737 (5e-22)	NCU09813 (7e-22)	AN1191 (1e-23)
	Skt5	MG10646 (4e-75)	NCU09322 (2e-74)	AN1554 (2e-75)
	Spr28	MG03087 (3e-49)	NCU02464 (8e-50)	AN4667 (2e-49)
	Spr3	MG07466 (1e-62)	NCU03795 (4e-65)	AN8182 (1e-62)
	Utr2	MG01134 (2e-89)	NCU05686 (1e-94)	AN4515 (9e-91)
Septin ring assembly	Bni5	No significant hit	No significant hit	No significant hit
	Dma1	MG05257 (4e-42)	NCU06269 (2e-58)	AN6908 (7e-32)
	Dma2	MG05257 (7e-40)	NCU06269 (5e-56)	AN6908 (4e-31)
	Gin4	MG02810 (5e-75)	NCU09064 (2e-83)	AN8693 (3e-90)
	Kcc4	MG02810 (5e-76)	NCU09064 (4e-79)	AN8693 (1e-85)
Cytokinesis, completion	Chs1	MG09551 (0)	Chs-3 (0)	AN4566 (0)
of separation	Cts1	MG10333 (2e-34)	NCU04500 (1e-50)	AN8241 (2e-40)
	Dse2	No significant hit	No significant hit	No significant hit
	Dse4	MG01001 (3e-126)	NCU07076 (7e-116)	AN0472 (8e-141)
	Hym1	MG03219 (4e-35)	NCU03576 (3e-39)	AN3095 (3e-39)
	Scw11	MG04582 (1e-33)	NCU09326 (2e-38)	AN1551 (2e-36)
Cell separation during cytokinesis	Skm1	*CHM1* (7e-86)	NCU00406 (1e-86)	AN8836 (3e-88)
Cytokinesis contractile ring contraction	Mlc2	No significant hit	No significant hit	No significant hit

[a]S. *cerevisiae* genes were categorized according to Gene Ontology at SGD (www.yeastgenome. org). For genes involved in bipolar and axial bud site selection in yeast, Atc1, Bud3, Gic1, Gic2, Pea2, and Sph1 have no significant homologue in A. *nidulans*, M. *oryzae*, and N. *crassa*. All other 60 genes (including Bud2, Bud4, Bud5, Cdc3, Cdc10, Cdc11, and Cdc12) are conserved in filamentous fungi. Septin genes Cdc3, Cdc10, Cdc11, and Cdc12 were listed under A. *nidulans*.

recognize hydrophobic surfaces and form subapical swollen bodies, but fail to form appressoria. The expression of *PMK1* is enhanced in developing appressoria and young conidia (Bruno *et al.*, 2004). Nuclear localization of a GFP-Pmk1 fusion is observed during appressorium formation. Site-directed mutagenesis of *PMK1* indicated that the kinase activity of Pmk1 is required for appressorium formation (Bruno *et al.*, 2004). In M. *oryzae*, appressorium formation is sensitive to several Ca^{2+} modulators and calmodulin (CaM) antagonists (Lee and Lee, 1998). A known activator of protein kinase C (PKC), diacylglycerol (DAG) induces appressorium formation, indicating that Ca^{2+}/CaM signaling also may be involved in appressorium formation. Although molecular data to support these pharmacological observations are lacking in M. *oryzae*, silencing of the calmodulin gene or disruption of a lipid-induced kinase gene prevents appressorium formation in C. *trifolii* (Dickman *et al.*, 2003). In M. *oryzae*, deletion of the cyclophilin gene *CYP1* results in a reduction in penetration and virulence (Viaud *et al.*, 2002), suggesting a role for calcineurin in infection-related morphogenesis. M. *oryzae* has one catalytic (MG07456) and one regulatory (MG06933) subunit of calcineurin and many other proteins involved in Ca^{2+} signaling, including one CaM, one CaM-dependent kinase, Ca^{2+}-permeable channels, Ca^{2+} pumps and transporters (Zelter *et al.*, 2004). At least nine predicted genes appear to be able to bind Ca^{2+}/CaM and many of them have putative kinase domains. However, similar to N. *crassa*, M. *oryzae* lacks significant homologues of InsP3 receptor, ADP ribosyl cyclase, spingosine kinase, or SCaMPER (Zelter *et al.*, 2004).

D. The Pmk1 MAP kinase pathway

In addition to its inability to form appressoria, the *pmk1* mutant is defective in infectious hyphal growth and fails to colonize through wounds. Studies of several phytopathogenic fungi, including C. *lagenarium*, F. *graminearum*, *Claviceps purpurea*, and *Cochliobolus heterostrophus*, have shown that *PMK1* homologues are well conserved for regulating appressorium formation and other plant infection processes. The M. *oryzae* genome contains distinct homologues of components of the yeast pheromone and filamentation pathways, including Ste2, Ste3, Ste20, Ste11, Ste7, Ste12, and Ste50 (Bardwell, 2004). The MAP kinase kinase (MEK) Mst7 and MEK kinase (MEKK) Mst11 (homologues of yeast Ste7 and Ste11, respectively) are required for appressorium formation and infectious growth in M. *oryzae* (Zhao *et al.*, 2005). Although there is no direct interaction between Pmk1 and Mst11 or Mst7, Mst50 (a Ste50 homologue) interacts with both Mst11 and Mst7. Mst50 may function as an adaptor protein interacting with upstream components and plays critical roles in activating the Pmk1 cascade (Fig. 5.2). Both Mst11 and Mst50 contain a sterile alpha motif (SAM) that is known to be involved in protein–protein interactions. SAM is essential for the Mst11–Mst50 interaction and the function of Mst11 or Mst50 (Park *et al.*, 2006). The function of another SAM-containing protein (MG06334) in *M. oryzae* is not clear. *F. graminearum* and *N. crassa* genomes all have only three SAM-containing proteins, suggesting that these genes are well conserved in fungi.

Mst50 and Mst11 also have a Ras-association (RA) domain. Two predicted Ras homologues in *M. oryzae*, Ras1 (MG09499) and Ras2 (MG06154), both interact directly with Mst50 in yeast two-hybrid assays. The *ras1* deletion mutant still forms appressoria and is pathogenic on rice plants. However, repeated attempts have failed to identify a *ras2* deletion mutant. Expressing a dominant active *RAS2* allele in the wild-type strain but not in the *pmk1* mutant stimulated abnormal appressorium formation on hydrophilic surfaces (Xue and Xu, unpublished data). These data indicate that *RAS1* and *RAS2* have distinct functions in *M. oryzae*. In *S. cerevisiae*, *CDC42*, *STE20*, *STE50*, and *STE11* are shared components of MAP kinase pathways regulating mating, filamentous growth, and osmotolerance (Bardwell, 2004; Ramezani-Rad, 2003). Cdc42 interacts with both Ste50 and the PAK kinase Ste20. The association with Cdc42 is responsible for bringing Ste20 together with the Ste11–Ste50 complex to activate Ste11 (Ramezani-Rad, 2003). In *M. oryzae*, however, deletion of *MST20* (*STE20* homologue) has no obvious effect on plant infection or responses to osmotic stresses (Li *et al.*, 2004). The *CDC42* homologue is also dispensable for appressorium formation and plant infection (S. Wu and Z. Wang, personal communication), indicating that MgCdc42 is not involved in linking a PAK kinase to the activation of Mst11. In *S. cerevisiae*, Ste5 is the scaffold protein that interacts with different components of the pheromone response pathway. *M. oryzae* and other filamentous ascomycetes lack a significant homologue of Ste5. The Mst11, Mst7, and Pmk1 proteins also lack homology with their yeast counterparts in the regions that are involved in interacting with Ste5. Therefore, *M. oryzae* and other fungi must deploy different strategies to regulate the specificity of this well-conserved MAP kinase pathway.

One putative downstream transcription factor of the Pmk1 pathway, Mst12, is homologous to yeast Ste12. The *mst12* deletion mutant forms melanized appressoria that are defective in penetration and fail to elicit plant defense responses in the underlying plant cells (Park *et al.*, 2002). The *mst12* mutant is nonpathogenic and fails to develop penetration pegs, probably due to cytoskeleton defects in mature appressoria (Park *et al.*, 2004). The C-terminal portion of Mst12 weakly interacts with Pmk1 in yeast two-hybrid assays, indicating that *MST12* functions downstream of *PMK1* to regulate genes involved in appressorial penetration and infectious growth. In *S. cerevisiae*, Dig1 and Dig2 are two regulatory proteins involved in Ste12 activation (Bardwell, 2004). The region interacting with Dig1 and Dig2 is not conserved in Mst12. The *M. oryzae* genome also lacks homologues of Dig1 and Dig2. In addition to *MST12*, *M. oryzae* has putative homologues of *S. cerevisiae TEC1*, *C. albicans EFG1*, and *N. crassa VIB1*. Preliminary data indicate that none of them is essential for appressorium formation and plant infection (Xu, unpublished data). Multiple transcription factors may form heterodimers or a transcription complex, similar to the Ste12–Tec1, Ste12–Mcm1, and Flo8–Sfl1 interactions in yeast or Efg1–Tec1 interaction in *C. albicans* (Lane *et al.*, 2001; Pan and Heitman, 2002), to regulate different processes of appressorium formation or plant infection processes in *M. oryzae*. The *MCM1* (MG02773) homologues are well conserved in *M. oryzae* and other filamentous fungi. However, *M. oryzae* lacks a significant homologue of Flo8 and has one putative Sfl1 homologue. One of the restriction enzyme-mediated integration (REMI) mutants is disrupted in a putative transcription factor gene *PTH12* (Sweigard *et al.*, 1998). However, the characterization of *PTH12* has not been published and its relationship with the cAMP signaling or *PMK1* pathway is not clear. In contrast to the *pmk1* mutant, the *pth12* deletion mutant generated in our laboratory still forms abnormal appressoria that are not melanized and can reestablish polarized growth without penetration.

    Overall, the *PMK1* pathway regulates appressorium formation and infectious growth, two processes that do not exist in *S. cerevisiae* (Fig. 5.2). Although the Mst11-Mst7-Pmk1 MAP kinase cascade is conserved, the upstream signal inputs and downstream transcription factors of the *PMK1* pathway must be different between *M. oryzae* and *S. cerevisiae*. To identify genes regulated by this MAP kinase pathway, subtraction libraries enriched with genes regulated by *PMK1* have been constructed and sequenced (Xue *et al.*, 2002). Two genes identified in this library, *GAS1* and *GAS2*, encode small proteins that are specific to filamentous fungi. Both are expressed specifically during appressorium formation. Mutants deleted of *GAS1*, *GAS2*, or both *GAS1* and *GAS2* have no defect in growth, conidiation, or appressoria formation, but are reduced in appressorial penetration and lesion development. Comparative analysis of expressed sequence tags (ESTs) sequenced from cDNA libraries of wild-type appressoria and *pmk1* germlings also have been used to identify genes that are differentially expressed in the *pmk1* mutant during appressorium formation (Ebbole *et al.*, 2004; Soanes and Talbot, 2005). Interestingly, some of these genes with enhanced expression in the *pmk1* germlings appear to be involved in protein or melanin synthesis (Soanes and Talbot, 2005). However, their expression and function need to be experimentally examined. When green fluorescent protein (GFP) was used as the reporter gene to examine the expression of 12 selected appressorium-stage ESTs, only 1 putative Mac1-interacting gene and 1 putative membrane-associated protein gene were found to be specifically expressed in appressoria. Targeted knockout mutants of either gene failed to show detectable phenotypic changes under laboratory conditions (Banno et al., 2003).

E. Genes expressed during appressorium formation

Various approaches have been used to identify genes highly or specifically expressed during appressorium formation. Over 2500 ESTs were sequenced from a cDNA library constructed with RNA isolated from appressoria (Ebbole et al., 2004). MPG1, UVI-1, GAS1, and GAS2 are among the most abundant ESTs in this library (Ebbole et al., 2004; Soanes and Talbot, 2005). By serial analysis of gene expression (SAGE) with mRNA isolated from conidia germinating in the presence and absence of cAMP, 57 and 53 genes have been found to be up- and downregulated by cAMP treatment, respectively (Irie et al., 2003). Many of these cAMP-induced genes have no known homologues in GenBank, but some of them are well-characterized pathogenicity factors such as MPG1, GAS2, and MAC1. A subtraction library was constructed and sequenced to isolate genes with enhanced expression levels in developing appressoria (Kamakura et al., 2002). One gene identified, CBP1, encodes a putative chitin-binding protein with signal peptide. Mutants deleted of CBP1 form abnormal appressoria on artificial surfaces but produce normal, functional appressoria on the leaf surface, indicating that CBP1 plays an important role in surface recognition (Kamakura et al., 2002). Sequencing 250 clones from a subtraction library enriched in genes expressed during appressorium maturation revealed 142 unique genes (Lu et al., 2005). RT-PCR analysis identified 71 of them specifically expressed in appressoria, including GAS1, GAS3, and PTH11. Among five proteins that are identified by two-dimensional (2D) gel analysis to be induced during appressorium formation, two were 20S proteasome alpha subunits (Mgp1 and Mgp4) and one was a scytalone dehydratase. The 20S proteasome components were confirmed by Western blot analysis to be highly expressed during appressorium formation and nitrogen starvation, indicating that the 20S proteasome may be involved in utilizing storage proteins under these conditions (Kim et al., 2004b).

III. MECHANISMS OF PENETRATION

A. Penetration peg formation and penetration forces

Mature appressoria tightly adhere to the plant surface by a ring of appressorium mucilage and develop the thin-walled appressorium pore area at the contact site (Bourett and Howard, 1992). The penetration peg emerging from the appressorial pore contains high concentrations of actin filaments that may be necessary to stabilize the tip of the penetration peg (Howard and Valent, 1996). The mst12 mutant is normal in appressorium pore development but incapable of penetration peg formation. The network of vertical microtubules observed in mature appressoria of the wild-type strain is absent in the *mst12* mutant (Park *et al.*, 2004), indicating that actin and cytoskeletal elements may be involved in the selection of peg emergence site and reestablishment of polarized growth. The M. *oryzae* genome has one actin gene (MG03982) and one predicted gene each for α-, β-, and γ-tubulins that are typical of filamentous fungi.

In M. *oryzae*, the elevated osmotic pressure within melanized appressoria is the primary force for direct penetration of plant cuticles and synthetic membranes. The appressorial turgor is estimated to be as high as 8 MPa or 80 bars by cytorrhysis assays. Up to 3-M glycerol has been observed in mature appressoria (de Jong *et al.*, 1997). However, the microfibrils of host cell walls around the penetration pegs are disorganized (Koga, 1995), indicating that hydrolytic enzymes also are involved in plant penetration. In the genome, several of the expanded gene families (Dean *et al.*, 2005) are predicted to encode hydrolytic enzymes that can erode the plant cuticle and degrade plant cell walls. Although deletion of a cutinase gene *CUT1* has no detectable effect on plant infection (Sweigard *et al.*, 1992), there are at least seven other putative cutinase genes in M. *oryzae*. *CUT1* was not among those cutinase genes that are significantly upregulated during infection-related development (Dean *et al.*, 2005). Among 10 putative xylanase genes, 3 of them, *XYL1*, *XYL2*, and *XYL6*, are expressed in axenic cultures and dispensable for plant infection in M. *oryzae* (Wu *et al.*, 1997, 2006).

B. Appressorium turgor generation

Carbon compounds stored in conidia, mainly glycogen, trehalose, and lipids, appear to be used to synthesize glycerol, which is responsible for the enormous turgor pressure. Deletion of the *TPS1* gene encoding a trehalose-6-phosphate synthase blocks trehalose synthesis and reduces the efficiency of appressorial penetration (Foster *et al.*, 2003). For two trehalase genes that have been characterized in M. *oryzae*, only *NTH1* but not *TRE1* is required for pathogenesis (Foster *et al.*, 2003; Sweigard *et al.*, 1998). Both glycerol-3-phosphate dehydrogenase and glycerol dehydrogenase are expressed in developing appressoria and may be responsible for glycerol synthesis from glycolytic intermediates (Thines *et al.*, 2000), but their enzymatic activities are not induced during appressorium maturation. In contrast, triacylglycerol lipase activity is strongly induced during appressorium maturation, and lipid droplets are completely degraded in appressoria formed on inert artificial surfaces (Weber *et al.*, 2001). In the M. *oryzae* genome, there are several putative acyl-CoA dehydrogenase genes but no acyl-CoA oxidase gene. This genome feature may allow β-oxidation to occur both in mitochondria and in glyoxysome-like bodies and confer flexibility in lipid metabolism (Dean *et al.*, 2005). Therefore,

M. *oryzae* has a versatile capacity to synthesize glycerol in the appressorium for turgor generation.

The M. *oryzae OSM1* gene encodes a MAP kinase homologous to yeast *HOG1*. In *S. cerevisiae*, *HOG1* regulates glycerol accumulation in response to hyperosmolarity. The *osm1* mutant of M. *oryzae* is hypersensitive to high osmotic stress during vegetative growth, but it has no defect in turgor generation and plant infection, indicating that glycerol accumulation in appressoria is not regulated by *OSM1* (Dixon *et al.*, 1999). Deletion of a histidine kinase gene (*HIK1*) homologous *Os-1/Nik-1* of N. *crassa* also has no effect on appressorial penetration and pathogenicity on rice plants (Motoyama *et al.*, 2005). The sensitivity of the *hik1* deletion mutant to high concentrations of sugars but not salts is elevated. Since the mobilization and degradation of glycogen and lipid bodies are affected in both *pmk1* and *cpkA* mutants (Thines *et al.*, 2000), it is likely that the cAMP signaling and *PMK1* pathways may coordinate with each other in the utilization of conidial storage products and generation of appressorial turgor.

C. Appressorial penetration involves many coordinated processes

To reestablish polarized growth, the elaboration of a penetration peg by the mature appressorium likely depends on controlled cell wall and membrane modifications. *MPS1* is highly homologous to the *S. cerevisiae* MAP kinase *SLT2* that regulates nutrient sensing and cell wall integrity (Xu *et al.*, 1998). The *mps1* deletion mutant is defective in cell wall integrity and is nonpathogenic on healthy rice leaves. Appressoria formed by the *mps1* mutant are melanized but fail to penetrate underlying plant cells and fail to form infectious hyphae. Functional characterization of the *MPS1* homologues in *C. purpurea* and *F. graminearum* has shown that this MAP kinase may also be well conserved among fungal pathogens and play important roles during plant infection (Hou *et al.*, 2002; Mey *et al.*, 2002). In *F. graminearum*, the *mgv1* mutant is self-incompatible, indicating that Mgv1 MAP kinase is essential for hyphal fusion and heterokaryon formation, another process involving controlled cell wall modification. Predicted genes highly homologous to several components of the yeast *SLT2* MAP kinase pathway have been identified in the M. *oryzae* genome (Dean *et al.*, 2005). Preliminary data indicate that the *BCK1* and *MMK2* homologues also are essential for appressorial penetration and plant infection (Xu, unpublished data).

The *PLS1* gene was identified by insertional mutagenesis to be essential for appressorial penetration. The *pls1* deletion mutant, similar to the *mps1* mutant, forms melanized appressoria but fails to penetrate and infect rice plants (Clergeot *et al.*, 2001). *PLS1* encodes a putative membrane protein structurally related to the tetraspanins in animals. It is specifically expressed in appressoria and localized in plasma membranes and vacuoles. Although its exact function is not clear, Pls1 may be involved in reestablishing polarized growth (Veneault-Fourrey *et al.*, 2006b) and its orthologues appear to be conserved in other fungal pathogens for regulating plant infection processes (Gourgues *et al.*, 2004). Pls1 is the only tetraspanin in M. *oryzae* and a single copy of an orthologous tetraspanin gene also exists in other fungi, including N. *crassa* and F. *graminearum*. Phylogenetic analysis indicated that Pls1 and its orthologues form a fungal-specific family of tetraspanins (Gourgues *et al.*, 2002).

Several other genes, including *PDE1*, *MMT1*, and *EMP1*, are known to be important for penetration in M. *oryzae*. *PDE1* encodes a P-type ATPase that is expressed in germinating conidia and developing appressoria. The *pde1* mutant is reduced in virulence on rice but normal on barley leaves (Balhadere and Talbot, 2001). *MMT1* encodes a metallothionein protein that is highly expressed throughout the growth and development. The *mmt1* mutant is defective in penetration (Tucker *et al.*, 2004) but it grows faster than the wild-type strain on complete medium (CM) or CM supplemented with hydrogen peroxide. The *emp1* deletion mutant also is reduced in appressorium formation and virulence (Ahn *et al.*, 2004). The *EMP1* gene encodes a putative extracellular matrix protein that is orthologous to the F. *oxysporum* Fem1. Mitosis and autophagic cell death have been linked with appressorium formation and penetration in M. *oryzae* (Veneault-Fourrey *et al.*, 2006a). Interfering with mitosis by hydroxyurea treatment or expressing the MgNimA^{E37G} allele at restrictive temperature adversely affects appressorium formation. Deletion of the MgATG8 gene, a homologue of yeast ATG8, blocks autophagy, conidial cell death, and appressorial penetration (Veneault-Fourrey *et al.*, 2006a).

IV. INVASIVE GROWTH AND HOST–PATHOGEN INTERACTIONS

A. Infectious hyphal growth

After penetration, the penetration peg develops into bulbous, lobed infectious hyphae that are morphologically distinct from germ tubes and somatic hyphae. M. *oryzae* has been described as a hemibiotrophic pathogen, and the infectious hyphae can grow intra- and intercellularly without damaging plant cells during the initial biotrophic stage (Heath *et al.*, 1992). However, there are contradictory reports on whether an intact plant cytoplasmic membrane surrounds the infectious hyphae. Studies using the FM4–64 fluorescent dye indicate that the invasive fungal hyphae are surrounded by the plant cytoplasmic membrane (Kankanala and Valent, 2005). In hemibiotrophic pathogens, the switch of infectious hyphae from the biotrophic phase to the necrotrophic phase in later infection stages results in plant cell death. However, to date, there is no evidence to show when and how this biotrophic–necrotrophic switch occurs in M. oryzae.

Like other fungal pathogens, M. oryzae must overcome constitutive and induced plant defense responses during infectious hyphal growth. Some of the enzymes secreted by M. oryzae may degrade or detoxify antimicrobial compounds produced by rice (Koga et al., 1998). The M. oryzae genome contains many secreted enzymes, such as 31 putative cytochrome P450 and 13 putative laccase genes, that may have overlapping roles in detoxifying plant defense molecules (Pedras and Ahiahonu, 2005). M. oryzae also may simply use transporters to pump out toxic or inhibitory molecules produced by plant cells. Interestingly, ABC1 encodes a putative membrane transporter with all the characteristics of the ATP-binding cassette (ABC) superfamily (Urban et al., 1999). The abc1 deletion mutant is normal in penetration but is dramatically reduced in lesion formation. After penetration, the abc1 mutant is restricted in its infectious hyphal growth and soon dies. The transcription level of ABC1 is induced by several toxic compounds and a rice phytoalexin (Urban et al., 1999). It is likely that ABC1 is involved in the tolerance of M. oryzae to certain toxic plant defense compounds or exporting fungal metabolites that suppress to host defense responses. Another ABC transporter gene has been characterized in M. oryzae. The abc2 disruption mutant has no obvious defect in plant infection but displays enhanced sensitivity to several antifungal compounds, including tebuconazole and cycloheximide (Lee et al., 2005b). The M. grisea genome contains 35 predicted genes with the ABC transporter domain (pfam00005). It is likely that additional ABC transporter genes play important roles in different plant infection processes and fungicide resistance in M. oryzae.

B. Genes involved in race-specific interactions

The rice–M. oryzae pathosystem is governed by specific interactions between fungal avirulence (AVR) genes and their corresponding plant resistance (R) genes. Several avirulence genes have been cloned in M. oryzae, including PWL2, AVR-Pita, AVR-CO39, and ACE1. AVR-Pita is a putative neutral zinc metalloprotease gene and PWL2 encodes a glycine-rich, hydrophilic protein. Both have signal peptides and likely are secreted proteins. Single amino acid mutations in the Pi-ta leucine-rich domain or in the AVR-Pita protease motif disrupt the physical interaction between Pi-ta and Avr-Pita proteins and result in loss of resistance responses in rice plants (Orbach et al., 2000), indicating that the direct interaction between Avr-Pita and Pi-ta inside plant cells is necessary for the race-specific recognition (Jia et al., 2000). Several PWL2 homologues with varying degrees of sequence homology and different chromosome locations, including PWL1, PWL3, and PWL4, have been identified (Kang et al., 1995).

The *AVR1-CO39* gene conferring avirulence on the rice cultivar CO39 was cloned from a weeping lovegrass pathogen. It is not present in M. *oryzae* strains Guy11, 70–15, and most of the 45 rice-infecting isolates examined. Ancestral rearrangements may have resulted in the loss of the *AVR1-CO39* locus during early evolution of the Oryza-specific subgroup of M. *oryzae* (Farman *et al.*, 2002; Tosa *et al.*, 2005). Another *AVR* gene, *ACE1*, encodes a protein with combined PKS and nonribosomal peptide synthetase (NRPS) domains (Bohnert *et al.*, 2004). *ACE1* is specifically expressed in penetrating appressoria but not in infectious hyphae. Sequence analysis of the genomic region adjacent to *ACE1* has revealed a cluster of genes potentially involved in secondary metabolism. It is likely that an unidentified secondary metabolite synthesized by Ace1 in appressoria is responsible for specific interaction with the resistance gene *Pi33* (Bohnert *et al.*, 2004).

The sequenced M. *oryzae* strain 70–15 contains four known avirulence genes: *AVR-Pita*, *ACE1*, *PWL2*, and *PWL3*. Three other known avirulence genes, *PWL1*, *PWL4*, and *AVR1-CO39*, are absent. For over 40 known major rice blast resistance genes, many corresponding avirulence genes have been mapped in M. *oryzae* (Dioh *et al.*, 2000; Valent and Chumley, 1991). However, it is impossible to systematically search for *AVR* genes in the genome sequence because of the lack of common structural features or conserved domains among fungal *AVR* genes. The M. *oryzae* genome does not have orthologues of well-characterized *AVR* genes from other pathogenic fungi, including *Avr2*, *Avr4*, *Avr9*, *ECP2*, *ECP3*, and *ECP5* from *Cladosporium fulvum* and *NIP1* from *Rhynchosporium secalis* (Dean *et al.*, 2005). Similarly, *F. graminearum* lacks distinct orthologues of known M. *oryzae* AVR genes *PWL2* and *AVR-CO39*. The lack of sequence similarity or conservation in fungal *AVR* genes may indicate that they are not important virulence factors conserved in many plant pathogenic fungi, or their roles in plant infection are host specific.

C. Physiological activities of infectious hyphae

As a hemibiotrophic pathogen, M. *oryzae* must adapt to the plant environment and produce enzymes necessary for nutrient absorption from plant tissues in a balanced manner during infectious growth. Unfortunately, only a few M. *oryzae* mutants are known to be defective in certain metabolic activities. The sulfate nonutilizing (*sub*) mutant has no obvious defects in plant infection (Harp and Correll, 1998). The M. *oryzae* genome contains orthologues of N. *crassa* genes involved in sulfur acquisition and processing (Borkovich *et al.*, 2004). The arginine auxotrophic mutant *argB* is fully pathogenic (Sweigard *et al.*, 1992) but the histidine and methionine auxotrophic REMI mutants are significantly reduced in plant colonization and lesion development (Balhadere *et al.*, 1999; Sweigard *et al.*, 1998). In *F. graminearum* and the human pathogen

Cryptococcus neoformans, methionine auxotrophic mutants are reduced in virulence (Seong *et al.*, 2005). Failure to complement the defects of the methionine synthase deletion mutant with methionine supplements is due to the accumulation of a toxic intermediate (Pascon *et al.*, 2004). Therefore, the phenotypes of auxotrophic mutants in plant infection should be interpreted with caution.

In M. *oryzae*, nitrogen starvation may mimic the *in planta* growth condition of infectious hyphae because nitrogen limitation induces the expression of many infection-related genes and stimulates the production of phytotoxic compounds in cultures (Talbot *et al.*, 1997). However, deletion of the major nitrogen regulatory gene *NUT1* has no effect on virulence in M. *oryzae* (Froeliger and Carpenter, 1996). *NUT1* is an orthologue of A. *nidulans areA* and N. *crassa NIT-2*. The *nut1* mutant is defective in growth on a variety of nitrogen sources but it can use proline and alanine. The *areA* and *nit-2* mutants fail to utilize nitrogen sources other than ammonium and glutamine (Marzluf, 1997). M. *grisea* also has orthologues of the *NIT-4*, *NMR*, and *PCO-1* genes that regulate nitrate utilization metabolism in N. *crassa* (Borkovich *et al.*, 2004), but none of them has been functionally characterized. Two nonallelic mutants, *npr1* and *npr2*, defective in utilization of a wide range of nitrogen sources (e.g., nitrate or amino acids) have been shown to be defective in plant infection (Lau and Hamer, 1996). However, genes responsible for observed phenotypes in these two mutants have not been identified.

It is not known what carbon sources are taken up by infectious hyphae in M. *oryzae*. Characterization of the glycerol-3-phosphate dehydrogenase gene in C. *gloeosporioides* has indicated that glycerol may be one of the major carbon sources available from plant cells (Wei *et al.*, 2004). Interestingly, the most abundant fungal clone identified in ESTs from a cDNA library of infected rice leaves is a homologue of the N. *crassa* glucose-repressible gene *GRG1* (Kim *et al.*, 2001). However, the function of this *GRG1* homologue in plant infection is not clear. The expression of the isocitrate lyase gene *ICL1* is also upregulated during plant infection (Wang *et al.*, 2003). The *icl1* deletion mutant is delayed in conidium germination and appressorium formation and reduced in lesion development. The isocitrate lyase gene also is important for virulence in *Leptosphaeria maculans* (Idnurm and Howlett, 2002) and C. *albicans*. Therefore, the glyoxylate cycle may play an important, conserved role in fungal pathogenesis. In addition, the *pth1* and *pth2* mutants of M. *oryzae* that are disrupted in homologues of yeast *GRR1* and *CAT2*, respectively, are reduced in virulence (Sweigard *et al.*, 1998). *GRR1* encodes a glucose repression resistant protein and the yeast *grr1* mutant is defective in high-affinity glucose transport and glucose repression. *PTH2* encodes a putative carnitine acetyltransferase associated with peroxisomes and mitochondria. In A. *nidulans*, CreA-dependent carbon catabolite repression has been well characterized (Ruijter and Visser, 1997).

It will be interesting to functionally characterize CreA-like genes in M. *oryzae* to determine the role of carbon catabolite repression in pathogenesis. Interestingly, orthologues of CreA are well conserved in N. *crassa* and F. *graminearum* but is not found in the available M. *oryzae* genome sequence by BLASTP and TBLASTN searches.

D. Genes involved in infectious growth

Many recent molecular studies have focused on the early infection stages. Only a few mutants such as *abc1*, *pth8*, *orp1*, *pal144*, and *pat531* (Balhadere *et al.*, 1999; Fujimoto *et al.*, 2002; Sweigard *et al.*, 1998) are known to be defective in plant infection but normal in appressorium morphogenesis or penetration. *PTH8* encodes a putative UDP-glucose:sterol glucosyltransferase homologous to *UGT51*. The *ORP1* disruption mutant generated by transposon *impala*-mediated insertional mutagenesis also is defective in plant infection (Villalba *et al.*, 2001). *ORP1* is a novel gene that is unique to filamentous ascomycetes. A P-type ATPase gene (*MgAPT2*) involved in exocytosis and rapid induction of host defense responses has been identified (Gilbert *et al.*, 2006). MgApt2-dependent exocytotic processes may play an important role in protein secretion and infectious hyphal growth during plant infection. Mutants blocked in the key components of the *PMK1* pathway and the *mac1* mutant also are defective in infectious growth and fail to colonize abraded leaves (Xu, 2000). A GFP-PMK1 fusion construct is expressed in infectious hyphae and complements the *pmk1* mutant for infectious growth (Bruno *et al.*, 2004). Therefore, the cAMP signaling and *PMK1* MAP kinase pathways may have overlapping or coordinated functions in regulating different infection processes.

Using a promoter-trapping approach, we have transformed Guy11 with a promoter-less EGFP construct and identified several transformants expressing GFP specifically in infectious hyphae. In one of these transformants, GFP signals were strong in infectious hyphae and localized to nuclei (Li and Xu, unpublished data). Further analyses indicated that EGFP was inserted in the ORF of a predicted gene *MIR1* (MG00129), which encodes a low-complexity protein with no known motif or domain. *MIR1* has no homologue in GenBank or other sequenced fungal genomes but it is highly conserved among field isolates. Its induced expression in infectious hyphae and localization to nuclei suggest that this M. *oryzae*-specific gene may play a role in infectious hyphal growth under field conditions. Differential screen and subtraction approaches also have been used to identify genes expressed in infectious hyphae. Similar to *MPG1*, the polyubiquitin gene *PUB4* and the ubiquitin extension protein genes *UEP1* and *UEP3* were isolated by differential screening of a rice infection library (McCafferty and Talbot, 1998). Of the ESTs sequenced from cDNA libraries of infected rice leaves, less than 15% (a total of ~140 clones) are putative fungal genes (Kim *et al.*, 2001; Rauyaree *et al.*, 2001), and only the *MHP1* gene has been shown to be involved in plant infection (Kim *et al.*, 2005). Due to the low abundance of fungal biomass in leaf samples, it is likely that only a few highly expressed fungal genes will be identified in these ESTs. In two independent 2D gel studies with infected rice leaves, no M. *oryzae* protein was identified (Kim *et al.*, 2004a; Konishi *et al.*, 2001).

V. GENES AND GENOME FEATURES

A. Genome sequence and annotation

The M. *oryzae* genome was sequenced with a whole-genome shotgun (WGS) approach at the Broad Institute. Approximately sevenfold coverage of strain 70–15 the genome was generated by sequencing ends of random clones from plasmid, fosmid, and BAC libraries of strain 70–15. The draft sequence was assembled with the Arachne package (Batzoglou *et al.*, 2002). The resulting assembly (V2) consists of 2273 sequence contigs longer than 2 kb, ordered and oriented within 159 supercontigs (scaffolds) with sequences of paired ends. The total length of all sequence contigs is 38.8 Mb (Dean *et al.*, 2005). The assembly displays considerable long-range continuity with 50% of all bases residing in scaffolds longer than 1.6 Mb. To verify the assembly, 33 scaffolds representing 32.8 Mb or 85% of the draft assembly were ordered on the genetic map and assigned to each of the seven linkage groups (Dean *et al.*, 2005).

The M. *oryzae* genome was predicted to contain 11,109 genes longer than 100 amino acids and a gene density of 1 gene per 3.5 kb. Genic sequences account for about 48% of the genome. On average, the amino acid sequences of M. *oryzae* genes share 47% and 46% identity, respectively, with their orthologues in N. *crassa* and A. *nidulans* (Dean *et al.*, 2005). Over 95% of the 28,682 ESTs sequenced from nine different libraries aligned to the M. *oryzae* sequence (Ebbole *et al.*, 2004). The 8177-unigene set of clusters and singletons identified from these ESTs by Ebbole *et al.* (2004) represents about 73% of M. *oryzae* genes from automated prediction. However, many genes represented by ESTs are not identified by automated annotation. Therefore, gene calling algorithms need improvement or better training to predict fungal genes (Galagan *et al.*, 2005b). In the latest release (V2.2) available in 2006, the Broad Institute changed the annotation of 3605 predicted genes by modifying the gene prediction guidelines. The revised genome has a total assembly length of 39.4 Mb (739 contigs) and 12841 predicted genes. For some genes located near or in the sequence gaps, it is likely that only incomplete sequences are available. Additional sequence

information may be necessary to fill the gaps before these genes can be
annotated properly.

B. Gene families and secreted proteins

Single linkage clustering of protein sequences resulted in 348 families containing
five or more genes (Dean *et al.*, 2005). In total, 1266 predicted proteins from
M. *oryzae* were classified into these families. The two largest families have 166
(including MG00001 and MG07835) and 105 (including MG00008) members
that contain DDE (pfam03184) or rvt (pfam00078) domains. The abundance of
transposase-like and retrotransposon proteins is directly related to highly repetitive sequences in M. *oryzae*. In comparison with N. *crassa* and A. *nidulans* (Dean
et al., 2005), several gene families are larger in M. *oryzae*. For example, M. *oryzae*
has 8 putative cutinase, 31 putative cytochrome P450 monooxygenase, and 96
subtilisin-like protease genes. Many of these genes are likely to be involved in
fungal pathogenesis. Interestingly, the M. *oryzae* genome has only three pectate
lyase genes and lacks recognizable pectin lyase. Considering that pectin is an
essential component of plant cell walls and pectate lyases have been implicated in
fungal pathogenesis, this observation may reflect the hemibiotrophic life style of
M. *oryzae*. The wheat scab fungus F. *graminearum* possesses at least 13 pectate
lyase and 4 pectin lyase genes.

In the M. *oryzae* genome, a total of over 700 proteins are predicted to
be secreted (D. Ebbole, personal communication), considerably more than
secreted proteins encoded by the N. *crassa* and A. *nidulans* genomes. In addition
to cell wall degrading enzymes, many families of secreted proteins have an
expansion in M. *oryzae*. Twenty-one secreted proteins in M. *oryzae* contain
the novel variant cysteine pattern CX_7CCX_5C, which exists only eight times in
A. *nidulans*, four times in N. *crassa*, and not at all in S. *cerevisiae* (Dean *et al.*,
2005). The M. *oryzae* CBP1 gene contains two CX_7CCX_5C patterns. It is
specifically expressed in germ tubes and involved in the recognition of physical
factors on the host surface (Kamakura *et al.*, 2002). In the tomato pathogen
C. *fulvum*, the avirulence gene Avr4 is a chitin-binding protein with similar
cysteine patterns that may protect the fungus from plant chitin-degrading
enzymes. It is possible that these putative chitin-binding proteins play various
roles in fungal–plant interactions. Among the predicted secreted M. *oryzae*
proteins, we find that 117 of them have nuclear localization signals. Some of
these proteins are unique to M. *oryzae* and may be secreted into plant cells and
localized to nuclei. There are three families of cysteine-rich polypeptides (clusters 180, 360, and 641) that may function like similar proteins in C. *fulvum* as
pathogen-associated molecular patterns (PAMPs). Another protein family is
similar to the necrosis-inducing peptide of *Phytophthora infestans* (NPP1,

pfam05630) and may function as putative effector proteins in M. *oryzae* (Dean *et al.*, 2005).

C. Chromosome organization

M. *oryzae* has seven chromosomes that terminate in 20–30 repeats of the telomere repeat sequence $(TTAGGG)_n$. The current assembly contains one telomere repeat in contig 1538, but all 14 M. *oryzae* telomeres are represented by 538 excluded reads containing at least six units of the CCCTAA repeat sequences (Li *et al.*, 2005). By identifying and sequencing fosmid clones with telomeric repeats, Farman and colleagues (Rehmeyer *et al.*, 2006) have generated the sequences of all 14 chromosome ends. The distal portions of 11 chromosome ends have the same basic organization, with each containing a telomere-linked RecQ helicase (*TLH*) gene (Gao *et al.*, 2002) surrounded by blocks of tandem repeats (Rehmeyer *et al.*, 2006). Interestingly, the *TLH1* family is ubiquitous among M. *oryzae* field strains isolated from rice but not from other host plants, indicating *TLH1* genes may be unique to rice pathogens (Gao *et al.*, 2002). Similar helicase gene families have been found in chromosome ends of *Metarhizium anisopliae*, S. *cerevisiae* and U. *maydis* but not in F. *graminearum* or N. *crassa*.

Telomeres are among the most variable regions of the M. *oryzae* genome, and novel telomeres arise frequently during vegetative culture (Farman and Kim, 2005) and during meiosis (Farman and Leong, 1995). The subtelomeric regions of M. *oryzae* contain large numbers of full-length and truncated transposons (Rehmeyer *et al.*, 2006). The dynamic environment of chromosome ends is also suitable for generating variation in genes critical for fungal–plant interactions. A total of 113 putative nontransposon-related genes were identified within 100 kb of the telomeres (Rehmeyer *et al.*, 2006). Thirty of them are not represented in the automated annotation of the genome. Among 23 putative secreted proteins, two contain a cellulose-binding motif and a glycosyl hydrolase domain, and may have cell wall-degrading capabilities. In M. *oryzae*, both *Avr-Pita* and *Avr-Tsuy* genes map to telomeric locations and are highly unstable (Valent and Chumley, 1994). Sequence analysis of the *Avr-Pita* gene revealed that its 3'-UTR is only 48 base pairs (bp) proximal to the telomeric repeat (Orbach *et al.*, 2000). However, in the sequenced strain 70–15, *PWL2*, *PWL3*, and *AVR-Pita* all occupy internal locations away from telomeres.

In the assembled M. *oryzae* genome sequence, centromere sequences have not been identified. It is likely that M. *oryzae* centromeres are composed of repetitive sequences derived from transposable elements. The centromere sequences of N. *crassa* and pericentromeric sequences in L. *maculans* contain clusters of inactive transposons, which have been truncated or heavily RIPped (Attard *et al.*, 2005; Cambareri *et al.*, 1998; Galagan and Selker, 2004). However, some of these highly repetitive sequences may have been excluded from the current version of genome assembly. To date, centromeres have not been well characterized in filamentous fungi, including *N. crassa*. On the basis of the cloned *CEN-VII* and *CEN* sequences identified in the genome assembly, cen-tromeres of *N. crassa* are 200–400 kb in length, larger than those of *S. cerevisiae* (Borkovich *et al.*, 2004), but it is not clear whether there is any conserved core centromeric sequence among the chromosomes. In *C. albicans*, there is no consensus centromeric sequence and each chromosome has a unique centromer-ic sequence (Sanyal *et al.*, 2004). *M. oryzae* and other filamentous fungi may have similar degrees of variation in centromeric sequences or structures. Some *M. oryzae* field isolates are known to contain conditionally dispensable chromo-somes. The mitotic and meiotic apparatus must be able to adjust to the elastic nature of fungal genomes. Most of the centromere-binding proteins annotated in *N. crassa* (Borkovich *et al.*, 2004) have orthologues in *M. grisea*, including Cbf1 helix-loop-helix transcription factor and Mif2/*CENP-C*. However, the *HsCENP-B* (NCU00996) and *S. cerevisiae OKP1* (NCU00367) genes have no significant homologues in *M. oryzae* and *F. graminearum*.

D. Colinearity

Colinearity of chromosome segments (synteny) is widely reported for plant and animal genomes but is largely unknown in filamentous ascomycete fungi. Analysis of orthologous pairs of genes in *M. oryzae* and *N. crassa* revealed no evidence for extensive regions of conserved synteny, although linkage group assignments were often conserved between the two species. Only 113 regions containing 4 or more genes are colinear between *M. oryzae* and *N. crassa*. One example, also conserved in several other filamentous fungi, is the quinate/shikimate (Qa) metabolic pathway gene cluster. This seven-gene cluster spanning about 20 kb is on chromosome 3 in *M. oryzae* and has syntenic regions in *N. crassa*, *F. graminearum*, *A. nidulans*, and other filamentous ascomycetes (Dean *et al.*, 2005). This cluster, involved in quinate metabolism and aromatic amino acid catabolism, is not present in *S. cerevisiae*, *S. pombe*, and other yeasts.

A draft sequence of 38 BAC clones spanning chromosome 7 from *M. oryzae* was completed, which when combined with the *M. oryzae* WGS sequence, yielded a sequence assembly that was 4 Mb in length and contained only 50 gaps (Thon *et al.*, 2006). Using the FISH software package, 21 syntenic blocks ($p < 0.001$) between chromosome 7 and the *N. crassa* genome ranging from 5 to 16 orthologous gene pairs were identified. Interestingly, all of the blocks were found on *N. crassa* chromosome 1, though the relative order of the syntenic blocks was not retained between the two chromosomes. Similar patterns of conserved synteny were found when the same analytical methods were used to compare chromosome 7 to two other filamentous fungi. Seventeen blocks were identified between chromosome 7 and the *F. graminearum* genome, 14 of which were found on chromosome 2. The remaining three were found on chromosome 4, suggesting either a translocation of a large chromosomal segment in the *F. graminearum* lineage or an error in the genetic map. Only two syntenic blocks were identified in *A. nidulans*, reflecting its greater evolutionary distance from *M. oryzae* (Thon et al., 2006).

E. Repetitive sequences

Approximately 9.7% of the *M. oryzae* genome assembly comprises repetitive DNA sequences longer than 200 bp and with greater than 65% similarity. Most repetitive sequences in the assembly are repetitive elements comprising eight major families (Dean et al., 2005). Five are retroelements and three are DNA transposons. Four previously unknown repeats have been discovered as well as alternative forms for three previously described transposons. The genome sequence also reveals full-length sequences of two elements for which only incomplete sequences were previously available. These repetitive elements are not uniformly distributed in the genome assembly, but form discrete clusters. Further examination has revealed many examples of transposable elements inserted into copies of themselves or other repetitive elements. On chromosome 7, transposable elements are largely restricted to three clusters located in chromosomal segments that have a high recombination rate (Thon et al., 2006). These clusters are marked by more frequent gene duplications, and genes within the clusters have greater sequence diversity to orthologous genes from other fungi. In the shotgun sequences of the 1.6-Mb minichromosome from a Japanese field isolate (9439009), there are at least three new classes of retrotransposons (Sone et al., 2005), indicating that the field isolates may have more active transposable elements. The prevalence of intact and essentially identical repeated DNA elements in the genome suggests that *M. oryzae* has been unable to stop their proliferation.

Given the prevalence of repetitive elements and their ability to participate in recombination, it is perhaps surprising that an organism could tolerate such substantial genomic changes. However, in nature, rice pathogenic strains of *M. oryzae* propagate asexually and, as such, genome organization is rarely, if ever, subject to the potentially catastrophic effects of meiotic recombination involving homologous chromosomes with radically different structures. Thus, rearrangements that normally would have been purged by meiosis appear to have been maintained in the absence of deleterious effects on vegetative fitness. Some rearrangements are expected to have positive fitness benefits, especially those that result in loss of genes whose products would normally trigger defense responses in potential hosts such as the insertion of a 1.9-kb MINE retrotransposon in the *ACE1* gene of a virulent isolate (Fudal et al., 2005). Novel chimeric genes generated by transposition events of MAGGY have been identified (Powell *et al.*, 2005). These observations suggest that transposable elements have a profound impact on the M. *oryzae* genome organization and genetic variability.

F. Receptors

During different infection stages, M. *oryzae* must be able to use different receptors to recognize various signal molecules or ligands and regulate plant penetration and infectious growth. Among three major classes of receptors known in eukaryotes, the G-protein–coupled receptors (GPCRs) are the biggest group involved in recognizing diverse external signals and regulating different cellular processes by association with heterotrimeric G-proteins (Liebmann, 2004). In S. *cerevisiae*, three known GPCRs, Ste2, Ste3, and Gpr1, are important for pheromone and carbon source receptions and play critical roles in mating and filamentous growth. The M. *oryzae* genome contains a large number of GPCR-like genes (Kulkarni *et al.*, 2005), including putative homologues of known fungal GPCRs, such as GprD and Pre-1, and the cAMP receptors from D. *discoideum* cAMP receptors (Kim and Borkovich, 2004; Verkerke-Van Wijk *et al.*, 1998). Twelve of these putative GPCR genes form a subfamily and contain an N-terminal extracellular membrane-spanning domain (CFEM) that is unique to fungi. A member of this new class, *PTH11*, is known to be involved in surface recognition during appressorium formation and is required for pathogenesis (DeZwaan *et al.*, 1999). Microarray analysis revealed that all 12 CFEM-GPCR genes are expressed, and two of them, MG09863 and MG05871, are upregulated during appressorium formation (Dean *et al.*, 2005). The CFEM-GPCRs are novel and unique to filamentous ascomycetes in the Pezizomycotina. Homologues of Pth11 have not been identified in yeast and basidiomycete genomes that have been sequenced. These putative GPCRs may be involved in recognizing different environmental and physiological signals/conditions for M. *oryzae* to adjust to *in planta* growth and be a successful plant pathogen. However, predicting GPCRs is not reliable and it is not clear whether these genes are true orphan GPCRs.

G. Secondary metabolism and phytotoxic compounds

Like many other plant pathogenic fungi, M. *oryzae* produces a variety of phytotoxic compounds, including dihydropyriculol, pyrichalasin H, terrestric acid, pyriculol, pyriculariol, pyriculone, pyricuol, and other pyriculol-related metabolites (Nukina, 1999). In comparison with N. *crassa*, the M. *oryzae* genome has many more genes that may be involved in secondary metabolism. There are 23 putative PKSs genes in M. *oryzae* but only 7 in N. *crassa*. Three of them, MG10072, MG04775, and MG07219, are upregulated during appressorium formation (Dean *et al.*, 2005). Although *ALB1*, the PKS involved in melanin biosynthesis in M. *oryzae*, has not been cloned yet, MG07219 is highly similar to the *PKS1* gene of C. *lagenarium*. Other recently sequenced plant pathogens, including F. *graminearum*, *Botrytis cinerea*, and *Gibberella moniliformis*, all have over 15 putative PKS genes, which are absent in the ascomycetous yeasts (Kroken *et al.*, 2003). Phylogenetic analysis indicated that fungal PKS genes vary significantly, and even closely related fungal genomes share only a few putative orthologous *PKS* genes. Phytopathogenic fungi may be able to synthesize a variety of polyketide metabolites because of the diverse sequence and domain structure of PKS genes. Interestingly, most of these putative PKS genes in M. *oryzae* appear to occur in gene clusters with neighboring genes that encode enzymes such as cytochrome P450 and monooxygenases (Dean *et al.*, 2005). The M. *oryzae* genome also contains six predicted NRPS genes and eight putative hybrid PKS-NRPS genes (NRPS fused to PKS). More NRPS genes have been identified in F. *graminearum*, but the N. *crassa* genome has only two predicted NRPS genes and one NRPS-related gene. Small peptides synthesized by NRPS and polyketide compounds synthesized by PKS are known to play roles in fungal survival and pathogenesis (Keller *et al.*, 2005; Wolpert *et al.*, 2002). However, there are also many more PKS and NRPS genes in three sequenced saprophytic *Aspergillus* species, particularly A. *oryzae*, than in N. *crassa* (Machida *et al.*, 2005; Nierman *et al.*, 2005). Therefore, experimental data are necessary to determine the role of individual PKS and NRPS genes in M. *oryzae*.

Besides *ALB1*, the avirulence gene *ACE1* is the only other secondary metabolism gene that is known to be involved in M. *oryzae* pathogenesis. *ACE1* is specifically expressed in late stages of appressorium formation (Bohnert *et al.*, 2004), suggesting that the secondary metabolite(s) synthesized by this hybrid PKS-NRPS gene must be able to enter plant cells for the race-specific interaction. In C. *heterostrophus* and F. *graminearum*, one orthologous NRPS gene *NPS6* is important for plant infection (Lee *et al.*, 2005a). Similarly, *CPS1* and its orthologues are likely involved in secondary metabolism and have been shown to be an important virulence factor in C. *heterostrophus* and other fungi (Lu *et al.*, 2003). *CPS1* and *NPS6* orthologues exist in M. *oryzae*, but their functions in pathogenesis are not clear. The sirodesmin biosynthesis cluster of L. *maculans* also is conserved in M. *oryzae* and other filamentous fungi (Gardiner *et al.*, 2004). However, the product(s) synthesized by this two-module NRPS gene cluster and its function in rice infection have not characterized.

In fungi, genes involved in the biosynthesis of secondary metabolites are usually controlled by a complex regulatory network, including pH regulation, carbon repression, and nitrogen limitation (Keller *et al.*, 2005). For regulatory genes that are known to be involved in secondary metabolism in A. *flavus* and A. *nidulans* (Yu and Keller, 2005), the M. *oryzae* genome has well-conserved orthologues of *pacC* and components involved in PKA and trimeric G-protein signaling. M. *oryzae* contains 11 predicted proteins that have limited homology with LaeA, which is a global regulator of several secondary metabolite synthesis pathways (Bok and Keller, 2004). It appears that LaeA is not well conserved between Pezizamycetes and Eurotiomycetes because *F. graminearum* and *N. crassa* also lack a well-defined LaeA orthologue. AflR is a transcription factor regulating the expression of the aflatoxin synthesis gene cluster (Yu and Keller, 2005). The cysteine-rich motif of AflR is conserved among several fungal proteins for DNA-binding and has been identified as InterPro IPR002409. The *M. oryzae* and *F. graminearum* genomes have 7 and 20 predicted proteins containing this AflR motif, respectively, considerably more than 4 in *N. crassa* and 2 in *S. cerevisiae* (Table 5.2).

H. Repeat-induced mutation and silencing

Repeat-induced point (RIP) mutation is a phenomenon first described in *N. crassa* and may be widely conserved in filamentous ascomycetes for protecting fungal genomes from invading sequences or amplification of selfish DNA (Galagan and Selker, 2004). In M. *oryzae*, RIP occurs in duplicated sequences introduced by transformation but at a weaker level than *N. crassa* based on the number of point mutations induced (Ikeda *et al.*, 2002). Sequence analysis of repetitive elements also indicates that RIP may be responsible for sequence variations (Dean *et al.*, 2005). There is one M. *oryzae* predicted gene (MG02795) encoding a DNA methyltransferase homologous to RID, which is required for RIP in *N. crassa* (Freitag *et al.*, 2002). However, the *M. oryzae* genome contains many intact, highly similar elements (>90% nucleotide identity) of some repeat families such as Pyret and Pot2 (Dean *et al.*, 2005). Therefore, RIP in M. *oryzae* may be a less efficient or a less frequent event than in *N. crassa*. Although RIP-like transitions were found in all the field isolates tested (Ikeda *et al.*, 2002), RIP is a meiosis-related event, and sexual reproduction in M. *oryzae* in nature has not been observed. The M. *oryzae* genome also contains orthologues of *N. crassa DIM-2* (MG00889), a DNA methyltransferase responsible for *de novo* and maintenance methylation (Kouzminova and Selker, 2001), and *DIM-5* (MG06852), a histone H3 methyltransferase controlling DNA methylation (Tamaru and Selker, 2001).

Different from RIP, quelling of repeated sequences in *N. crassa* does not involve point mutations and only occurs to specific genes (Pickford *et al.*, 2002). Three genes essential for quelling have been identified in *N. crassa*. *QDE-1*, *QDE-2*, and *QDE-3* encoding a putative RNA-dependent RNA polymerase (RdRP), an eIF2C-like protein, and an RecQ DNA helicase, respectively, and are homologous to genes known to be involved in posttranscriptional gene silencing (PTGS) or RNAi in plants and animals (Nakayashiki, 2005).

Table 5.2. Putative Transcription Factor Genes in M. *oryzae* and other Fungi

InterPro ID	Description	Mo[a]	Fg	Nc	An	Um	Sc
IPR009057	Homeodomain-like	140	50	45	67	31	33
IPR001138	Fungal transcriptional regulatory protein, N-terminal	133	317	114	243	103	57
IPR007087	Zn-finger, C2H2 type	94	114	96	81	44	52
IPR001878	Zn-finger, CCHC type	78	13	9	12	15	11
IPR007219	Fungal-specific transcription factor	67	164	57	158	32	25
IPR001841	Zn-finger, RING	50	62	57	40	48	35
IPR004827	Basic-leucine zipper (bZIP)	22	32	24	20	14	14
IPR008917	Eukaryotic transcription factor, DNA-binding	15	15	14	9	7	5
IPR001965	Zn-finger-like, PHD finger	16	21	17	18	17	16
IPR001789	Response regulator receiver	13	20	14	19	11	4
IPR001092	Basic helix-loop-helix (bHLH) domain	12	16	15	13	13	8
IPR000637	HMG-I and HMG-Y DNA-binding domain (A + T-hook)	12	13	6	2	8	1
IPR000910	HMG1/2 (high mobility group) box	11	10	10	7	8	8
IPR002197	Helix-turn-helix, Fis-type	10	8	14	13	5	5
IPR000679	Zn-finger, GATA type	10	8	6	7	11	10
IPR000571	Zn-finger, C-x8-C-x5-C-x3-H type	8	13	10	10	4	6
IPR001356	Homeobox	8	14	9	8	10	9
IPR002409	Aflatoxin biosynthesis regulatory protein	7	20	4	21	1	2
IPR003347	Transcription factor jumonji, jmjC	7	9	7	5	8	5
IPR002893	Zn-finger, MYND type	7	8	6	3	2	1
IPR003604	Zn-finger, U1-like	7	10	7	8	8	6
IPR001594	Zn-finger, DHHC type	6	5	5	6	7	7

[a]Based on data of intergenomic comparison of InterPro domains available at the MIPS Web site. Mo for M. *oryzae*; Fg for F. *graminearum*; Nc for N. *crassa*; An for A. *nidulans*, Um for U. *maydis*; Sc for S. *cerevisiae*. Only InterPro domains with more than six copies in M. *oryzae* are listed in this table.

Orthologues of Qde-1, Qde-2, and Qde-3 are conserved in M. *oryzae* and F. *graminearum*. Although it has not been reported, some transforming genes with multiple copies of integration or overexpression constructs under the control of a strong promoter in M. *oryzae* are likely to be subjected to quelling or RNAi silencing. Introduced MAGGY elements are methylated during proliferation but the transpositional frequency is independent of DNA methylation (Nakayashiki *et al.*, 2001), indicating that both pre- and posttranscriptional gene suppression mechanisms are involved in the repression of MAGGY transposition. RNAi silencing with introduced hairpin RNA molecules has been reported in M. *oryzae* (Kadotani *et al.*, 2003). During RNAi silencing, dsRNA is processed into short interfering RNA (siRNA) by an RNase III enzyme called Dicer. There are two Dicer-like (DCL) genes in M. *oryzae* named *MDL1* and *MDL2* (Kadotani *et al.*, 2004). The *mdl2* but not the *mdl1* deletion mutant is defective in silencing of EGFP by hairpin RNA and has a reduced level of related siRNA accumulation, indicating that only *MDL2* is responsible for siRNA production. In N. *crassa*, two Dicer-like proteins are redundantly involved in quelling of transgenes (Catalanotto *et al.*, 2004). The lack of functional redundancy between *MDL1* and *MDL2* in M. *oryzae*, however, may be specific to dsRNA silencing constructs because the N. *crassa* quelling experiments dealt with simple repeats of transgenes.

VI. FUNCTIONAL GENOMICS

The M. *oryzae* genome sequence has provided basic information on the content and organization of the genome and is valuable for comparative analysis with closely related fungi. To understand molecular mechanisms of fungal pathogenesis, however, it is necessary to determine the function of individual genes and genome-wide networks. A variety of advanced functional genomics tools and resources have been developed in S. *cerevisiae* over the past few years (Hughes *et al.*, 2004). Although functional genomics research in M. *oryzae* and filamentous fungi in general is still in its infancy, resources are being developed and several large-scale genome-wide analyses have already been initiated.

A. Large-scale mutagenesis

Due to low efficiency of homologous recombination, flanking sequences longer than 0.5 kb are required to obtain 5–10% gene replacement transformants in M. *oryzae*. Therefore, the direct PCR approach developed for yeast knockout analysis is not applicable. To date, conventional strategies and a modified ligation-PCR approach (Zhao *et al.*, 2004) have been used to generate gene replacement constructs for over 100 genes in M. *oryzae*. However, more efficient approaches must be developed to systematically characterize all the predicted genes. The split-marker approach (Catlett *et al.*, 2003) worked efficiently in F. *graminearum* in our laboratory but it only succeeded for one of the five genes tested in M. *oryzae* (Xu, unpublished data). RNAi silencing has been used to silence EGFP and two other fungal genes by expressing hairpin RNA constructs (Kadotani *et al.*, 2004) but whether it is suitable for large-scale mutagenesis remains to be tested. In our experiments with five M. *oryzae* genes, only two were silenced by ds-RNA constructs expressed with a strong constitutive promoter (Xue and Xu, unpublished data). The M. *grisea* genome contains orthologues of Ku70 and Ku80. Using deletion mutants of Ku70 or Ku80 may increase the homologous recombination frequency (Ninomiya *et al.*, 2004). In N. *crassa*, there is an ongoing project to generate a large collection of deletion mutants with the Ku70 or Ku80 deletion mutant (www.dartmouth.edu/~neurosporagenome). However, due to low female fertility of most laboratory M. *oryzae* strains, particularly 70–15, the removal of the Ku70 or Ku80 deletion background will be technically challenging.

   Another approach is to generate a collection of random insertional mutants that is large enough to allow each M. *oryzae* gene having the probability of being disrupted in at least one transformant. In the past few years, several laboratories have generated random insertion transformants by REMI or *Agrobacterium tumefaciens*-mediated transformation (ATMT) or transposon hopping approaches to screen for novel pathogenicity factors (Balhadere *et al.*, 1999; Sweigard *et al.*, 1998; Villalba *et al.*, 2001). Recently, over 55,000 insertion mutants have been created as a part of a project funded by NSF, the majority (>40,000) by ATMT and the remainder by protoplast transformation. All the data related to these transformants were recorded with a Web-based open source software system, PACLIMS (Donofrio *et al.*, 2005). All the transformants were analyzed for growth, pigmentation, and sporulation defects and deposited at the Fungal Genetic Stock Center (www.fgsc.net) for distribution. Southern analysis of over 200 randomly selected transformants indicate that most (>86%) ATMT transformants have single DNA insertions in contrast to the apparent tandem insertions in the protoplast transformants. Sequence analysis with flanking DNA recovered from over 200 ATMT transformants indicates that the vast majority (>90%) of transformants have no deletion or other rearrangement at the integration site. Among more than half of the transformants tested, 0.3% of them are auxotrophic. About 1% of 14,000 ATMT transformants examined have defects in appressorium formation, and of these, half are nonpathogenic or reduced in virulence on rice. Genes disrupted in several of these appressorium mutants have been identified, including a novel G-protein (MG10193), a STAT-signal protein (MG00455), and a membrane protein (MG07535). In infection assays with young rice seedlings, 81 out of 29,000 transformants (0.3%) tested to date are defective in plant infection. About one-third of them cause no symptoms or only small dark flecks on rice leaves. Flanking sequences have been isolated from about 30 pathogenicity mutants. Insertions have occurred in several genes involved in regulating transcription and development, including kinesin, as well as genes encoding proteins containing SET and homeobox domains such as MG01588, MG06515, MG09255, and MG02425 (Dean, unpublished data).

Similar large-scale mutagenesis projects are being carried out by other members of the International Rice Blast Genomics Consortium, mainly scientists in China and in South Korea. Over 50,000 insertional mutants have been generated by REMI and ATMT with a Japanese and a Chinese field isolate (Y. Peng, personal communication). Similar numbers of ATMT mutants have been generated with a Korean field isolate (Y. Lee, personal communication). In total, ~150,000–200,000 M. *oryzae* random insertional transformants have been created. If all these existing transformants are combined and pools of genomic DNA of these transformants become available for mutant screening, that will be a valuable functional genomics resource. However, M. *oryzae* is a plant pathogen and appropriate permits are necessary for exchanging strains among laboratories in the United States and other countries.

Proteins secreted or present in the extracellular matrix are crucial components of the ability of fungi to perceive and respond to the environment. Some phytotoxic factors of M. *oryzae* are known to be proteinacious, such as the 30-kDa rice protoplast-disrupting protein, leaf senescence-promoting factors in filtrates of cultures starved for nitrogen, and heat-labile molecules (Bucheli *et al.*, 1990; Talbot *et al.*, 1997). Another on-going functional project funded by NSF is on secreted proteins. Over 100 candidate genes predicted to encode proteins with signal peptides were over-expressed in M. *oryzae* and purified by immunoprecipitation. Several purified proteins, including CWDEs and proteinases, caused severe necrosis and hydrogen peroxide generation when placed on wound sites. Others induced hydrogen peroxide without causing a detectable necrotic response (D. Ebbole, personal communication). Further functional characterization of these genes is under the way to verify their roles in plant infection. Results from this project on secreted proteins will provide necessary tools and procedures for the systematic functional analysis of the M. *oryzae* secretome.

B. Genome-wide transcriptional profiling

A 22,000-element oligo-based microarray was designed in collaboration with Agilent Technologies that contains the Broad Institute predicted gene set, plus additional predicted features from other gene models (including GenScan) as well as ESTs for genes not predicted, yielding a total of 13,666 elements. The array also contains elements for ~7000 rice genes and is available from Agilent. With this array, detailed transcription profiling experiments have been conducted to identify genes differentially expressed during rice infection, appressorium formation, under various nutritional stresses, and in mutants with specific genetic backgrounds. Additional information can be found at MGOS (www.mgosdb.org). As anticipated, many hundreds of genes were differentially expressed (>twofold change with $p < 0.05$). For example, ~4% of the total predicted gene set were upregulated during spore germination (Dean, unpublished data). Similar numbers of genes were downregulated. Intriguingly, a number of transposable elements including MAGGY were found to be highly expressed in conidia. These findings were confirmed by RT-PCR. It is not known whether or not these elements are actively transposing, but transposon activity in germline cells is well documented (Prak et al., 2003). In addition, about 2% and 4% of genes were significantly differentially regulated in immature (7 h) and mature appressoria (12 h), respectively, compared to spores germinated on a noninducing surface. In addition to genes expected, such as those involved in melanin biosynthesis, genes encoding transporters and secondary metabolites were found to be upregulated (Dean, unpublished data).

As described above, cAMP signaling is a key regulator of appressorium formation. Comparison of whole genome expression data from appressoria stimulated by an inductive hydrophobic surface versus by cAMP identified a common core of 226 upregulated and 110 downregulated genes, respectively. Functional categorization using Genome Ontology (GO) revealed that many upregulated genes are involved in lipid metabolism, amino acid metabolism, and proteolysis. Gene knockout of a subtilisin-like protease Spm1 (MG03670) (Fukiya et al., 2002) demonstrated its requirement for pathogenicity (Dean, unpublished data). Among the core set of downregulated genes, a significant fraction were grouped into translation, including translation factors and genes encoding ribosomal proteins. It is noteworthy that cAMP-PKA pathways are known in a number of organisms to regulate ribosome biogenesis and responses to stress, including nutritional stress. Microarray experiments have shown that several hundred genes were differentially regulated during growth in carbon- and nitrogen-limited environments. Approximately 25% of the upregulated genes were also upregulated during appressorium formation. For example, an NAD + glumatate dehydrogenase gene MGD1 (MG05247) was found to be upregulated during both nitrogen starvation and appressorium formation. Targeted deletion of this gene revealed that it is required for appressorium formation and pathogenicity. These data suggest that proteolysis and recycling of amino acids is essential for successful infection by M. oryzae and provide further evidence for a link between nutrient stress and infection-related development (Talbot et al., 1997). Other transcriptional profiling experiments have focused on in planta gene expression and selected mutants blocked in plant infection (e.g., the pmk1 and mst12 mutant). At 48 hours postinoculation (hpi), the expression of 17 fungal genes could be detected in infected barley leaves, which rose to 348 genes at 96 hpi. Twelve genes, including CYP1 (Viaud et al., 2002), were in common to all the time points assayed. Analysis across all microarray experiments to date revealed that about 60 genes were specifically expressed *in planta* (Dean, unpublished data), including several transporters (MG09973, MG0735), and proteins involved in xylan (MG07016) and lipid metabolism (MG07868).

SAGE and MPSS analyses also have been used to identify M. *oryzae* genes differentially expressed during different growth and development stages. Among 12,119 SuperSAGE tags obtained from M. *oryzae*-infected rice leaves, 74 (0.6%) are derived from fungal genes. The most abundantly expressed gene is *MPG1*, which accounts for 38 tags. *PUB4* and a nucleoside-diphosphate kinase gene are among the other M. *oryzae* genes represented by the SAGE tags (Matsumura *et al.*, 2003). The mycelium and appressorium transcriptomes have been analyzed by MPSS and RL-SAGE methods (Wang and Dean, unpublished data). A large number of unique MPSS tags were identified from the mycelium (20,299) and appressorium (16,427) libraries. Nearly 85% of the significant MPSS tags common in both libraries matched the M. *oryzae* genome. Comparing genome-matched tags with current EST collections, about 40% of the MPSS tags and 55% of the RL-SAGE tags apparently represent novel transcripts. Interestingly, about 50% of the annotated genes contained antisense transcript tags (Wang and Dean, unpublished data).

VII. CONCLUDING REMARKS

In the past few years, genetic and cell biological studies of M. *oryzae* have advanced our understanding of many aspects of rice–rice blast interaction and disease development, including characterization of avirulence genes, identification of signal transduction pathways regulating infection processes, and mechanisms involved in appressorium turgor generation and infectious hyphal growth. As the first plant pathogenic fungus with its genome sequence published, M. *oryzae* has great potential to be further developed as a model to illustrate molecular mechanisms of fungal pathogenesis. Large-scale insertional mutagenesis projects and genome-wide transcriptional profiling and proteomic analyses are in progress. In addition, we believe more and more functional genomics resources and tools will become available for both M. *oryzae* and rice. Recently, we were funded to develop and use a M. *oryzae* protein chip to identify transcription factors regulated by conserved MAP kinase and cAMP signaling pathways. In the same project, we will use ChIP-chip to identify binding motifs and target genes of transcription factors identified in protein chip experiments. We expect proteomics tools will be improved and more proteomics resources will be available to rice and M. *oryzae*. Therefore, the M. *oryzae*–rice pathosystem is uniquely positioned for applying a genome-wide, systems biology approach to functionally dissect and characterize the processes by which a fungus causes an important crop disease (Dean *et al.*, 2005; Veneault-Fourrey and Talbot, 2005).

Analysis of the M. *oryzae* genome has provided useful information about fungal genes that may be important for plant infection and colonization such as genes encoding secreted proteins or involved in secondary metabolism. Genome information and resources will also be valuable for comparative analyses with other model fungi, such as N. *crassa* and A. *nidulans* (Galagan *et al.*, 2003, 2005a), to identify common features involved in growth and development of filamentous fungi. However, the phylogenetic distance between M. *oryzae* and F. *graminearum* or N. *crassa* or A. *nidulans* is too large for the type of comparative analyses that have been done with *Saccharomyces* species (Cliften *et al.*, 2003; Kellis *et al.*, 2003). In three recently published *Aspergillus* species, comparative analysis has been used to identify conserved syntenic blocks, novel regulatory elements, and subsets of genes that are unique to each species, possibly for expansion in secondary metabolism or human infection (Galagan *et al.*, 2005a; Machida *et al.*, 2005; Nierman *et al.*, 2005). Sequencing field isolates of M. *oryzae* and other closely related *Magnaporthe* species will be very helpful to improve the genome annotation and provide important information about genome structures and adaptations to different environments or host plants.

Acknowledgments

We thank Drs. Daniel Ebbole, Mark Farman, Marc Orbach, Yong-Hwan Lee, You-Liang Peng, Michael Thon, Zonghua Wang, and Sheng-Cheng Wu for sharing unpublished data. We also thank Drs. Larry Dunkle and Steve Goodwin for critical reading of this chapter. This work was supported by a grant from the USDA National Research Initiative to J.-R.X. and a grant from NSF Plant Genome Program (R.D. and J.-R.X.).

References

Adachi, K., and Hamer, J. E. (1998). Divergent cAMP signaling pathways regulate growth and pathogenesis in the rice blast fungus *Magnaporthe grisea*. *Plant Cell* **10**, 1361–1373.

Ahn, N., Kim, S., Choi, W., Im, K. H., and Lee, Y. H. (2004). Extracellular matrix protein gene, *EMP1*, is required for appressorium formation and pathogenicity of the rice blast fungus, *Magnaporthe grisea*. *Mol. Cells* **17**, 166–173.

Asiegbu, F. O., Choi, W., Jeong, J. S., and Dean, R. A. (2004). Cloning, sequencing and functional analysis of *Magnaporthe grisea MVP1* gene, a hex-1 homolog encoding a putative 'woronin body' protein. *FEMS Microbiol. Lett.* **230**, 85–90.

Atkinson, H. A., Daniels, A., and Read, N. D. (2002). Live-cell imaging of endocytosis during conidial germination in the rice blast fungus, *Magnaporthe grisea*. *Fungal Genet. Biol.* **37**, 233–244.

Attard, A., Gout, L., Ross, S., Parlange, F., Cattolico, L., Balesdent, M. H., and Rouxel, T. (2005). Truncated and RIP-degenerated copies of the LTR retrotransposon Pholy are clustered in a pericentromeric region of the *Leptosphaeria maculans* genome. *Fungal Genet. Biol.* **42**, 30–41.

Balhadere, P. V., and Talbot, N. J. (2001). *PDE1* encodes a P-type ATPase involved in appressorium-mediated plant infection by the rice blast fungus *Magnaporthe grisea*. *Plant Cell* **13**, 1987–2004.

Balhadere, P. V., Foster, A. J., and Talbot, N. J. (1999). Identification of pathogenicity mutants of the rice blast fungus *Magnaporthe grisea* by insertional mutagenesis. *Mol. Plant Microbe Interact.* **12**, 129–142.

Banno, S., Kimura, M., Tokai, T., Kasahara, S., Higa-Nishiyama, A., Takahashi-Ando, N., Hamamoto, H., Fujimura, M., Staskawicz, B. J., and Yamaguchi, I. (2003). Cloning and characterization of genes specifically expressed during infection stages in the rice blast fungus. *FEMS Microbiol. Lett.* **222**, 221–227.

Bardwell, L. (2004). A walk-through of the yeast mating pheromone response pathway. *Peptides* **25**, 1465–1476.

Barhoom, S., and Sharon, A. (2004). cAMP regulation of 'pathogenic' and 'saprophytic' fungal spore germination. *Fungal Genet. Biol.* **41**, 317–326.

Batzoglou, S., Jaffe, D. B., Stanley, K., Butler, J., Gnerre, S., Mauceli, E., Berger, B., Mesirov, J. P., and Lander, E. S. (2002). ARACHNE: A whole-genome shotgun assembler. *Genome Res.* **12**, 177–189.

Bohnert, H. U., Fudal, I., Dioh, W., Tharreau, D., Notteghem, J. L., and Lebrun, M. H. (2004). A putative polyketide synthase peptide synthetase from *Magnaporthe grisea* signals pathogen attack to resistant rice. *Plant Cell* **16**, 2499–2513.

Bok, J. W., and Keller, N. P. (2004). LaeA, a regulator of secondary metabolism in *Aspergillus* spp. *Eukaryot. Cell* **3**, 527–535.

Borkovich, K. A., Alex, L. A., Yarden, O., Freitag, M., Turner, G. E., Read, N. D., Seiler, S., Bell-Pedersen, D., Paietta, J., Plesofsky, N., Plamann, M., Goodrich-Tanrikulu, M., *et al.* (2004). Lessons from the genome sequence of *Neurospora crassa*: Tracing the path from genomic blueprint to multicellular organism. *Microbiol. Mol. Biol. Rev.* **68**, 1–108.

Bourett, T. M., and Howard, R. J. (1992). Actin in penetration pegs of the fungal rice blast pathogen *Magnaporthe grisea*. *Protoplasma* **168**, 20–26.

Bruno, K. S., Tenjo, F., Li, L., Hamer, J. E., and Xu, J. R. (2004). Cellular localization and role of kinase activity of *PMK1* in *Magnaporthe grisea*. *Eukaryot. Cell* **3**, 1525–1532.

Bucheli, P., Doares, S. H., Albersheim, P., and Darvill, A. (1990). Host pathogen interactions 36. Partial purification and characterization of heat-labile molecules secreted by the rice blast pathogen that solubilize plant cell wall fragments that kill plant cells. *Physiol. Mol. Plant Pathol.* **36**, 159–173.

Cambareri, E. B., Aisner, R., and Carbon, J. (1998). Structure of the chromosome VII centromere region in *Neurospora crassa*: Degenerate transposons and simple repeats. *Mol. Cell. Biol.* **18**, 5465–5477.

Catalanotto, C., Pallotta, M., ReFalo, P., Sachs, M. S., Vayssie, L., Macino, G., and Cogoni, C. (2004). Redundancy of the two Dicer genes in transgene-induced posttranscriptional gene silencing in *Neurospora crassa*. *Mol. Cell. Biol.* **24**, 2536–2545.

Catlett, N. L., Lee, B., Yoder, O. C., and Turgeon, B. G. (2003). Split-marker recombination for efficient targeted deletion of fungal genes. *Fungal Genet. Newsl.* **50**, 9–11.

Choi, W. B., and Dean, R. A. (1997). The adenylate cyclase gene *MAC1* of *Magnaporthe grisea* controls appressorium formation and other aspects of growth and development. *Plant Cell* **9**, 1973–1983.

Clergeot, P. H., Gourgues, M., Cots, J., Laurans, F., Latorse, M. P., Pepin, R., Tharreau, D., Notteghem, J. L., and Lebrun, M. H. (2001). *PLS1*, a gene encoding a tetraspanin-like protein, is required for penetration of rice leaf by the fungal pathogen *Magnaporthe grisea*. *Proc. Natl. Acad. Sci. USA* **98**, 6963–6968.

Cliften, P., Sudarsanam, P., Desikan, A., Fulton, L., Fulton, B., Majors, J., Waterston, R., Cohen, B. A., and Johnston, M. (2003). Finding functional features in *Saccharomyces* genomes by phylogenetic footprinting. *Science* **301**, 71–76.

Correa, A., Staples, R. C., and Hoch, H. C. (1996). Inhibition of thigmostimulated cell differentiation with RGD-peptides in *Uromyces* germlings. *Protoplasma* **194,** 91–102.

Couch, B. C., and Kohn, L. M. (2002). A multilocus gene genealogy concordant with host preference indicates segregation of a new species, *Magnaporthe oryzae,* from M. *grisea. Mycologia* **94,** 683–693.

de Jong, J. C., McCormack, B. J., Smirnoff, N., and Talbot, N. J. (1997). Glycerol generates turgor in rice blast. *Nature* **389,** 244–245.

Dean, R. A., Talbot, N. J., Ebbole, D. J., Farman, M. L., Mitchell, T. K., Orbach, M. J., Thon, M., Kulkarni, R., Xu, J. R., Pan, H., Read, N. D., Lee, Y. H., *et al.* (2005). The genome sequence of the rice blast fungus *Magnaporthe grisea. Nature* **434,** 980–986.

DeZwaan, T. M., Carroll, A. M., Valent, B., and Sweigard, J. A. (1999). *Magnaporthe grisea* Pth11p is a novel plasma membrane protein that mediates appressorium differentiation in response to inductive substrate cues. *Plant Cell* **11,** 2013–2030.

Dickman, M. B., Ha, Y. S., Yang, Z., Adams, B., and Huang, C. (2003). A protein kinase from *Colletotrichum trifolii* is induced by plant cutin and is required for appressorium formation. *Mol. Plant Microbe Interact.* **16,** 411–421.

Dioh, W., Tharreau, D., Notteghem, J. L., Orbach, M., and Lebrun, M. H. (2000). Mapping of avirulence genes in the rice blast fungus, *Magnaporthe grisea,* with RFLP and RAPD markers. *Mol. Plant Microbe Interact.* **13,** 217–227.

Dixon, K. P., Xu, J. R., Smirnoff, N., and Talbot, N. J. (1999). Independent signaling pathways regulate cellular turgor during hyperosmotic stress and appressorium-mediated plant infection by *Magnaporthe grisea. Plant Cell* **11,** 2045–2058.

Donofrio, N., Rajagopalon, R., Brown, D., Diener, S., Windham, D., Nolin, S., Floyd, A., Mitchell, T., Galadima, N., Tucker, S., Orbach, M. J., Patel, G., *et al.* (2005). 'PACLIMS': A component LIM system for high-throughput functional genomic analysis. *BMC Bioinformatics* **6,** 94.

Dufresne, M., and Osbourn, A. E. (2001). Definition of tissue-specific and general requirements for plant infection in a phytopathogenic fungus. *Mol. Plant Microbe Interact.* **14,** 300–307.

Durrenberger, F., Wong, K., and Kronstad, J. W. (1998). Identification of a cAMP-dependent protein kinase catalytic subunit required for virulence and morphogenesis in *Ustilago maydis. Proc. Natl. Acad. Sci. USA* **95,** 5684–5689.

Ebbole, D. J., Jin, Y., Thon, M., Pan, H. Q., Bhattarai, E., Thomas, T., and Dean, R. (2004). Gene discovery and gene expression in the rice blast fungus, *Magnaporthe grisea:* Analysis of expressed sequence tags. *Mol. Plant Microbe Interact.* **17,** 1337–1347.

Fang, E. G. C., and Dean, R. A. (2000). Site-directed mutagenesis of the *MAGB* gene affects growth and development in *Magnaporthe grisea. Mol. Plant Microbe Interact.* **13,** 1214–1227.

Farman, M. L., and Kim, Y.-S. (2005). Telomere hypervariability in *Magnaporthe oryzae. Mol. Plant Pathol.* **6,** 287–298.

Farman, M. L., and Leong, S. A. (1995). Genetic and physical mapping of telomeres in the rice blast fungus, *Magnaporthe grisea. Genetics* **140,** 479–492.

Farman, M. L., Eto, Y., Nakao, T., Tosa, Y., Nakayashiki, H., Mayama, S., and Leong, S. A. (2002). Analysis of the structure of the *AVR1-CO39* avirulence locus in virulent rice-infecting isolates of *Magnaporthe grisea. Mol. Plant Microbe Interact.* **15,** 6–16.

Foster, A. J., Jenkinson, J. M., and Talbot, N. J. (2003). Trehalose synthesis and metabolism are required at different stages of plant infection by *Magnaporthe grisea. EMBO J.* **22,** 225–235.

Freitag, M., Williams, R. L., Kothe, G. O., and Selker, E. U. (2002). A cytosine methyltransferase homologue is essential for repeat-induced point mutation in *Neurospora crassa. Proc. Natl. Acad. Sci. USA* **99,** 8802–8807.

Froeliger, E. H., and Carpenter, B. E. (1996). *NUT1,* a major nitrogen regulatory gene in *Magnaporthe grisea,* is dispensable for pathogenicity. *Mol. Gen. Genet.* **251,** 647–656.

Fudal, I., Bohnert, H. U., Tharreau, D., and Lebrun, M. H. (2005). Transposition of MINE, a composite retrotransposon, in the avirulence gene *ACE1* of the rice blast fungus *Magnaporthe grisea*. *Fungal Genet. Biol.* **42**, 761–772.

Fujimoto, D., Shi, Y., Christian, D., Mantanguihan, J. B., and Leung, H. (2002). Tagging quantitative loci controlling pathogenicity in *Magnaporthe grisea* by insertional mutagenesis. *Physiol. Mol. Plant Pathol.* **61**, 77–88.

Fukiya, S., Kuge, T., Tanishima, T., Sone, T., Kamakura, T., Yamaguchi, I., and Tomita, F. (2002). Identification of a putative vacuolar serine protease gene in the rice blast fungus *Magnaporthe grisea*. *Biosci. Biotechnol. Biochem.* **66**, 663–666.

Galagan, J. E., and Selker, E. U. (2004). RIP: The evolutionary cost of genome defense. *Trends Genet.* **20**, 417–423.

Galagan, J. E., Calvo, S. E., Borkovich, K. A., Selker, E. U., Read, N. D., Jaffe, D., FitzHugh, W., Ma, L. J., Smirnov, S., Purcell, S., Rehman, B., Elkins, T., *et al.* (2003). The genome sequence of the filamentous fungus *Neurospora crassa*. *Nature* **422**, 859–868.

Galagan, J. E., Calvo, S. E., Cuomo, C., Ma, L. J., Wortman, J. R., Batzoglou, S., Lee, S. I., Basturkmen, M., Spevak, C. C., Clutterbuck, J., Kapitonov, V., Jurka, J., *et al.* (2005a). Sequencing of *Aspergillus nidulans* and comparative analysis with A. *fumigatus* and A. *oryzae*. *Nature* **438**, 1105–1115.

Galagan, J. E., Henn, M. R., Ma, L. J., Cuomo, C. A., and Birren, B. (2005b). Genomics of the fungal kingdom: Insights into eukaryotic biology. *Genome Res.* **15**, 1620–1631.

Gale, C., Finkel, D., Tao, N. J., Meinke, M., McClellan, M., Olson, J., Kendrick, K., and Hostetter, M. (1996). Cloning and expression of a gene encoding an integrin-like protein in *Candida albicans*. *Proc. Natl. Acad. Sci. USA* **93**, 357–361.

Gale, C., Gerami-Nejad, M., McClellan, M., Vandoninck, S., Longtine, M. S., and Berman, J. (2001). *Candida albicans* Int1p interacts with the septin ring in yeast and hyphal cells. *Mol. Biol. Cell* **12**, 3538–3549.

Gao, W. M., Khang, C. H., Park, S. Y., Lee, Y. H., and Kang, S. C. (2002). Evolution and organization of a highly dynamic, subtelomeric helicase gene family in the rice blast fungus *Magnaporthe grisea*. *Genetics* **162**, 103–112.

Gardiner, D. M., Cozijnsen, A. J., Wilson, L. M., Pedras, M. S. C., and Howlett, B. J. (2004). The sirodesmin biosynthetic gene cluster of the plant pathogenic fungus *Leptosphaeria maculans*. *Mol. Microbiol.* **53**, 1307–1318.

Gilbert, M. J., Thornton, C. R., Wakley, G. E., and Talbot, N. J. (2006). A P-type ATPase required for rice blast disease and induction of host resistance. *Nature* **440**, 535–539.

Gourgues, M., Clergeot, P. H., Veneault, C., Cots, J., Sibuet, S., Brunet-Simon, A., Levis, C., Langin, T., and Lebrun, M. H. (2002). A new class of tetraspanins in fungi. *Biochem. Biophys. Res. Commun.* **297**, 1197–1204.

Gourgues, M., Brunet-Simon, A., Lebrun, M. H., and Levis, C. (2004). The tetraspanin BcPls1 is required for appressorium-mediated penetration of *Botrytis cinerea* into host plant leaves. *Mol. Microbiol.* **51**, 619–629.

Harp, T. L., and Correll, J. C. (1998). Recovery and characterization of spontaneous, selenate-resistant mutants of *Magnaporthe grisea*, the rice blast pathogen. *Mycologia* **90**, 954–963.

Harris, S. D. (2001). Septum formation in *Aspergillus nidulans*. *Curr. Opin. Microbiol.* **4**, 736–739.

Heath, M. C., Howard, R. J., Valent, B., and Chumley, F. G. (1992). Ultrastructural interactions of one strain of *Magnaporthe grisea* with goosegrass and weeping lovegrass. *Can. J. Bot.* **70**, 779–787.

Hou, Z. M., Xue, C. Y., Peng, Y. L., Katan, T., Kistler, H. C., and Xu, J. R. (2002). A mitogen-activated protein kinase gene (*MGV1*) in *Fusarium graminearum* is required for female fertility, heterokaryon formation, and plant infection. *Mol. Plant Microbe Interact.* **15**, 1119–1127.

Howard, R. J., and Valent, B. (1996). Breaking and entering: Host penetration by the fungal rice blast pathogen *Magnaporthe grisea*. *Ann. Rev. Microbiol.* **50**, 491–512.

Hughes, T. R., Robinson, M. D., Mitsakakis, N., and Johnston, M. (2004). The promise of functional genomics: Completing the encyclopedia of a cell. *Curr. Opin. Microbiol.* **7,** 546–554.

Idnurm, A., and Howlett, B. J. (2002). Isocitrate lyase is essential for pathogenicity of the fungus leptosphaeria maculans to canola (*Brassica napus*). *Eukaryot. Cell* **1,** 719–724.

Ikeda, K., Nakayashiki, H., Kataoka, T., Tamba, H., Hashimoto, Y., Tosa, Y., and Mayama, S. (2002). Repeat-induced point mutation (RIP) in *Magnaporthe grisea*: Implications for its sexual cycle in the natural field context. *Mol. Microbiol.* **45,** 1355–1364.

Irie, T., Matsumura, H., Terauchi, R., and Saitoh, H. (2003). Serial Analysis of Gene Expression (SAGE) of *Magnaporthe grisea*: Genes involved in appressorium formation. *Mol. Genet. Genomics* **270,** 181–189.

Jia, Y., McAdams, S. A., Bryan, G. T., Hershey, H. P., and Valent, B. (2000). Direct interaction of resistance gene and avirulence gene products confers rice blast resistance. *EMBO J.* **19,** 4004–4014.

Kadotani, N., Nakayashiki, H., Tosa, Y., and Mayama, S. (2003). RNA silencing in the phytopathogenic fungus *Magnaporthe oryzae*. *Mol. Plant Microbe Interact.* **16,** 769–776.

Kadotani, N., Nakayashiki, H., Tosa, Y., and Mayama, S. (2004). One of the two Dicer-like proteins in the filamentous fungi *Magnaporthe oryzae* genome is responsible for hairpin RNA-triggered RNA silencing and related small interfering RNA accumulation. *J. Biol. Chem.* **279,** 44467–44474.

Kamakura, T., Yamaguchi, S., Saitoh, K., Teraoka, T., and Yamaguchi, I. (2002). A novel gene, *CBP1*, encoding a putative extracellular chitin-binding protein, may play an important role in the hydrophobic surface sensing of *Magnaporthe grisea* during appressorium differentiation. *Mol. Plant Microbe Interact.* **15,** 437–444.

Kang, S. C., Sweigard, J. A., and Valent, B. (1995). The *PWL* host specificity gene family in the blast fungus *Magnaporthe grisea*. *Mol. Plant Microbe Interact.* **8,** 939–948.

Kankanala, P., and Valent, B. (2005). Molecular and cellular biology of biotrophic interactions in rice blast disease. *Fungal Genet. Newsl.* **52S,** 286.

Kasahara, S., Wang, P., and Nuss, D. L. (2000). Identification of *bdm-1*, a gene involved in G protein beta- subunit function and alpha-subunit accumulation. *Proc. Natl. Acad. Sci. USA* **97,** 412–417.

Kawamura, C., Moriwaki, J., Kimura, N., Fujita, Y., Fuji, S., Hirano, T., Koizumi, S., and Tsuge, T. (1997). The melanin biosynthesis genes of *Alternaria alternata* can restore pathogenicity of the melanin-deficient mutants of *Magnaporthe grisea*. *Mol. Plant Microbe Interact.* **10,** 446–453.

Kays, A. M., and Borkovich, K. A. (2004). Severe impairment of growth and differentiation in a *Neurospora crassa* mutant lacking all heterotrimeric G alpha proteins. *Genetics* **166,** 1229–1240.

Keller, N. P., Turner, G., and Bennett, J. W. (2005). Fungal secondary metabolism—from biochemistry to genomics. *Nat. Rev. Microbiol.* **3,** 937–947.

Kellis, M., Patterson, N., Endrizzi, M., Birren, B., and Lander, E. S. (2003). Sequencing and comparison of yeast species to identify genes and regulatory elements. *Nature* **423,** 241–254.

Kershaw, M. J., Wakley, G., and Talbot, N. J. (1998). Complementation of the *mpg1* mutant phenotype in *Magnaporthe grisea* reveals functional relationships between fungal hydrophobins. *EMBO J.* **17,** 3838–3849.

Kershaw, M. J., Thornton, C. R., Wakley, G. E., and Talbot, N. J. (2005). Four conserved intramolecular disulphide linkages are required for secretion and cell wall localization of a hydrophobin during fungal morphogenesis. *Mol. Microbiol.* **56,** 117–125.

Kim, H., and Borkovich, K. A. (2004). A pheromone receptor gene, *pre-1*, is essential for mating type-specific directional growth and fusion of trichogynes and female fertility in *Neurospora crassa*. *Mol. Microbiol.* **52,** 1781–1798.

Kim, S., Ahn, I. P., and Lee, Y. H. (2001). Analysis of genes expressed during rice–*Magnaporthe grisea* interactions. *Mol. Plant Microbe Interact.* **14**, 1340–1346.

Kim, S. T., Kim, S. G., Hwang, D. H., Kang, S. Y., Kim, H. J., Lee, B. H., Lee, J. J., and Kang, K. Y. (2004a). Proteomic analysis of pathogen-responsive proteins from rice leaves induced by rice blast fungus *Magnaporthe grisea*. *Proteomics* **4**, 3569–3578.

Kim, S. T., Yu, S., Kim, S. G., Kim, H. J., Kang, S. Y., Hwang, D. H., Jang, Y. S., and Kang, K. Y. (2004b). Proteome analysis of rice blast fungus (*Magnaporthe grisea*) proteome during appressorium formation. *Proteomics* **4**, 3579–3587.

Kim, S., Ahn, I. P., Rho, H. S., and Lee, Y. H. (2005). *MHP1*, a *Magnaporthe grisea* hydrophobin gene, is required for fungal development and plant colonization. *Mol. Microbiol.* **57**, 1224–1237.

Koga, H. (1995). An electron-microscopic study of the infection of spikelets of rice by *Pyricularia oryzae*. *J. Phytopathol.* **143**, 439–445.

Koga, J., Yamauchi, T., Shimura, M., Ogawa, N., Oshima, K., Umemura, K., Kikuchi, M., and Ogasawara, N. (1998). Cerebrosides A and C, sphingolipid elicitors of hypersensitive cell death and phytoalexin accumulation in rice plants. *J. Biol. Chem.* **273**, 31985–31991.

Konishi, H., Ishiguro, K., and Komatsu, S. (2001). A proteomics approach towards understanding blast fungus infection of rice grown under different levels of nitrogen fertilization. *Proteomics* **1**, 1162–1171.

Kouzminova, E., and Selker, E. U. (2001). DIM-2 encodes a DNA methyltransferase responsible for all known cytosine methylation in *Neurospora*. *EMBO J.* **20**, 4309–4323.

Kroken, S., Glass, N. L., Taylor, J. W., Yoder, O. C., and Turgeon, B. G. (2003). Phylogenomic analysis of type I polyketide synthase genes in pathogenic and saprobic ascomycetes. *Proc. Natl. Acad. Sci. USA* **100**, 15670–15675.

Kulkarni, R. D., and Dean, R. A. (2004). Identification of proteins that interact with two regulators of appressorium development, adenylate cyclase and cAMP-dependent protein kinase A, in the rice blast fungus *Magnaporthe grisea*. *Mol. Genet. Genomics* **270**, 497–508.

Kulkarni, R. D., Thon, M. R., Pan, H. Q., and Dean, R. A. (2005). Novel G-protein-coupled receptor-like proteins in the plant pathogenic fungus *Magnaporthe grisea*. *Genome Biol.* **6**(3), R24.

Lane, S., Birse, C., Zhou, S., Matson, R., and Liu, H. P. (2001). DNA array studies demonstrate convergent regulation of virulence factors by Cph1, Cph2, and Efg1 in *Candida albicans*. *J. Biol. Chem.* **276**, 48988–48996.

Lau, G., and Hamer, J. E. (1996). Regulatory genes controlling *MPG1* expression and pathogenicity in the rice blast fungus *Magnaporthe grisea*. *Plant Cell* **8**, 771–781.

Lee, S. C., and Lee, Y. H. (1998). Calcium/calmodulin-dependent signaling for appressorium formation in the plant pathogenic fungus *Magnaporthe grisea*. *Mol. Cells* **8**, 698–704.

Lee, B. N., Kroken, S., Chou, D. Y. T., Robbertse, B., Yoder, O. C., and Turgeon, B. G. (2005a). Functional analysis of all nonribosomal peptide synthetases in *Cochliobolus heterostrophus* reveals a factor, *NPS6*, involved in virulence and resistance to oxidative stress. *Eukaryot. Cell* **4**, 545–555.

Lee, Y. J., Yamamoto, K., Hamamoto, H., Nakaune, R., and Hibi, T. (2005b). A novel ABC transporter gene *ABC2* involved in multidrug susceptibility but not pathogenicity in rice blast fungus *Magnaporthe grisea*. *Pest. Biochem. Physiol.* **81**, 13–23.

Li, L., Xue, C. Y., Bruno, K., Nishimura, M., and Xu, J. R. (2004). Two PAK kinase genes, *CHM1* and *MST20*, have distinct functions in *Magnaporthe grisea*. *Mol. Plant Microbe Interact.* **17**, 547–556.

Li, W. X., Rehmeyer, C. J., Staben, C., and Farman, M. L. (2005). TERMINUS-Telomeric end-read mining in unassembled sequences. *Bioinformatics* **21**, 1695–1698.

Liebmann, C. (2004). G protein-coupled receptors and their signaling pathways: Classical therapeutical targets susceptible to novel therapeutic concepts. *Curr. Pharma. Design* **10**, 1937–1958.

Liu, S. H., and Dean, R. A. (1997). G protein alpha subunit genes control growth, development, and pathogenicity of *Magnaporthe grisea*. *Mol. Plant Microbe Interact.* **10**, 1075–1086.

Lu, J. P., Liu, T. B., and Lin, F. C. (2005). Identification of mature appressorium-enriched transcripts in *Magnaporthe grisea*, the rice blast fungus, using suppression subtractive hybridization. *FEMS Microbiol. Lett.* **245,** 131–137.

Lu, S. W., Kroken, S., Lee, B. N., Robbertse, B., Churchill, A. C. L., Yoder, O. C., and Turgeon, B. G. (2003). A novel class of gene controlling virulence in plant pathogenic ascomycete fungi. *Proc. Natl. Acad. Sci. USA* **100,** 5980–5985.

Machida, M., Asai, K., Sano, M., Tanaka, T., Kumagai, T., Terai, G., Kusumoto, K., Arima, T., Akita, O., Kashiwagi, Y., Abe, K., Gomi, K., *et al.* (2005). Genome sequencing and analysis of *Aspergillus oryzae*. *Nature* **438,** 1157–1161.

Marzluf, G. A. (1997). Genetic regulation of nitrogen metabolism in the fungi. *Microbiol. Mol. Biol. Rev.* **61,** 17–32.

Matsumura, H., Reich, S., Ito, A., Saitoh, H., Kamoun, S., Winter, P., Kahl, G., Reuter, M., Kruger, D. H., and Terauchi, R. (2003). Gene expression analysis of plant host-pathogen interactions by SuperSAGE. *Proc. Natl. Acad. Sci. USA* **100,** 15718–15723.

McCafferty, H. R. K., and Talbot, N. J. (1998). Identification of three ubiquitin genes of the rice blast fungus *Magnaporthe grisea*, one of which is highly expressed during initial stages of plant colonisation. *Curr. Genet.* **33,** 352–361.

Mey, G., Oeser, B., Lebrun, M. H., and Tudzynski, P. (2002). The biotrophic, non-appressorium-forming grass pathogen *Claviceps purpurea* needs a *FUS3/PMK1* homologous mitogen-activated protein kinase for colonization of rye ovarian tissue. *Mol. Plant Microbe Interact.* **15,** 303–312.

Mitchell, T. K., and Dean, R. A. (1995). The cAMP-dependent protein kinase catalytic subunit is required for appressorium formation and pathogenesis by the rice blast pathogen *Magnaporthe grisea*. *Plant Cell* **7,** 1869–1878.

Motoyama, T., Kadokura, K., Ohira, T., Ichiishi, A., Fujimura, M., Yamaguchi, I., and Kudo, T. (2005). A two-component histidine kinase of the rice blast fungus is involved in osmotic stress response and fungicide action. *Fungal Genet. Biol.* **42,** 200–212.

Nakayashiki, H. (2005). RNA silencing in fungi: Mechanisms and applications. *FEBS Lett.* **579,** 5950–5957.

Nakayashiki, H., Ikeda, K., Hashimoto, Y., Tosa, Y., and Mayama, S. (2001). Methylation is not the main force repressing the retrotransposon MAGGY in *Magnaporthe grisea*. *Nucleic Acids Res.* **29,** 1278–1284.

Ni, M., Rierson, S., Seo, J. A., and Yu, J. H. (2005). The *pkaB* gene encoding the secondary protein kinase a catalytic subunit has a synthetic lethal interaction with pkaA and plays overlapping and opposite roles in *Aspergillus nidulans*. *Eukaryot. Cell* **4,** 1465–1476.

Nierman, W. C., Pain, A., Anderson, M. J., Wortman, J. R., Kim, H. S., Arroyo, J., Berriman, M., Abe, K., Archer, D. B., Bermejo, C., Bennett, J., Bowyer, P., *et al.* (2005). Genomic sequence of the pathogenic and allergenic filamentous fungus *Aspergillus fumigatus*. *Nature* **438,** 1151–1156.

Ninomiya, Y., Suzuki, K., Ishii, C., and Inoue, H. (2004). Highly efficient gene replacements in *Neurospora* strains deficient for nonhomologous end-joining. *Proc. Natl. Acad. Sci. USA* **101,** 12248–12253.

Nishimura, M., Park, G., and Xu, J. R. (2003). The G-beta subunit *MGB1* is involved in regulating multiple steps of infection-related morphogenesis in *Magnaporthe grisea*. *Mol. Microbiol.* **50,** 231–243.

Nukina, M. (1999). The blast disease fungi and their metabolic products. *J. Pest. Sci.* **24,** 293–298.

Orbach, M. J., Farrall, L., Sweigard, J. A., Chumley, F. G., and Valent, B. (2000). A telomeric avirulence gene determines efficacy for the rice blast resistance gene Pi-ta. *Plant Cell* **12,** 2019–2032.

Osherov, N., and May, G. S. (2001). The molecular mechanisms of conidial germination. *FEMS Microbiol. Lett.* **199,** 153–160.

Pan, X. W., and Heitman, J. (2002). Protein kinase A operates a molecular switch that governs yeast pseudohyphal differentiation. *Mol. Cell. Biol.* **22,** 3981–3993.

Park, G., Xue, G. Y., Zheng, L., Lam, S., and Xu, J. R. (2002). *MST12* regulates infectious growth but not appressorium formation in the rice blast fungus *Magnaporthe grisea*. *Mol. Plant Microbe Interact.* **15**, 183–192.

Park, G., Bruno, K. S., Staiger, C. J., Talbot, N. J., and Xu, J. R. (2004). Independent genetic mechanisms mediate turgor generation and penetration peg formation during plant infection in the rice blast fungus. *Mol. Microbiol.* **53**, 1695–1707.

Park, G., Xue, C., Zhao, X., Kim, Y., Orbach, M., and Xu, J.-R. (2006). Multiple upstream signals converge on an adaptor protein Mst50 to activate the *PMK1* pathway in *Magnaporthe grisea*. *Plant cell* **18**, 2822–2835.

Pascon, R. C., Ganous, T. M., Kingsburry, J. M., Cox, G. M., and McCusker, J. H. (2004). *Cryptococcus neoformans* methionine synthase: Expression analysis and requirement for virulence. *Microbiology* **150**, 3013–3023.

Pedras, M. S. C., and Ahiahonu, P. W. K. (2005). Metabolism and detoxification of phytoalexins and analogs by phytopathogenic fungi. *Phytochemistry* **66**, 391–411.

Pickford, A. S., Catalanotto, C., Cogoni, C., and Macino, G. (2002). Quelling in *Neurospora crassa*. *In* "Advances in Genetics: Homology Effects" (C. Wu and J. C. Dunlap, eds.), Vol. 46, pp. 277–303. Academic Press, San Diego, CA.

Powell, A. J., Pan, H. Q., Diener, S. E., and Dean, R. A. (2005). Transposition, recombination and gene genesis in *Magnaporthe grisea*. *Fungal Genet. Newsl.* **52S**, 106.

Prak, E. T., Dodson, A. W., Farkash, E. A., and Kazazian, H. H. (2003). Tracking an embryonic L1 retrotransposition event. *Proc. Natl. Acad. Sci. USA* **100**, 1832–1837.

Ramezani-Rad, M. (2003). The role of adaptor protein Ste50-dependent regulation of the MAPKKK Ste11 in multiple signaling pathways of yeast. *Curr. Genet.* **43**, 161–170.

Rauyaree, R., Choi, W., Fang, E., Blackmon, B., and Dean, R. A. (2001). Genes expressed during early stages of rice infection with the rice blast fungus *Magnaporthe grisea*. *Mol. Plant Pathol.* **2**, 347–354.

Rehmeyer, C., Li, W., Kusaba, M., Kim, Y. K., Brown, D., Staben, C., Dean, R. A., and Farman, M. (2006). Organization of chromosome ends in the rice blast fungus *Magnaporthe oryzae*. *Nucleic Acids Res.* **34**, 4685–4701.

Ruijter, G. J. G., and Visser, J. (1997). Carbon repression in *Aspergilli*. *FEMS Microbiol. Lett.* **151**, 103–114.

Sanyal, K., Baum, M., and Carbon, J. (2004). Centromeric DNA sequences in the pathogenic yeast *Candida albicans* are all different and unique. *Proc. Natl. Acad. Sci. USA* **101**, 11374–11379.

Seong, K., Hou, Z. M., Tracy, M., Kistler, H. C., and Xu, J. R. (2005). Random insertional mutagenesis identifies genes associated with virulence in the wheat scab fungus *Fusarium graminearum*. *Phytopathology* **95**, 744–750.

Sesma, A., and Osbourn, A. E. (2004). The rice leaf blast pathogen undergoes developmental processes typical of root-infecting fungi. *Nature* **431**, 582–586.

Soanes, D. M., and Talbot, N. J. (2005). A bioinformatic tool for analysis of EST transcript abundance during infection-related development by *Magnaporthe grisea*. *Mol. Plant Pathol.* **6**, 503–512.

Sone, T., Oguchi, A., Kikuchi, H., Senoh, A., Nakagawa, S., and Tomita, F. (2005). Three novel retrotransposons from *Magnaporthe grisea* mini-chromosome. *Fungal Genet. Newsl.* **52S**, 347.

Soundararajan, S., Jedd, G., Li, X. L., Ramos-Pamplona, M., Chua, N. H., and Naqvi, N. I. (2004). Woronin body function in *Magnaporthe grisea* is essential for efficient pathogenesis and for survival during nitrogen starvation stress. *Plant Cell* **16**, 1564–1574.

Sweigard, J. A., Chumley, F. G., and Valent, B. (1992). Disruption of a *Magnaporthe grisea* cutinase gene. *Mol. Gen. Genet.* **232**, 183–190.

Sweigard, J. A., Carroll, A. M., Farrall, L., Chumley, F. G., and Valent, B. (1998). *Magnaporthe grisea* pathogenicity genes obtained through insertional mutagenesis. *Mol. Plant Microbe Interact.* **11**, 404–412.

Talbot, N. J., McCafferty, H. R. K., Ma, M., Moore, K., and Hamer, J. E. (1997). Nitrogen starvation of the rice blast fungus *Magnaporthe grisea* may act as an environmental cue for disease symptom expression. *Physiol. Mol. Plant Pathol.* **50,** 179–195.

Tamaru, H., and Selker, E. U. (2001). A histone H3 methyltransferase controls DNA methylation in *Neurospora crassa*. *Nature* **414,** 277–283.

Tenney, K., Hunt, I., Sweigard, J., Pounder, J. I., McClain, C., Bowman, E. J., and Bowman, B. J. (2000). *HEX-1*, a gene unique to filamentous fungi, encodes the major protein of the Woronin body and functions as a plug for septal pores. *Fungal Genet. Biol.* **31,** 205–217.

Thines, E., Weber, R. W. S., and Talbot, N. J. (2000). MAP kinase and protein kinase A–dependent mobilization of triacylglycerol and glycogen during appressorium tugor generation by *Magnaporthe grisea*. *Plant Cell* **12,** 1703–1718.

Thon, M. R., Pan, H., Diener, S., Papalas, J., Taro, T., Mitchell, T., and Dean, R. A. (2006). The role of transposable element clusters in genome evolution and loss of synteny in the rice blast fungus *Magnaporthe oryzae*. *Genome Biol.* **7,** R16.

Tosa, Y., Osue, J., Eto, Y., Oh, H. S., Nakayashiki, H., Mayama, S., and Leong, S. A. (2005). Evolution of an avirulence gene, *AVR1-CO39*, concomitant with the evolution and differentiation of *Magnaporthe oryzae*. *Mol. Plant Microbe Interact.* **18,** 1148–1160.

Tsuji, G., Kenmochi, Y., Takano, Y., Sweigard, J., Farrall, L., Furusawa, I., Horino, O., and Kubo, Y. (2000). Novel fungal transcriptional activators, Cmr1p of Colletotrichum lagenarium and Pig1p of *Magnaporthe grisea*, contain Cys2His2 zinc finger and Zn(II)2Cys6 binuclear cluster DNA-binding motifs and regulate transcription of melanin biosynthesis genes in a developmentally specific manner. *Mol. Microbiol.* **38,** 940–954.

Tucker, S. L., and Talbot, N. J. (2001). Surface attachment and pre-penetration stage development by plant pathogenic fungi. *Annu. Rev. Phytopathol.* **39,** 385–419.

Tucker, S. L., Thornton, C. R., Tasker, K., Jacob, C., Giles, G., Egan, M., and Talbot, N. J. (2004). A fungal metallothionein is required for pathogenicity of *Magnaporthe grisea*. *Plant Cell* **16,** 1575–1588.

Urban, M., Bhargava, T., and Hamer, J. E. (1999). An ATP-driven efflux pump is a novel pathogenicity factor in rice blast disease. *EMBO J.* **18,** 512–521.

Valent, B., and Chumley, F. G. (1991). Molecular genetic analysis of the rice blast fungus *Magnaporthe grisea*. *Annu. Rev. Phytopathol.* **29,** 443–467.

Valent, B., and Chumley, F. G. (1994). Avirulence genes and mechanisms of genetic instability in the rice blast fungus. *In* "Rice Blast Disease" (R. S. Zeigler, S. A. Leong, and P. S. Teng, eds.), pp. 111–153. CAB International, Wallingford, United Kingdom.

Veneault-Fourrey, C., and Talbot, N. J. (2005). Moving toward a systems biology approach to the study of fungal pathogenesis in the rice blast fungus *Magnaporthe grisea*. *Adv. Appl. Microbiol.* **57,** 177–215.

Veneault-Fourrey, C., Barooah, M., Egan, M., Wakley, G., and Talbot, N. J. (2006a). Autophagic fungal cell death is necessary for infection by the rice blast fungus. *Science* **312,** 580–583.

Veneault-Fourrey, C., Lambou, K., and Lebrun, M. H. (2006b). Fungal Pls1 tetraspanins as key factors of penetration into host plants: A role in re-establishing polarized growth in the appressorium. *FEMS Microbiol. Lett.* **256,** 179–184.

Verkerke-Van Wijk, I., Kim, J. Y., Brandt, R., Devreotes, P. N., and Schaap, P. (1998). Functional promiscuity of gene regulation by serpentine receptors in *Dictyostelium discoideum*. *Mol. Cell. Biol.* **18,** 5744–5749.

Viaud, M. C., Balhadere, P. V., and Talbot, N. J. (2002). A *Magnaporthe grisea* cyclophilin acts as a virulence determinant during plant infection. *Plant Cell* **14,** 917–930.

Villalba, F., Lebrun, M. H., Hua-Van, A., Daboussi, M. J., and Grosjean-Cournoyer, M. C. (2001). Transposon impala, a novel tool for gene tagging in the rice blast fungus *Magnaporthe grisea*. *Mol. Plant Microbe Interact.* **14,** 308–315.

Walther, A., and Wendland, J. (2003). Septation and cytokinesis in fungi. *Fungal Genet. Biol.* **40,** 187–196.

Wang, Z. Y., Thornton, C. R., Kershaw, M. J., Li, D. B., and Talbot, N. J. (2003). The glyoxylate cycle is required for temporal regulation of virulence by the plant pathogenic fungus *Magnaporthe grisea. Mol. Microbiol.* **47,** 1601–1612.

Weber, R. W. S., Wakley, G. E., Thines, E., and Talbot, N. J. (2001). The vacuole as central element of the lytic system and sink for lipid droplets in maturing appressoria of *Magnaporthe grisea. Protoplasma* **216,** 101–112.

Wei, Y. D., Shen, W. Y., Dauk, M., Wang, F., Selvaraj, G., and Zou, J. T. (2004). Targeted gene disruption of glycerol-3-phosphate dehydrogenase in *Colletotrichum gloeosporioides* reveals evidence that glycerol is a significant transferred nutrient from host plant to fungal pathogen. *J. Biol. Chem.* **279,** 429–435.

Wolpert, T. J., Dunkle, L. D., and Ciuffetti, L. M. (2002). Host-selective toxins and avirulence determinants: What's in a name? *Annu. Rev. Phytopathol.* **40,** 251–285.

Wu, S. C., Ham, K. S., Darvill, A. G., and Albersheim, P. (1997). Deletion of two endo-beta-1, 4-xylanase genes reveals additional isozymes secreted by the rice blast fungus. *Mol. Plant Microbe Interact.* **10,** 700–708.

Wu, S. C., Halley, J. E., Luttig, C., Fernekes, L. M., Gutierrez-Sanchez, G., Darvill, A. G., and Albersheim, P. (2006). Identification of an endo-beta-1,4-D-xylanase from *Magnaporthe grisea* by gene knockout analysis, purification, and heterologous expression. *Appl. Environ. Microbiol.* **72,** 986–993.

Xu, J. R. (2000). MAP kinases in fungal pathogens. *Fungal Genet. Biol.* **31,** 137–152.

Xu, J. R., and Hamer, J. E. (1996). MAP kinase and cAMP signaling regulate infection structure formation and pathogenic growth in the rice blast fungus *Magnaporthe grisea. Genes Dev.* **10,** 2696–2706.

Xu, J. R., Urban, M., Sweigard, J. A., and Hamer, J. E. (1997). The *CPKA* gene of *Magnaporthe grisea* is essential for appressorial penetration. *Mol. Plant Microbe Interact.* **10,** 187–194.

Xu, J. R., Staiger, C. J., and Hamer, J. E. (1998). Inactivation of the mitogen-activated protein kinase Mps1 from the rice blast fungus prevents penetration of host cells but allows activation of plant defense responses. *Proc. Natl. Acad. Sci. USA* **95,** 12713–12718.

Xue, C. Y., Park, G., Choi, W. B., Zheng, L., Dean, R. A., and Xu, J. R. (2002). Two novel fungal virulence genes specifically expressed in appressoria of the rice blast fungus. *Plant Cell* **14,** 2107–2119.

Yamauchi, J., Takayanagi, N., Komeda, K., Takano, Y., and Okuno, T. (2004). cAMP-PKA signaling regulates multiple steps of fungal infection cooperatively with Cmk1 MAP kinase in *Colletotrichum lagenarium. Mol. Plant Microbe Interact.* **17,** 1355–1365.

Yu, J. H., and Keller, N. (2005). Regulation of secondary metabolism in filamentous fungi. *Annu. Rev. Phytopathol.* **43,** 437–458.

Zelter, A., Bencina, M., Bowman, B. J., and Read, N. D. (2004). A comparative genomic analysis of the calcium signaling machinery in *Neurospora crassa, Magnaporthe grisea,* and *Saccharomyces cerevisiae. Fungal Genet. Biol.* **41,** 827–841.

Zhao, X., Xue, C., Kim, Y., and Xu, J. R. (2004). A ligation-PCR approach for generating gene replacement constructs in *Magnaporthe grisea. Fungal Genet. Newsl.* **51,** 17–18.

Zhao, X. H., Kim, Y., Park, G., and Xu, J. R. (2005). A mitogen-activated protein kinase cascade regulating infection-related morphogenesis in *Magnaporthe grisea. Plant Cell* **17,** 1317–1329.

6

Genetic and Genomic Dissection of the *Cochliobolus heterostrophus Tox1* Locus Controlling Biosynthesis of the Polyketide Virulence Factor T-toxin

B. Gillian Turgeon* and Scott E. Baker[†]

*Department of Plant Pathology, Cornell University
Ithaca, New York 14853
[†]Fungal Biotechnology Team, Chemical and Biological Process
Development, Environmental Technology Directorate, Pacific Northwest
National Laboratory, Richland, Washington 99352

Advances in Genetics, Vol. 57
0065-2660/07 $35.00
DOI: 10.1016/S0065-2660(06)57006-3

ABSTRACT

Fungal pathogenesis to plants is an intricate developmental process requiring
biological components found in most fungi, as well as factors that are unique to
fungal taxa that participate in particular fungus–plant interactions. The host-
selective polyketide toxin known as T-toxin produced by *Cochliobolus heterostrophus* race T, a highly virulent pathogen of maize, is an intriguing example of
the latter type of virulence determinant. The *Tox1* locus, which controls biosynthesis of T-toxin, originally defined as a single genetic locus, it is, in fact, two
exceedingly complex loci on two chromosomes that are reciprocally translocated
with respect to their counterparts in weakly pathogenic race O. Race O lacks the
Tox1 locus and does not produce T-toxin. Highly virulent race T was first
recognized when it caused an epidemic of Southern Corn Leaf Blight, which
devastated the US corn crop in 1970. The evolutionary origin of the *Tox1* locus
remains unknown.   © 2007, Elsevier Inc.

I. INTRODUCTION

Fungal pathogenesis to plants is a multifaceted developmental process. Molecular mechanisms underlying each distinct fungus–plant interaction necessitate
orchestration of both universal and specific factors. Herein is an historical
tracing of several decades of research on one specific factor important in one
particular pathosystem, that is, the filamentous Ascomycete *Cochliobolus heterostrophus* on its host plant maize. The focus is on the genetic and genomic pursuit
of genes required for production of the host-selective toxin (HST), T-toxin, a
family of linear polyketides that renders the fungus highly and specifically
virulent to maize carrying Texas male sterile cytoplasm (T-cms). Our narrative
of this necrotrophic interaction begins with research in the laboratory of O. C.
Yoder at Cornell University in the early 1970s and ends with the present
state of knowledge. In 1982, one of us, Turgeon, joined the endeavor, as did
C. R. Bronson at Iowa State University. The Yoder/Turgeon research partner-
ship continued until the year 2003 and included a leave of absence from

Cornell in San Diego at the Novartis Agricultural Discovery Institute (NADII), subsequently the Torrey Mesa Research Institute (TMRI). During this period, the *C. heterostrophus* genome was sequenced and the quest to understand fungal pathogenic development was expanded to the genome-wide level. Issuing from this effort was the now commonplace discovery that, while determinants of pathogenicity include factors that appear to be or are unique to one pathogen or to pathogens in general, many genes whose products serve the pathogenicity niche are found in saprobes as well as in pathogens. With respect to the genetics of T-toxin production, genome data confirmed that the genetic and physical nature of the genetic locus required for T-toxin production is remarkably complex, as described below.

Why *C. heterostrophus*? The fungus was known as a mild pathogen of corn until 1969–1970 when a superpathogenic race caused the Southern Corn Leaf Blight (SCLB) epidemic, which devastated the US corn crop along the eastern seaboard. Since then, *C. heterostrophus* has emerged as a model for the study of plant pathogenesis. Research focused initially on physiological and genetic characterization of T-toxin, then broadened to include other factors required for pathogenicity and virulence. The appeal of *C. heterostrophus* as a model necrotrophic fungus is due in large part to the tractability of the organism for classical and molecular genetic analyses.

A. *C. heterostrophus* biology

The genus *Cochliobolus*, related genera [*Lewia* (*Alternaria*), *Pyrenophora*, *Setosphaeria*], and their asexual relatives are a species rich collection of taxa in the Ascomycetes. *Cochliobolus* includes both saprobic and pathogenic species that are significant monocot pathogens worldwide; individual species/races attack corn, rice, barley, sugarcane, wheat, and oats, all major cereal crops. Phylogenetic relationships among these genera have been well resolved using a number of molecular characters (Berbee, 2001; Berbee *et al.*, 1996, 1999; Turgeon, 1998; Turgeon and Berbee, 1998) such that it is possible to choose and compare closely related pairs of *Cochliobolus* spp. taxa that differ in lifestyle to extract key determinants of a particular lifestyle and also to explore genome evolution. For example, one can examine *C. heterostrophus* (sexual) versus *Bipolaris sacchari* (asexual) for studies of sexual development, *C. heterostrophus* (heterothallic) versus *C. luttrellii* (homothallic) for studies of sexual reproductive mode, *C. heterostrophus* (pathogen of maize) versus *C. miyabeanus* (pathogen of rice) for cereal host range determinants, *C. heterostrophus* (pathogenic) versus *C. luttrellii* (nonpathogenic) for evolution of pathogenicity, and more broadly, *Cochliobolus* spp. (pathogens of monocots) versus *Alternaria* spp. (pathogens of dicots), for monocot/dicot host range determinants.

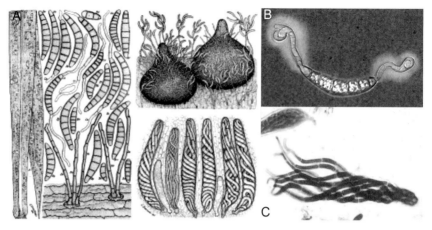

Figure 6.1. (A) Original description of C. *heterostrophus* by Charles Drechsler (1925). (B) A C. *heterostrophus* conidium (asexual spore) that has germinated at both ends (hence the anamorph "*Bipolaris*"), negatively stained with India ink to delineate extra-cellular polysaccharide capsule around germ tubes and appressoria. (C) A C. *heterostrophus* tetrad consisting of eight filamentous, entwined ascospores (sexual spores).

C. *heterostrophus* is the most widely distributed species in the genus *Cochliobolus*. The fungus was first identified (Fig. 6.1A) as *Ophiobolus hetero-strophus*, based on its teleomorph (Drechsler, 1925, 1927), but later chosen as the type species of a newly erected genus *Cochliobolus* in the family Pleospor-aceae which included those *Helminthosporium* (a Deuteromycete classification) species with ascospores arranged in a helicoid pattern in the ascus (Drechsler, 1934). The proper teleomorph designation, C. *heterostrophus*, was not widely used until nomenclature conventions for plant pathogenic fungi were proposed (Yoder et al., 1986). In most of the older literature, the fungus is frequently referred to by one of its anamorphic designations, *Helminthosporium maydis* (Nisikado and Miyake), *Drechslera maydis* (Nisikado and Miyake) Subramanian and Jain, or *Bipolaris maydis* (Nisikado and Miyake) Shoemaker.

As a natural pathogen of corn, C. *heterostrophus* can be found in many tropical and subtropical areas of the world (Drechsler, 1925, 1934; Orillo, 1952; Yu, 1933). In the United States, the fungus is usually found in the warmer southern states, thus, the disease it causes is commonly known as SCLB (Hooker, 1974). The life cycle includes both asexual and sexual stages (Figs. 6.1B, C and 6.2). The fungus overwinters as conidia or mycelia on debris of dead corn plants. The sexual stage has rarely, if ever, been observed in the field, but can be induced easily in the laboratory. The fungus is heterothallic: only strains of opposite mating type can cross and undergo sexual reproduction (Debuchy and Turgeon, 2006; Nelson, 1957; Sivanesan, 1984; Turgeon et al., 1993). Sexual reproduction in

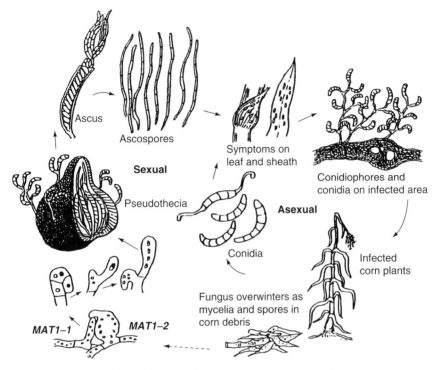

Figure 6.2. Life and disease cycles of C. *heterostrophus*. Drawn by S.-W. Lu.

C. heterostrophus is controlled by a single mating-type locus (*MAT1*). For all self incompatible Dothideomycetes examined to date, *MAT1* in each haploid genome contains one of two different sequences known as *MAT1–1* and *MAT1–2* idiomorphs (Debuchy and Turgeon, 2006).

Asexual reproduction can occur repeatedly in a single growing season; conidia infect corn leaves predominantly by direct penetration (Wheeler, 1977) and cause small, tan lesions which, in epidemic conditions, cover the entire leaf, killing the plant (Miller *et al.*, 1970; Smith *et al.*, 1970). Although appressoria are usually associated with penetration sites, they may not be necessary for infection (Horwitz *et al.*, 1999).

B. *Cochliobolus* and disease

C. heterostrophus and related taxa are notorious for their ability to produce HSTs, a group of chemically diverse, low-molecular weight compounds that serve as virulence or pathogenicity factors (Walton, 1996; Wolpert *et al.*, 2002;

Yoder, 1980; Yoder et al., 1997). Each HST is necessary for development of a particular disease; for example, T-toxin, a family of linear polyketides, is required by C. heterostrophus for high virulence to T-cytoplasm maize; HC-toxin and victorin, small cyclic peptide toxins (synthesized by nonribosomal peptide synthetases) are required by C. carbonum and C. victoriae, respectively, for pathogenicity to corn and oats, respectively. Phylogenetic analysis has revealed that Cochliobolus spp. fall into two distinct groups (Berbee et al., 1999). Group 2 contains 18 species, none of which is notable as a pathogen. In contrast, all members of the genus that are known to cause serious crop diseases, for example, C. heterostrophus, C. carbonum, C. victoriae, C. sativus, C. miyabeanus, are found among the 13 species in Group 1. This observation suggests that a progenitor within the genus Cochliobolus gave rise, over a relatively short period, to a series of distinct biotypes, each distinguished by having unique pathogenic capability to a particular type of plant. This radiation may be associated with acquisition of ability to produce HSTs, although the data are incomplete on this issue. It appears that the concentration of "heavy duty" pathogens in Cochliobolus Group 1 resulted from a shared origin of traits, predisposing members of this group to evolve high levels of virulence on susceptible host genotypes (Berbee et al., 1999).

On two occasions, Cochliobolus species have caused devastating losses to US agriculture. In the 1940s, C. victoriae caused widespread destruction (spreading to 20 states) of oat varieties containing the recently introduced Pc-2 gene for general crown rust resistance. In 1970, a previously unseen race (race T) of C. heterostrophus caused the worst epidemic (SCLB) in US agricultural history, which destroyed more than 15% of the US maize crop (Hooker, 1974). T-cms corn, widely planted at that time, is exquisitely sensitive to race T (Section III.B). The epidemic swept all corn-growing areas and in a short period (May–September), the disease spread from Florida, north to Maine and finally into southern Canada (Moore, 1970) (Fig. 6.3). More than 30 states reported serious damage to corn (ranging from 50% reduced yields to complete loss) valued at more than 1 billion dollars. The disease was still prominent in 1971, but the overall severity was much reduced because of cooler weather, unfavorable for the fungus (Ullstrup, 1972).

In fact, increased susceptibility of T-cms corn to certain isolates of C. heterostrophus had been reported in the Philippines as early as 1961 (Mercado and Lantican, 1961) and also in the United States (southern Iowa and Illinois) in 1968 (Scheifele et al., 1970; Ullstrup, 1970, 1972). The presence of a new race of the fungus was not seriously considered until the epidemic in 1970, when Smith et al. (1970) and Hooker and colleagues (Hooker et al., 1970b; Lim and Hooker, 1971, 1972a,b) published studies on a C. heterostrophus isolate obtained from severely diseased corn leaves collected in central Illinois in 1969. The isolate was designated "race T" for its high virulence on T-cms corn

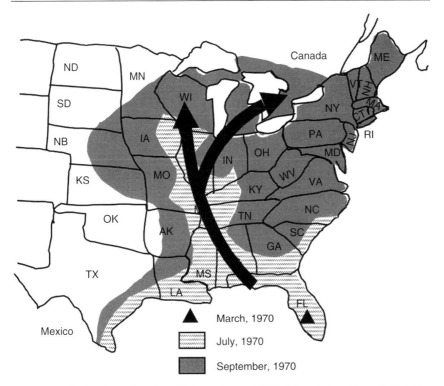

Figure 6.3. The Southern Corn Leaf Blight epidemic of 1970. Adapted from Moore (1970). The epidemic was first reported in Belle Glade, FL (black triangle), then spread over the growing season to more than 30 states and Canada.

and for its ability to produce T-toxin, a substance that specifically affects T-corn (Fig. 6.4A and B). Prior to 1970, *C. heterostrophus* was known as an endemic pathogen of minor economic importance in the United States. As noted above, it was first reported in Florida and the Philippines in 1925 (Drechsler, 1925) as a corn leaf spot that was distinct from leaf blights caused by several known species of fungi. The race of the fungus that had been known since 1925 was named "race O." Race O does not produce T-toxin and is mildly virulent on both T- and N-cytoplasm (Hooker *et al.*, 1970a; Smith *et al.*, 1970). Note that both races O and T are pathogenic, but race T is highly and specifically virulent on T-cms corn (Fig. 6.4A). The concept of two races of *C. heterostrophus* was further formalized by subsequent genetic and biochemical analyses (Lim and Hooker, 1971; Yoder, 1980; Yoder and Gracen, 1975; Yoder *et al.*, 1976). A detailed history and commentary on the origins of race T can be found in Wise *et al.* (1999).

Figure 6.4. (A) Race T, inoculated on T-cms corn (right), has a much more devastating effect than when inoculated on N-cytoplasm corn (left). (B) Culture filtrate of Race T, injected into T-cms corn (left), causes massive yellowing of the leaves, in contrast to the same filtrate injected into N-cytoplasm corn (right). (See Color Insert.)

The SCLB epidemic of 1970 generated enormous concern not just about the short term effects on growers' incomes and corn market prices, but also about the overall "genetic vulnerability" of our essential crop plants. Excerpts from an article by Doyle (1985) serve to illustrate the SCLB impact:

> ..."In the summer of 1968, ... the first signs of trouble went almost unnoticed. Out in the heartland, on a few isolated seed farms in Illinois and Iowa, a mysterious disease was producing "ear rot" on corn plants."...
>
> ..."Reproducing rapidly in the unusually warm and moist weather of 1970, its spores carried on the wind, the new disease began moving northward toward a full-scale invasion of America's vast corn empire. Later to be

*identified as "race T" of the fungus Helminthosporium maydis, it soon
became known as the Southern Corn Leaf Blight.". . .*

*. . ."The new fungus moved like wildfire through one corn field after another.
In some cases it would wipe out an entire stand of corn in ten days. . . . In
extreme infections, whole ears of corn would fall to the ground and crumble at
the touch.". . .*

*. . ."Nevertheless, with the opening of business that Monday, panic gripped
the commodity and futures markets. Estimates that the blight might wipe out
half the nation's corn crop fueled frantic trading and speculation. At the
Chicago Board of Trade, the nation's largest commodities market, 193
million bushels of corn changed hands in one day, smashing a trading record
that had stood for 122 years. The Dow Jones index for commodity futures hit
145.27, and had its highest one-day advance in nineteen years.". . .*

The stories of the SCLB and the Victoria blight epidemics are dramatic
examples of interactions between crop plants, whose "evolution" is driven by
human intervention, and their pathogens, which evolve naturally to exploit new
genetic susceptibilities. The genetic loci responsible for T-cms and resistance to
crown rust (*Pc-2*) were introduced into corn and oats, respectively, by breeders
fewer than 20 years before the epidemic outbreaks. Specifically, T-cms was
discovered in the 1950s, then incorporated into elite corn inbred lines increas-
ingly throughout the 1960s, and was present in almost all of the hybrid corn in
the United States by 1970. Race T likely appeared in the Philippines in 1963,
but was not recognized as a new race until the late-1960/1970s (Turgeon and Lu,
2000). It is currently unclear whether *C. heterostrophus* race T evolved around
the time of the epidemic or had been "lurking" in the field for a long time.
Regardless of the answer, the vast monoculture of T-cms maize was the perfect
host for this previously unknown race.

Concurrent with the epidemic of SCLB, a novel disease, Yellow Leaf
Blight (Arny and Nelson, 1971), caused by *Didymella zeae-maydis* (Mukunya &
Boothroyd) von Arx, formerly *Mycosphaerella zeae-maydis* Mukunya & Boothroyd
(Mukunya and Boothroyd, 1973; von Arx, 1987), also known as *Phyllosticta
maydis* Arny & Nelson (Arny and Nelson, 1971) was recognized. *D. zeae-maydis*
is also specific to T-cms and produces PM-toxin, a family of polyketides with
structural and functional similarity to T-toxin (Danko *et al.*, 1984).

Beyond US borders, an outbreak of *C. miyabeanus* contributed to the
Bengal rice famine of 1942/1943, which resulted in starvation of more than
2 million people (Dasgupta, 1984). Species of *Cochliobolus* clearly have proven
their ability to cause extraordinary crop losses. Perhaps it is only a matter of time

before some *Cochliobolus* pathogen, either a new species or a new race, again causes widespread destruction of one of our cereal crops.

II. TOOLS FOR GENETIC ANALYSIS

Key to the rise of *C. heterostrophus* as a model for fungal plant pathogenesis has been the development of tools that enable functional and molecular genetic analyses of biological processes.

A. Classical genetics

The 1970s epidemic led to the genetic domestication of *C. heterostrophus* and to its development as a model eukaryotic plant pathogen for rigorous analysis of fungal pathogenesis to plants by both conventional genetic and molecular genetic approaches. Lines of inbred (6–12 backcrosses) near-isogenic laboratory strains with paired genetic traits of interest called C (Leach *et al.*, 1982; Tegtmeier *et al.*, 1982) and K (Klittich and Bronson, 1986) strains are available, allowing functional analysis of candidate genes in a defined genetic background (Table 6.1). As noted above, *C. heterostrophus* is heterothallic and mating between isolates of opposite mating type (*MAT1–1* and *MAT1–2*) can be achieved readily in the laboratory (Leach *et al.*, 1982); ascospore progeny can be collected in 18–21 days for segregation analysis. Complete tetrads are easily isolated when necessary (e.g., for double or triple mutants, epistatic analysis, and so on). Collections of auxotrophic, morphological, and drug-resistant mutants were generated in the early 1980s (Leach *et al.*, 1982), using ethyl methanesulphonate (EMS). Cross data have been recorded in the Yoder/Turgeon laboratories since 1973. More than 4600 strains are stored in glycerol at −80 °C. Heterokaryon incompatibility exists in the field population, as demonstrated by failure of auxotrophs carrying different alleles at the same heterokaryon incompatibility locus to form heterokaryons. The actual number of *Het* loci is unknown.

B. Molecular genetics

The following is a brief summary of molecular technological milestones in the development of this particular fungal pathosystem.

1. Transformation

C. heterostrophus was the first phytopathogenic fungus to be transformed (Turgeon *et al.*, 1985). The technique used, $CaCl_2$-PEG-mediated protoplast transformation, remains the procedure of choice for this fungus. Stable integration of

Table 6.1. Genotypes of C. *heterostrophus* Laboratory Strains

Isolate[a]	Tox1	MAT1[b]	Alb1[c]	Cyh[d]
C1	+	1	+	S
C2	+	1	−	S
C3	−	2	+	S
C4	+	2	+	S
C5	−	1	+	S
C6	+	1	−	S
C7	−	1	−	S
C8	−	2	+	S
C9	+	1	+	S
C10	+	2	+	R
C11	+	1	+	R
C12	+	1	+	S
C13	+	2	+	S
C14	+	2	−	S
C15	−	2	−	S
C16	+	2	alb2	S
CB1	+	1	+	S
CB2	+	2	+	S
CB3	−	1	+	S
CB4	−	2	+	S
CB5	−	2	−	S
CB6	+	2	+	R
CB7	+	1	−	S
CB8	−	1	+	R
CB9	−	1	−	S
CB10	+	2	−	R
CB11	+	2	−	S
CB12	−	2	−	R
CB13	−	1	−	R
CB15	+	1	+	R
CB16	+	1	−	R

[a]C strains = result of six generations of backcrosses (Leach et al., 1982), CB strains = further backcrosses of C strains (Klittich and Bronson, 1986).

[b]MAT1 = mating-type locus. Mating type of strain = MAT1–1 or MAT1–2.

[c]Alb1 = locus for dark green pigment production. Strains lacking pigment are albino. C16 is alb2; phenotype is an off-white colony with black pseudothecia when mated.

[d]Cyh = resistance (R) or susceptibility (S) to cycloheximide.

transforming DNA into the chromosome by homologous recombination occurs at high frequency (usually close to 100%) with small amounts of homologous sequence (250–500 bp) (Turgeon et al., 1987; Wirsel et al., 1996). A variety of dominant selectable markers can be used, including the *Aspergillus nidulans amdS* gene for growth on acetamide as the sole carbon source (Turgeon *et al.*, 1985), the *Escherichia coli hygB* gene for resistance to hygromycin B (Turgeon *et al.*, 1987), the *Streptomyces hygroscopicus bar* gene for resistance to bialaphos (Straubinger *et al.*, 1992), the *Aspergillus terreus BSD* gene for resistance to blasticidin S (Kimura *et al.*, 1994), and the *Streptomyces noursei gene* (*NAT*) for resistance to nourseothricin (Schoch *et al.*, unpublished data). This versatile system makes it possible to clone genes and manipulate fungal genomic DNA by targeted gene disruption or gene replacement. Transformants, which are occasionally hetero-karyons containing both transformed and wild-type nuclei, are quickly purified by isolating single conidia.

2. Mutagenesis

a. Reverse genetics

As noted, *C. heterostrophus* undergoes high-frequency homologous recombination when small amounts of homologous DNA are included on the transforming DNA (Catlett *et al.*, 2003; Wirsel *et al.*, 1996). A PCR-based gene transformation strategy founded on approaches originally developed for *Saccharomyces cerevisiae* (Amberg *et al.*, 1995; Fairhead *et al.*, 1998) has been developed (Catlett *et al.*, 2003) and is used almost exclusively for gene manipulation. Successful-targeted deletions occur in >90% of *C. heterostrophus* transformants, for nonessential genes. The split-marker PCR strategy is simple and quick, eliminating the need to subclone of target sequences (Fig. 6.5).

b. Forward genetics

C. heterostrophus was the first filamentous fungus for which gene tagging by restriction enzyme-mediated integration (REMI) was reported to work (Lu *et al.*, 1994). Like transposon tagging in prokaryotes, REMI had a profound impact on molecular investigation of pathogenesis in many fungal species, particularly before whole-genome sequences became available (Akamatsu *et al.*, 1997; Boelker *et al.*, 1995; Sweigard, 1996; Wirsel *et al.*, 1996; Yun *et al.*, 1998). The Turgeon laboratory has a library consisting of over 3000 purified REMI transformants; analysis of a subset indicates at least 60% of those with an altered phenotype results from tagged insertions (Lu, 1998). These tagged mutations represent different genetic loci controlling diverse phenotypes, including ability to produce T-toxin, pathogenicity, mating ability, auxotrophy, pigmentation, ability to conidiate, and growth habit. This suggests that most, if not all, genes of *C. heterostrophus* can be mutated and simultaneously tagged by this procedure. DNA flanking the REMI insertion site can be recovered by plasmid rescue (Yang *et al.*, 1996) or PCR and used to query nucleotide sequence databases to help identify the disrupted genes. With the availability of an organism's genome sequence, flanking DNA

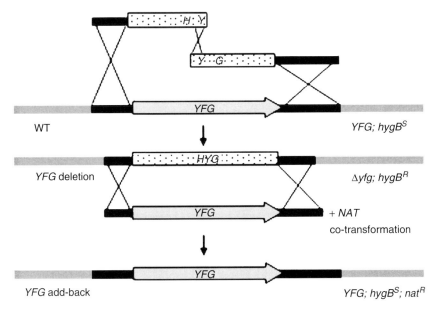

Figure 6.5. Split marker gene knockout strategy (Catlett *et al.*, 2003). Top line: wild-type (WT) chromosome (hygromycin B sensitive, *hygB^S*) carrying your favorite gene (YFG). Targeted integration of two PCR fragments, carrying (1) the left flank of YFG and the 5′ two thirds (HY) of the selectable marker *hygB* and (2) the right flank of YFG and the 3′ two thirds (YG) of the selectable marker *hygB*, would result in genotype *yfg-*; *hygB^R*. Middle and bottom lines: Add-back of a wild-type copy of YFG and replacement of the selectable marker by cotransformation with a plasmid carrying a second selectable marker, such as the gene (NAT), for resistance to nourseothricin would result in genotype YFG+;*hygB^S*;*nat^K*.

sequence information can be coupled to the genome sequence to quickly map the gene and recover the entire open reading frame (ORF). Molecular characterization suggests that the vector (single copy) usually integrates into the same site as that recognized by the restriction enzyme used for REMI; mutations are stable and chromosome rearrangements occur at low frequency (one out of six).

c. Chemical mutagenesis

Chemical mutagenesis of *Cochliobolus* was used in the early 1980s to develop a large collection of auxotrophic and morphological mutants for genetic analyses. For this, conidia were treated with chemical mutagens such as EMS and N-methyl-N′-nitro-N-nitrosoguanidine (NTG) (Leach *et al.*, 1982). In the mid-1990s, this procedure was combined with transformation technology in an attempt to enrich for Tox– mutants. In this case, conidia of a Tox+ strain that had been transformed with a version of the corn mitochondrial T-urf13 gene (conditionally expressed and shown to confer T-toxin sensitivity to the fungus itself, Section III.B) were first mutagenized with EMS, then grown on a medium that induced expression of the T-urf13 gene. The assumption was that survivors, if any, should include those that sustained mutations in the gene(s) for T-toxin production. Several leaky and one tight mutant (ctm45) were recovered; the latter completely failed to produce T-toxin as determined by the microbial assay (Yang et al., 1994). This was the first, laboratory produced, Tox– mutant of a Tox+ strain available for mutational analysis of the locus, known as *Tox1* (Section IV.A), responsible for T-toxin production. Although valuable, chemically induced mutants are not ideal in terms of molecular cloning because the mutation sites are not tagged.

C. Electrophoretic karyotype analysis

The *Cochliobolus* genome consists of 15 easily resolved chromosomes, ranging in size from about 1.3 to 3.7 Mb (Chang and Bronson, 1996; Kodama et al., 1999; Tasma and Bronson, 1998; Tzeng et al., 1992). Electrophoretic separation of these chromosomes, combined with subsequent blotting and probing of gels with markers or genes of interest, was a key experimental approach to unraveling and documenting the fact that *Tox1* is on two chromosomes in race T (Section IV.A.4), translocated with respect to a pair in race O (Chang and Bronson, 1996; Kodama et al., 1999; Tzeng et al., 1992).

D. Restriction fragment length polymorphism mapping

Although over 50 loci had been identified in early studies, C. heterostrophus genetic linkage groups were not established until 1992 when a combined genetic and restriction fragment length polymorphism (RFLP) map was developed (Tasma and Bronson, 1998; Tzeng et al., 1992). This map was based on the segregation pattern of known phenotypic or RFLP markers among progeny of a cross between *Tox1–;MAT1-1* field isolate, Hm540, and *Tox1+;MAT1-2* laboratory strain, B30.A3.R.45, combined with physical placement of these markers by probing electrophoretically separated chromosomes of both parents. Fifteen chromosomes (numbered sequentially, largest to smallest based on their sizes in strain Hm540) and a dispensable chromosome were identified and the chromosomal locations of 125 markers determined; the total map length was estimated to be 1501 cM and the total genome size $\sim$35 Mb (kb/cM = $\sim$23) (Tzeng et al., 1992). The genome size is in agreement with that estimated from whole-genome sequencing (Section V). The map was updated to include additional markers and telomeres and the map length recalculated as 1598 cM (Tasma and Bronson, 1998). This map confirmed the earlier genetic evidence

(Section IV.A.3) that the *Tox1* locus is tightly linked to the break points of a reciprocal translocation.

III. *C. HETEROSTROPHUS* AND SCLB

The SCLB epidemic involved the appearance of new race of C. *heterostrophus* that biosynthesizes a previously undescribed phytotoxin (T-toxin), which renders the fungus highly virulent on T-cms corn. T-toxin is toxic only to the same variety of plant that the fungus is able to attack, that is, maize containing mitochondria carrying the Texas cytoplasmic male sterility (cms) gene (T-urf13). HSTs like T-toxin are known from about 30 different plant pathogens (Turgeon and Lu, 2000; Walton, 1996; Wolpert *et al.*, 2002; Yoder, 1980; Yoder *et al.*, 1997).

A. T-toxin

Host specificity of T-toxin was first studied using culture filtrates produced by the fungus and inbred corn lines with T- or normal (N)-cytoplasm. Seedling assays for inhibition of root growth and leaf injection tests confirmed T-toxin caused the same symptoms on T-cytoplasm corn, as did the fungus itself (Gracen *et al.*, 1971; Lim and Hooker, 1971; Yoder *et al.*, 1976). Later, it was reported that T-toxin is a mixture of several linear polyketols ranging in length from C_{35} to C_{49} with the C_{39} and C_{41} components predominating (60–90%) in the native toxin (Kono and Daly, 1979; Kono *et al.*, 1981) (Fig. 6.6). Each of these components has the same specific toxicity against T-cms corn. Chemically synthesized T-toxin analogues, available in the early 1980s (Suzuki *et al.*, 1982, 1983), displayed the same activity as fungal preparations in a bioassay of inhibition of dark CO_2 fixation by susceptible corn leaves (Yoder and Gracen, 1977), thereby validating the structure.

B. T-cytoplasm corn and URF13 protein

One year after the discovery of race T and T-toxin, Genggenbach *et al.* (1972) and Miller and Koeppe (1971) reported that the T-toxin target site is T-cytoplasm mitochondria. Subsequent studies demonstrated that T-toxin binds to a 13-kDa inner mitochondrial membrane protein, which is the product of the mitochondrial gene, T-urf13 (Forde *et al.*, 1978). Binding of T-toxin causes pores in the URF13 oligomeric complex; small molecules required for normal mitochondrial function leak out (Levings, 1990; Levings and Siedow, 1992; Levings *et al.*, 1995), resulting in the cessation of ATP synthesis and subsequent cell death. The specific interaction between T-toxin and URF13 protein was also proven by heterologous expression of *T-urf13* in other organisms, including

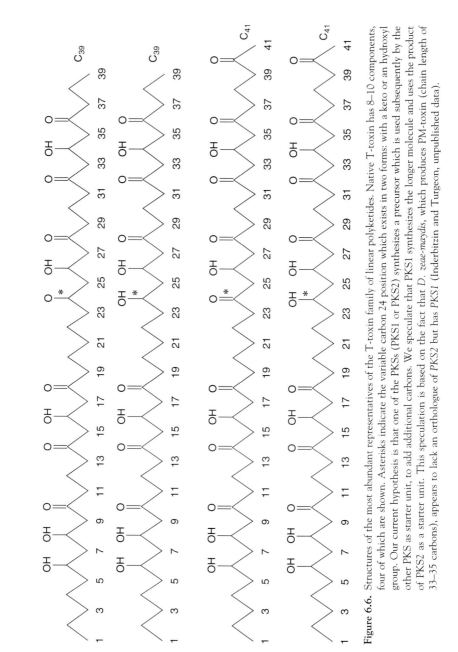

Figure 6.6. Structures of the most abundant representatives of the T-toxin family of linear polyketides. Native T-toxin has 8–10 components, four of which are shown. Asterisks indicate the variable carbon 24 position which exists in two forms: with a keto or an hydroxyl group. Our current hypothesis is that one of the PKSs (PKS1 or PKS2) synthesizes a precursor which is used subsequently by the other PKS as starter unit, to add additional carbons. We speculate that PKS1 synthesizes the longer molecule and uses the product of PKS2 as a starter unit. This speculation is based on the fact that *D. zeae-maydis*, which produces PM-toxin (chain length of 33–35 carbons), appears to lack an orthologue of *PKS2* but has *PKS1* (Inderbitzin and Turgeon, unpublished data).

E. coli (Dewey *et al.*, 1988; Huang *et al.*, 1990), tobacco (Vonallmen *et al.*, 1991), insects (Korth and Levings, 1993), and *C. heterostrophus* itself (Yang *et al.*, 1994). All transformants of these organisms that express *T-urf13* become sensitive to T-toxin. The URF13 protein not only confers sensitivity to T-toxin, it also confers a second phenotype, cms. The history of studies on T-cms in maize and the mechanisms of host susceptibility to T-toxin have been extensively reviewed (Wise *et al.*, 1999).

T-cms corn had been widely used for hybrid seed production and breeding since 1950s to avoid hand or mechanical emasculation. By 1970, 85% of hybrid corn produced in the United States was T-cms corn. The vast planting of the *C. heterostrophus* host undoubtedly facilitated the epidemic of SCLB in the United States. It is currently unclear whether race T originated at that time or had been in the field population at low levels all along but had no advantage over race O until a susceptible host was available.

C. Microbial bioassay for T-toxin production

Early evaluation of T-toxin action relied on plant assays; T-toxin activity was indicated by a typical race T symptom on susceptible corn plants, or by inhibition of root growth, dark CO_2 fixation in leaf discs, or death of protoplasts from the susceptible host (Bhullar *et al.*, 1975; Lim and Hooker, 1971; Yoder *et al.*, 1976, 1977). Although sensitive enough, these assays were laborious. In addition, certain chemical methods, such as a colorimetric assay, were developed (Karr and Hsu, 1975) but found unreliable (Yoder *et al.*, 1977).

The introduction of the *T-urf13* gene from mitochondrial DNA of T-cms corn (Dewey *et al.*, 1988) into *E. coli* cells provided a highly specific and efficient way to detect T-toxin (Fig. 6.7) (Ciuffetti *et al.*, 1992). Agar plugs bearing fungal mycelia are inoculated onto *E. coli* cells expressing the *T-urf13* gene. T-toxin-producing strains inhibit growth of the *E. coli* cells producing halos. Tox− mutants can be distinguished from wild type by failure to produce a halo and thus appear like wild-type race O, or by production of halos that are smaller (leaky) or larger (overproducing) than wild type.

IV. THE GENETICS OF T-TOXIN PRODUCTION

A. The *Tox1* locus

1. *Tox1* is defined as a single genetic locus

The ability to produce T-toxin is genetically inseparable from high virulence of the fungus on T-cms corn. In a cross between race O and race T, all progeny producing T-toxin are highly virulent and all progeny not producing T-toxin are

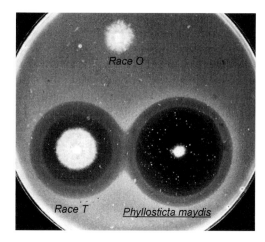

Figure 6.7. Microbial assay for T-toxin or PM-toxin production. The plate contains a lawn of
E. coli cells carrying the *URF13* gene from T-cms mitochondrial DNA. Agar plugs
bearing mycelium from Tox+ strains (race T, *Phyllosticta maydis*) kill the *E. coli* cells
around the plug, whereas mycelium from Tox− strains (e.g., race O) does not.
Phyllosticta maydis = D. *zeae-maydis*.

weakly virulent on T-cms corn; both races are pathogenic on N-cytoplasm
corn (Lim and Hooker, 1971). This 1:1 segregation of parental phenotypes is
observed in most crosses conducted in studies using either field isolates or inbred
laboratory strains (C-strains) (but see exceptions, Section IV.A.2; Yoder, 1976;
Yoder and Gracen, 1975), thus defining a single genetic locus, designated *Tox1*
(Leach *et al.*, 1982; Lim and Hooker, 1971; Tegtmeier *et al.*, 1982) that controls
T-toxin production and high virulence on T-cms corn of C. *heterostrophus*
race T (Fig. 6.8).

2. Odd ratios

Exceptions to 1:1 segregation of parental phenotypes (Tox+:Tox−) in crosses
between race O and race T strains were noticed (Yoder and Gracen, 1975).
Certain crosses between field isolates yielded an excess of Tox− progeny, which
might be explained if more than one locus were responsible for toxin production
or, alternatively, if nonrandom ascospore abortion were occurring. The former
explanation was ruled out by crossing field isolates to genetically defined laboratory strains and laboratory strains to each other (Bronson *et al.*, 1990). Nonrandom ascospore abortion was demonstrated to be due to the presence of a spore
killer gene in about 50% of race O field isolates (Taga *et al.*, 1985). Progeny of
crosses between a race O strain carrying this gene and a race T strain which does

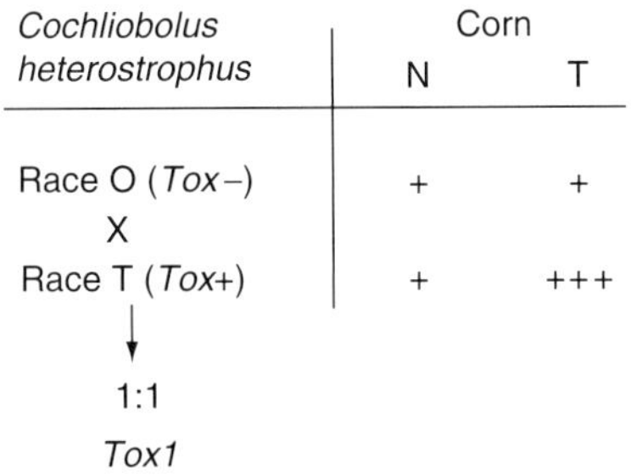

Cochliobolus heterostrophus	Corn	
	N	T
Race O (*Tox*–)	+	+
X		
Race T (*Tox*+)	+	+++
1:1		
Tox1		

Figure 6.8. When naturally occurring race T and O strains are crossed, *Tox1* appears to be a single genetic locus, as only parental types are recovered (ratio = 1:1). N, normal cytoplasm; T, T-cms cytoplasm. Both races are pathogenic (+), however race T is highly virulent on T-cms corn (+ + +).

not are nonviable if they lack the spore killer gene. Progeny of crosses between defined races O and T laboratory strains, lacking the spore killer factor, always segregate 1:1 for ability to produce toxin and for high virulence.

3. Races O and T differ by a reciprocal translocation at *Tox1*

Although early studies suggested a simple inheritance pattern, *Tox1* was subsequently found to be genetically inseparable from a reciprocal translocation break point (Fig. 6.9). This possibility was first proposed based on the comparison of nonviable ascospore frequencies in crosses homozygous versus heterozygous at *Tox1* (Bronson, 1988). The pattern of ascospore abortion suggested that races T and O differ by a reciprocal translocation and that *Tox1* is at or near the break point. Note that there were, in fact, two genetic complexities confounding *Tox1* analysis—the spore killer gene which, once recognized, could be eliminated by choosing strains which lacked it, and the reciprocal translocation which could not be eliminated until it was possible to make Tox– mutants of a Tox+ strain (Section II.B.2.c).

4. *Tox1* is two unlinked genetic loci

As noted above, in a cross between race O and race T, all progeny producing T-toxin are highly virulent and all progeny not producing T-toxin are weakly virulent on T-corn (Lim and Hooker, 1971; Yoder and Gracen, 1975). This 1:1 segregation of parental phenotypes identifies a single genetic locus, *Tox1*, that controls T-toxin production and high virulence. The 1988 observation in the Bronson laboratory, that *Tox1* is genetically inseparable from a reciprocal

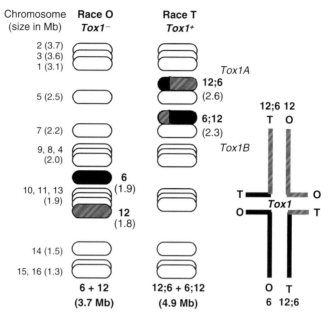

Figure 6.9. (A) The *Tox1* locus is actually two loci on two different chromosomes (6;12 and 12;6), reciprocally translocated with respect to their counterparts in race O (6 and 12). *Tox1A* is on chromosome 12;6, while *Tox1B* is on chromosome 6;12. In addition, the sum of the race T pair is 1.2 Mb greater than the sum of the race O pair, when chromosomes from isogenic strains are compared. (B) *Tox1* maps to the intersection of the four-armed linkage group made up of race T chromosomes 6;12 and 12;6 and race O chromosomes 6 and 12, and is genetically inseparable from the break points. When naturally occurring races T and O strains are crossed, it is impossible to tell if *Tox1* is on chromosome 6;12, or 12;6, or both. Crosses between race T and mutants of race T, lacking ability to produce T-toxin, revealed that *Tox1* is on both race T chromosomes.

translocation break point (Section IV.A.3), was supported by the construction, in the same laboratory, of an RFLP map (Tzeng *et al.*, 1992) that revealed a four-armed linkage group, diagnostic of a reciprocal translocation, with *Tox1* located at the intersection (Fig. 6.9B). The linkage group consists of a pair of race O chromosomes (6 and 12) and a pair of race T chromosomes (6;12 and 12;6) that are reciprocally translocated with respect to the race O pair (Tzeng *et al.*, 1992) (Fig. 6.9A). These findings demonstrated that *Tox1* was not a simple, single, Mendelian character and that the conclusions from the earlier genetic analyses were misleading. Although RFLP mapping placed *Tox1* at the intersection of the four-armed linkage group, it did not reveal its exact chromosomal location, that is, which translocated chromosome was it on, or was it on both?

The answer to this question was not available until it became possible to do genetic analysis of toxin production with tagged Tox− mutants of a progenitor *Tox1+* strain (Section V.B). Until this time, all genetic analyses relied on crosses between naturally occurring races O and T strains, which, as noted above, are distinguished by a reciprocal translocation intimately associated with toxin production. The translocation complicates genetic analysis since the pair of chromosomes carrying *Tox1* is heterozygous with respect to the pair in race O. Crosses between induced, tagged, Tox− mutants and a wild-type race O tester (the translocated chromosomes are heterozygous) showed that all induced Tox− mutations mapped at the previously defined *Tox1* locus since all progeny from these crosses were Tox−. Crosses between the Tox− mutants and a wild-type race T tester are between isogenic strains that differ only by the mutation at *Tox1* and, by necessity, at *MAT1*. This type of cross confirmed that each mutation was at a single site, since progeny segregated 1:1 for toxin production. Up to this point, these data led to the same conclusion as race O by race T crosses, that is, that *Tox1* is indeed a single locus. However, when mutants were crossed to each other (these crosses are also between isogenic strains differing only by the mutation at *Tox1* and at *MAT*), the major hypothesis that *Tox1* is a single locus collapsed. Progeny of a cross between one particular mutant and any of the other induced mutants were 25% Tox+, indicating that the mutation in this strain is not linked to the others. Furthermore, heterokaryons between an auxotroph of the unique mutant and different auxotrophs of any of the other induced mutants were all found to produce halos (produce T-toxin) in the microbial assay (Section III.C), about the same size as that of a control formed by two wild-type race T strains, indicating that the defects in the induced mutants could be complemented by the nucleus of the unique mutant (Kodama *et al.*, 1999).

Physical mapping of gel-separated chromosomes of the tagged Tox− mutants confirmed that *Tox1* is not a single locus (Fig. 6.9). The majority of the mutations were located on translocated chromosome 12;6; however, the unique mutation was located on chromosome 6;12, the other chromosome involved in the translocation (Kodama *et al.*, 1999). Thus, after about 25 years of data which suggested that *Tox1* was a single locus, it was demonstrated that *Tox1* is on two different chromosomes and that the two loci map to the chromosomes that are reciprocally translocated (Bronson, 1988; Tzeng *et al.*, 1992) in race T (i.e., chromosomes 6;12 and 12;6) with respect to the race O counterparts (chromosomes 6 and 12). *Tox1* behaves like a single Mendelian element because these two chromosomes must cosegregate during meiosis whenever a cross is heterozygous at *Tox1* (i.e., race O × race T) (Kodama *et al.*, 1999; Turgeon *et al.*, 1995). When they do not cosegregate, ascospore progeny are nonviable due to duplications and deletions. When crosses are homozygous (i.e., Tox− mutant of a *Tox1+* strain × another such mutant), the complexities of reciprocal translocation genetics are eliminated and the two translocated chromosomes segregate independently. To reflect the unique genetic organization of *Tox1*, the two loci were designated *Tox1*A and *Tox1*B (Fig. 6.9A) (Kodama *et al.*, 1999). Details regarding the genes encoded at these two *Tox1* loci are given below.

5. *Tox1* is associated with an insertion of DNA

Mendelian genetic analyses, RFLP mapping, and karyotype analysis demonstrated that the *Tox1* locus is complex. In addition to the translocation, *Tox1* is associated with a large insertion and highly repeated DNA (Chang and Bronson, 1996; Kodama *et al.*, 1999; Tzeng *et al.*, 1992). Evidence for this includes: (1) The total sum of the sizes of translocated chromosomes 6;12 and 12;6 in race T is about 1.2 Mb greater than the sum of the sizes of chromosomes 6 and 12 in near-isogenic race O (Fig. 6.9A). Since the strains used for this analysis are highly inbred, but heterozygous at *Tox1*, the inserted DNA is predicted to be located at or near *Tox1*. (2) Half of the RFLP probes mapping within 4 cM of *Tox1* are repetitive, in contrast to only ~4% repetitive probes in the remainder of race T genome (Chang and Bronson, 1996; Tzeng *et al.*, 1992; Yoder *et al.*, 1994). Cloning of genes from the two *Tox1* loci and the architecture of the genome sequence scaffolds carrying *Tox1*-associated genes in the genome assembly have confirmed that the *Tox1* genes are indeed embedded in highly repeated A + T-rich DNA (Section V). These genes are scattered on the scaffolds and are not closely linked or "clustered" as are many genes involved in production of secondary metabolites. (3) The genes are completely missing in race O (Rose *et al.*, 1996; Yang *et al.*, 1996; Baker *et al.*, 2006).

B. Identification of genes at *Tox1*

1. A polyketide synthase at *Tox1*A

The fact that T-toxin is a polyketide suggested the simple hypothesis that *Tox1* encodes a polyketide synthase (PKS)-encoding gene since biosynthesis of polyketides generally depends on PKS activity (Hopwood and Sherman, 1990). Indeed, the first gene identified at *Tox1* was found to encode this type of enzyme. This gene (designated *ChPKS1*) was cloned from the REMI-tagged Tox− mutant R.C4.350L (which maps at *Tox1*A), using the plasmid rescue procedure (Yang *et al.*, 1996). *ChPKS1* is a 7.6-kb ORF after splicing of four introns. The putative multifunctional protein has six enzymatic domains, including β-ketoacyl synthase (KS), acyltransferase (AT), dehydratase (DH), enoyl reductase (ER), β-ketoacyl reductase (KR), and acyl carrier protein (ACP, the REMI vector insertion site), plus a degenerate methyl transferase domain (MeT), all identified by the presence of conserved motifs found in known type I PKSs

(Yang *et al.*, 1996). *ChPKS1* lacks a chain-terminating thioesterase domain (TE) found at the C-terminus of some PKSs. Several features of *ChPKS1* helped in identification of additional genes at *Tox1*, most important of which are as follows: (1) *ChPKS1* is absent from the race O genome and from the genome of any other *Cochliobolus* species or related genus (Yang *et al.*, 1996), and (2) DNA on both flanks of the *ChPKS1* ORF is noncoding and A + T rich (~70%).

2. A decarboxylase (*DEC1*) and a reductase (*RED1*) are encoded at *Tox1B*

In contrast to the relatively straightforward cloning of *ChPKS1* from *Tox1A*, the path to cloning of genes at *Tox1B* was convoluted. A plasmid tagged, large deletion mutation mapping at *Tox1B*, which eliminated T-toxin production and high virulence on T-cms corn, was identified in a race T strain (Rose *et al.*, 2002). Comparisons among this mutant and wild-type races T and O DNA revealed that a diagnostic *Not*I restriction enzyme fragment associated with wild-type race T (not found in race O) was missing from the mutant and replaced by a smaller fragment. Screening a wild-type cDNA library with probes made from the polymorphic *Not*I restriction fragment from the deletion mutant and its wild-type *Tox1B* counterpart identified two single copy genes, a decarboxylase (*DEC1*) and a reductase (*RED1*), present in wild-type race T but absent in the mutant. Disruption of *DEC1* in wild-type race T eliminated T-toxin production and high virulence on T-cms corn. *DEC1* is similar to the acetoacetate decarboxylase gene of *Clostridium acetobutylicum*. RED1 is similar to members of the medium-chain dehydrogenase/reductase gene superfamily and disruption of *RED1* indicated it is not required for T-toxin production (but see Section V.B.3). DEC1 (46% G + C) and RED1 (44% G + C) are adjacent in the *C. heterostrophus* genome and divergently transcribed. DNA gel blot analysis indicates that they, like *PKS1*, are unique to race T. Like *PKS1*, these genes are flanked by A + T-rich (72%), highly repetitive, noncoding DNA (Rose *et al.*, 2002).

Both genetic and physical mapping confirmed that *PKS1* is at *Tox1A* on chromosome 12;6 and *DEC1/RED1* are at *Tox1B* on chromosome 6;12. PKS1 and DEC1 are required for T-toxin production. To determine if these genes were sufficient to make active T-toxin molecules, *PKS1* was introduced into Tox— mutant R.C4.186, which carries a deletion of ~700 kb at *Tox1A* (Kodama *et al.*, 1999). Transformants remained Tox— (Zhu, 1999). Similarly, DEC1 was introduced into Tox— mutant C4.PKS.13, which carries a deletion of ~75 kb at *Tox1B* (Kodama *et al.*, 1999; Rose *et al.*, 1996, 2002). Transformants also remained Tox— (Zhu, 1999). The failure of these single genes at *Tox1A* and *Tox1B*, respectively, to restore ability to produce T-toxin suggested that there were additional genes at both *Tox1A* and *Tox1B* necessary for T-toxin biosynthesis.

V. GENOMIC ANALYSIS OF THE *TOX1* LOCUS

The sojourn at TMRI provided the Turgeon/Yoder program with an opportunity to sequence the *C. heterostrophus* race T, strain C4 genome. Approximately, $2\times$ shotgun sequence coverage was generated, in-house, using paired end reads of 1- to 2-kb insert clones. Subsequently, an additional $3\times$ coverage was generated by Celera for TMRI, using paired end reads of 4-, 10-, and 14-kb insert libraries. Sequences contributing to the combined $5\times$ coverage were assembled into ~300 scaffolds. Additionally, several thousand ESTs were sequenced to aid in gene finding. The genome was estimated to be about 35 Mb. The complexity of *Tox1* was confirmed; contigs carrying known *Tox1* genes (*PKS1, DEC1,* and *RED1*) were located on the smallest scaffolds in the assembly (23–46 kb) and contigs comprising these scaffolds carried only a few ORFs embedded in highly A + T rich DNA.

A. cDNA profiling

Previous attempts to identify additional genes at *Tox1*, using positional-cloning procedures, failed; furthermore, it had not been possible to detect genes at *Tox1* that had already been identified (*PKS1, DEC1,* and *RED1*) in any of several dozen libraries (plasmid, lambda, BAC, and YAC), likely due to the A + T-rich nature of the regions. Thus, concurrent with the sequencing project, Ligation specificity-based Expression Analysis Display (LEAD), a comparative cDNA/AFLP/gel fractionation/capillary sequencing method coupled to software analysis (Li *et al.*, 2002) was undertaken. The LEAD approach was aimed at recovering genes expressed in race T, but not in near isogenic race O, strain C5 (Baker *et al.*, 2006) and was underpinned by the knowledge that the two known *Tox1* genes, *PKS1* and *DEC1*, are absent in race O. The cDNA subtraction procedure led to the identification of a second polyketide synthase-encoding gene, *PKS2*, which is unique to race T, maps at *Tox1A*, and, like *PKS1*, is required for both T-toxin biosynthesis and high virulence to maize (Baker *et al.*, 2006).

In all, 47 cDNA fragments were sequenced. Twelve of the sequences had similarity to polyketide synthases and three of the twelve fragments corresponded to *C. heterostrophus PKS1* (Yang *et al.*, 1996). The other nine fragments corresponded to an ORF designated *PKS2*. No other *PKS* fragments were identified. That *PKS2* is not in the genome of race O strain C5 was confirmed by lack of success in attempts to amplify corresponding DNA from race O genomic DNA using *PKS2*-specific primers and by lack of a PKS2 signal DNA in gel blots of race O DNA (Baker *et al.*, 2006).

The predicted PKS2 protein is 2144 amino acids and has domains corresponding to KS, AT, DH, ER, KR, and ACP. *PKS2* is distinct from *PKS1* in its sequence, domain structure, and phylogenetic placement, suggesting different functions of the two gene products in the assembly of the T-toxin molecule. The two PKS proteins are only 32% identical and 50% similar at the amino acid level in regions of alignment, and PKS2 is shorter than PKS1 (2144 cf 2528 amino acids) partly because it lacks the degenerate MeT domain, present in PKS1. This domain would not be predicted as a requirement for T-toxin biosynthesis since the molecule is not methylated. Phylogenetic analyses suggested that the product of PKS2 is a linear, variously reduced polyketide, as is the product of PKS1 (Kroken *et al.*, 2003).

B. *Tox1*-associated genome scaffolds

To determine if additional genes beyond *PKS1*, *PKS2*, and *DEC1* were required for T-toxin production, the sequences of the known genes at *Tox1* (*PKS1*, *PKS2*, *DEC1*, and *RED1*) were used to query the *C. heterostrophus* genome sequence database when the genome assembly became available. These genes mapped to three scaffolds: 4FP (carrying *PKS1*), 4LU (carrying *PKS2*), and 3PL (carrying *DEC1* and *RED1*) (Fig. 6.10). These scaffolds were the smallest in the *C. heterostrophus* genome assembly, confirming the complex and difficult to sequence nature of *Tox1*, which consists of genes embedded in highly repeated A + T rich DNA.

1. *Tox1A*

Two scaffolds (4FP and 4LU) mapped to *Tox1A* (Fig. 6.10). Scaffold 4FP is 23,099 kb in size, and composed of five contigs. Scaffold 4LU is 30,397 kb in size, and composed of seven contigs. *PKS1* maps to 4FP and was found to cover part of contig 1 and all of contig 2. *PKS2* maps to 4LU and was found to cover part of contig 4, all of contig 5, and part of contig 6. 4FP and 4LU sequences were examined using Artemis software (Rutherford *et al.*, 2000) and also blasted against GenBank using tbastx and manually annotated, but only one other putative ORF was found, on 4LU, contig 3, upstream of *PKS2*. Query of the putative protein sequence of this ORF against pfam database revealed high similarity to 3-hydroxyacyl-CoA dehydrogenase and to mouse lambda crystalline. This putative ORF is designated as *LAM1* (Asvarak, 2003). *LAM1* has two introns and a G + C content of 42.47%.

2. *Tox1B*

Scaffold 3PL is 43,533 kb in size, and composed of seven contigs (Fig. 6.10). *DEC1* was found to cover part of contig 2. Although *RED1* was not found, previous work had established that *RED1* is upstream of *DEC1* (Section V.B.2) and thus must lie within the gap between contig 1 and contig 2. Analysis of contigs 3 and 4 revealed two putative ORFs downstream of *RED1* (Fig. 6.10). Query of protein sequences of contigs 3 and 4 against the Pfam database showed

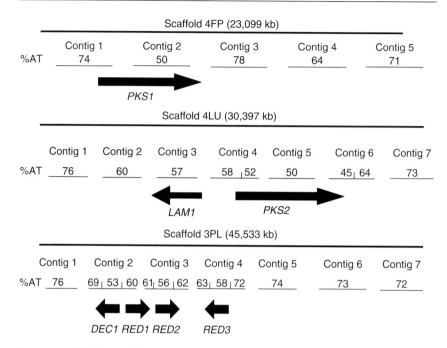

Figure 6.10. The three smallest scaffolds (4FP, 4LU, and 3PL) in the assembly of the C. *heterostrophus* genome. Known *Tox1* genes map to these scaffolds. Each scaffold consists of a number of "contigs," the sequences and order of which are known. Each contig is separated by a gap of predictable length. Numbers above thin lines representing contigs are the average percent A + T (or conversely G + C) content. Note, known genes are 50–58% A + T, while flanking regions are much higher average percent A + T. PKS1 and PKS2 encode polyketide synthases, *LAM1* encodes a lambda crystalline-like protein, while *DEC1* encodes a decarboxylase. All four are required for T-toxin production. Two reductase encoding genes (*RED2*, *RED3*) are required for toxin production, while deletion of *RED1* causes minimal reduction of toxin production, while deletion of *RED1* causes minimal reduction of toxin production.

high similarities to enzymes of the short chain dehydrogenase/reductase (SDR) superfamily. The two putative genes were designated as *RED2* (contig 3) and *RED3* (contig 4). *RED2* is 1133 nucleotides, including 2 putative introns; G + C content is 44.48%. *RED3* is 807 nucleotides, including a putative intron; G + C content is 41.57%.

3. Are the additional ORFs found on contigs associated with *Tox1* required for T-toxin production?

Targeted disruption of *LAM1* greatly reduced both T-toxin production and virulence to T-cytoplasm corn. Although Asvarak (2003) reported that simultaneous disruption of all three reductase genes *RED1*, *RED2*, and *RED3* had no apparent effect on T-toxin production or on virulence, recent analyses (Inderbitzin and Turgeon, unpublished data) indicate that deletion of *RED2* leads to drastically reduced ability to produce T-toxin while deletion of *RED3* leads to moderately reduced ability to produce T-toxin. Deletion of *RED1* slightly reduces production of T-toxin.

To date, seven genes on three scaffolds/metacontigs have been identified at *Tox1*, including *PKS1*, *PKS2*, and *LAM1* at *Tox1A*, and *DEC1*, *RED1*, *RED2*, and *RED3* at *Tox1B*. No gene identified to date at *Tox1* is present in race O.

4. *PKS2* and *PKS1* are unique to race T, whereas the remaining 23 *PKS* genes in the *C. heterostrophus* genome are common to both races

Kroken *et al.* (2003) described 25 PKS-encoding genes in the *C. heterostrophus* race T genome sequence. Except for reactions with *PKS1*- and *PKS2*-specific primers, products were amplified from both C4 (race T) and C5 (race O) genomic DNA, when DNAs from both races were tested for the presence of each *PKS*. Thus, none of the other 23 *PKS* genes in the *C. heterostrophus* genome is an obvious candidate for a role in T-toxin biosynthesis, since each is common to both races T and O. PCR products were of predicted size, and were the same size in both races (Table 6.2).

5. Phylogenetic relationship of *PKS1* and *PKS2* to other *PKS* genes

Phylogenomic analysis of fungal and certain bacterial *PKS*s (Kroken *et al.*, 2003) revealed that (1) fungal PKSs are common and numerous in filamentous Ascomycete genomes. *C. heterostrophus*, for example, encodes 25 of them; (2) fungal genomes share few orthologues among their abundant PKSs, even among species in the same genus; (3) fungal PKSs group separately, in most cases, from bacterial PKSs; and (4) the fungal PKSs diverged into two major clades, responsible for synthesis of reduced and unreduced polyketides, before the radiation of Euascomycetes some 400–700 Mya. Subsequent gene duplication, divergence, and loss can account for observed discontinuous distributions. Thus, horizontal gene transfer, a subject of much speculation, including our own regarding the origin of genes at *Tox1* (Walton, 2000; Yang *et al.*, 1996), need not be invoked (Kroken *et al.*, 2003).

The amino acid sequences of the KS and AT domains from all *C. heterostrophus* PKSs and from all fungal PKSs available in GenBank as of April, 2005 were used to construct a phylogenetic tree illustrating the relationship of

Table 6.2. *C. heterostrophus* PKS Genes

PKS[a]	Race T	Race O	Chromosome[b]
PKS1	+	−	12;6
PKS2	+	−	12;6
PKS3	+	+	6;12
PKS4	+	+	6;12
PKS5	+	+	
PKS6 (L14)	+	+	
PKS7	+	+	
PKS8	+	+	11
PKS9 (AT9)	+	+	
PKS10 (AT14)	+	+	
PKS11	+	+	
PKS12	+	+	12;6
PKS13	+	+	
PKS14	+	+	
PKS15	+	+	1
PKS16	+	+	6;12
PKS17	+	+	3
PKS18	+	+	
PKS19	+	+	4
PKS20	+	+	
PKS21	+	+	6;12
PKS22	+	+	3
PKS23	+	+	
PKS24	+	+	1
PKS25	+	+	

[a]L14, AT9, AT14 identified previously by Rose (1996).
[b]Located on chromosome indicated. Blank = unknown.

PKS1 and *PKS2* to each other, to other *C. heterostrophus PKSs*, and to known fungal PKSs (Baker *et al.*, 2006). Both PKS1 and PKS2 are found in a large clade of PKSs in which all of the characterized gene products make linear and variously reduced polyketide intermediates. This clade includes the majority of PKSs that make fungal toxins, and is subdivided into three subclades. One of these (fungal reducing PKS clade I, Fig. 6.11) contains *ChPKS1* (arrow) and its orthologue *MzmPKS1* (Fig. 6.11, double headed arrow). The latter is responsible for PM-toxin production (Yun, 1998; Yun *et al.*, 1998) by *D. zeae-maydis*. In addition, this clade contains *ChPKS2* (Fig. 6.11, arrow, above *ChPKS1*), and two orthologous genes whose PKS enzymes make the diketide component of lovastatin (*Aspergillus terreus lovF*) and compactin/citrinin (*Penicillium citrinum mlcB*), respectively (arrows above *ChPKS2*). *ChPKS1* and *ChPKS2* are not closely related within this subclade, as they are not united by any branch that received significant (>50%) bootstrap support.

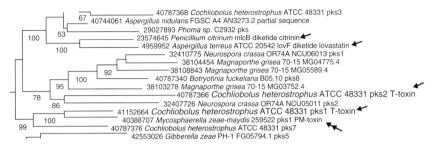

Figure 6.11. Portion of a phylogeny of fungal *PKS* genes inferred from amino acid sequence of their ketosynthase (KS) and acyl transferase (AT) domains (Baker *et al.*, 2006). The tree included all full-length PKS sequences available in GenBank as of April, 2005. Numbers below branches are bootstrap support values. Putative orthologues are in subclades supported by bootstrap values of 100% whose topology is consistent with presumed organismal phylogeny. The overall topology is similar to that reported in Kroken *et al.* (2003), with the exception that the number of major clades of reducing PKSs declined from four to three. Topology shown is one of two most parsimonious trees. Arrows point to *C. heterostrophus PKS1* and *PKS2* (involved in T-toxin production), double arrow indicates *D. zeae-maydis PKS1* (involved in PM-toxin production). Note *C. heterostrophus PKS1* and *D. zeae-maydis PKS1* are orthologues, while *C. heterostrophus PKS2* is not closely related. Arrows also indicate *PKS*s, *lovF* and *mclB*, involved in diketide production by *Aspergillus terreus* (producing lovastatin) and *Penicillium citrinum* (producing compactin/citrinin), respectively.

VI. THE PM-TOXIN GENE CLUSTER

D. zeae-maydis (Mukunya and Boothroyd, 1973), the causal agent of the Yellow Leaf Blight, appeared suddenly in the field in the late-1960s, early 1970s, as did race T. When first identified, it was described by its anamorph, *Phyllosticta maydis* (Arny and Nelson, 1971), later renamed *Phoma zeae-maydis* (Punithalingam, 1990). *D. zeae-maydis* has the same biological specificity as *C. heterostrophus* race T, showing high virulence to T-cytoplasm corn (Arny and Nelson, 1971; Arny *et al.*, 1970; Scheifele and Nelson, 1969). The discovery of T-toxin prompted researchers to screen *D. zeae-maydis* for a toxin that had the same specificity as T-toxin. Two years later, this was identified in culture filtrates and named PM-toxin based on the anamorph (Comstock *et al.*, 1972, 1973; Yoder, 1973; Yoder and Mukunya, 1972). Later, structural studies confirmed that PM-toxin is also a family of linear polyketides, similar, although not identical in structure, to T-toxin. The major difference is in chain length, C_{33}–C_{35} for PM-toxin versus C_{35}–C_{49} for T-toxin (Danko *et al.*, 1984; Kono *et al.*, 1983). Unlike *C. heterostrophus*, which occurs in two forms, race T which produces T-toxin and race O which does not, all known isolates of *D. zeae-maydis* produce PM-toxin.

The discovery of PM-toxin from *D. zeae-maydis*, with the same structure and biological activity as T-toxin, provided an additional comparative route to tracing the evolutionary origin of these toxins. Are the genes encoding the ability to produce toxin orthologues and were they horizontally transferred? If so, did one fungus acquire them from the other, or did they both get them from independent sources?

Because PM-toxin is structurally similar to T-toxin, a PCR-based strategy was used to clone *PKS* genes from *D. zeae-maydis*. Using primers corresponding to the conserved KS domain of known PKSs, a predicted 290-bp product was amplified and cloned from genomic DNA of wild type. The same PCR fragment yielded four different clones. One of them (MzPKS1) had 82% identity to the KS domain of *C. heterostrophus* PKS1, suggesting that it was a homologue.

Using the MzKS1 PCR product as a starting point, the entire MzPKS1 gene and its flanking sequences (totaling 23 kb) were cloned by a combination of TAIL-PCR (Liu and Whittier, 1995) and plasmid rescue (Yang *et al.*, 1996). Sequencing revealed that *MzPKS1* has 60% identity to *ChPKS1* over its entire length; identity in some enzymatic domain signature motifs is as high as 90%. The two genes have identical organization, each has four introns, three out of four are in conserved positions, and each is 7.6 kb after intron splicing (Yun, 1998).

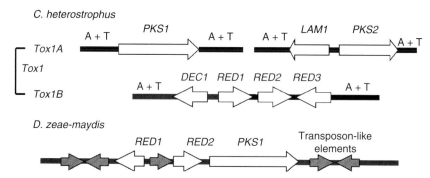

Figure 6.12. Organization of the *C. heterostrophus* Tox1A and Tox1B loci compared to the *Tox1* locus of *D. zeae-maydis*. Currently, two genome assembly scaffolds map to *C. heterostrophus* Tox1A, but their linkage is not known. One scaffold maps to Tox1B. The *D. zeae-maydis* Tox1 region is a tightly linked cluster of genes. For *C. heterostrophus* genes, see Fig. 6.10. None of the three *C. heterostrophus* RED genes is similar to each other and none is similar to the two *D. zeae-maydis* RED genes, which are also dissimilar from each other. All *D. zeae-maydis* genes shown (*PKS1*, *RED1*, and *RED2*) are required for PM-toxin production, while *C. heterostrophus* PKS1, PKS2, LAM1, DEC1, RED2, RED3, and possibly RED1 are required for T-toxin production.

In contrast to Ch*PKS1*, which is flanked mostly by A + T-rich, repeated, noncoding DNA, Mz*PKS1* is flanked by regularly spaced ORFs, including those with similarity to transposases. Two of the ORFs, designated Mz*RED1* and Mz*RED2*, have high similarity to ketoreductases involved in polyketide and fatty acid biosynthesis but are not similar to the Ch*RED1*, Ch*RED2*, and Ch*RED3* genes. When any one of Mz*RED1*, Mz*RED2*, or Mz*PKS1* is disrupted, PM-toxin production is lost, indicating they are indispensable for PM-toxin biosynthesis and that a gene cluster is involved (Yun, 1998) (Fig. 6.12).

VII. ARE ADDITIONAL *TOX* LOCI INVOLVED IN T-TOXIN PRODUCTION?

The working hypothesis is that there are two loci, one at *Tox1*A and one at *Tox1*B, each carrying genes (*PKS1*, *PKS2*, *LAM1*, *DEC1*, *RED2*, *RED3*, possibly *RED1* and others?) required for the T-toxin biosynthesis and that these genes include those involved in biosynthesis of precursors, regulation, secretion, and so on. Furthermore, since the ability to produce T-toxin segregates with *Tox1*, these two loci should contain all of the genes necessary for toxin production; genes at loci other than *Tox1* should not be required. In earlier studies, evidence for non 1:1 segregation of Tox+:Tox− progeny of crosses involving certain race O field isolates was presented (Section IV.A.2). These data, however, were attributable subsequently to the presence of a *Tox1*-linked spore killer gene and not to the requirement for products of genes at loci, other than *Tox1*, controlling T-toxin production. This conclusion was strengthened by a survey of the field population of the fungus; no locus other than *Tox1* was found to affect T-toxin production (Bronson *et al.*, 1990). Mutational analyses also supported this conclusion: 12 T-toxin production-deficient mutants generated by chemical or random mutagenesis, all map to the known *Tox1* locus. In addition, race O lacks genes required for T-toxin production implying that race O does not carry DNA associated with this trait. Therefore, it was a surprise when a REMI screen identified four loci, unlinked to *Tox1*, and present in race O, which when mutated in race T affect toxin production, as measured by the microbial assay (Lu, 1998).

Of the four strains carrying mutations-affecting ability to secrete toxin, three are tagged and one is not. One of these mutants shows a complete loss (tight, untagged, no halo produced in the microbial assay), two show a reduction (leaky, small halos), and one shows an increase (overproducing, halos larger than wild type) in ability to produce T-toxin. When these mutants were crossed to a race T tester, progeny segregated 1:1 (Tox+:Tox−), indicating a single gene mutation in each mutant. When crossed to a race O tester, 25% of the progeny were Tox+, indicating that none of these mutations is linked to the known *Tox1* locus. When the mutants were crossed to each other in all possible combinations, 25% of the progeny segregated as Tox+, indicating that none of these mutations is linked to any of the others. On the basis of these genetic analyses, four new loci that control toxin production in the genus *Cochliobolus* were proposed.

Site-specific disruption of the three tagged genes in the race T genome, using flanking DNA from the REMI insertion site, restored the original mutant phenotypes, confirming that these genes are involved in T-toxin production. All of the new *Tox* loci are present in the genomes of both races. Segregation analysis of crosses between the new Tox— mutants and wild-type race O testers yields 25% wild-type Tox+ recombinants, suggesting, as noted above, that none of these mutations maps at the known *Tox1* locus and that race O has genes that can complement these Tox— mutations. In addition, both races O and T DNA hybridize to genomic DNA fragments recovered from the insertion sites and both show the same hybridization patterns. This indicates that the new *Tox* loci are not T-toxin production specific, rather, they could be involved in both T-toxin biosynthesis and in other common, but dispensable, pathways. Indeed, linkage analyses revealed that two of the new *Tox* loci (R.C4.1771 and R.C4.1957) also contribute to a second mutant phenotype (reduced pigmentation) and a double mutation at these two loci is lethal (Lu, 1998).

VIII. MODEL FOR BIOSYNTHESIS OF T-TOXIN

The discovery that *PKS2* is required, along with *PKS1*, for T-toxin biosynthesis makes it likely that T-toxin is the combined product of two interacting polyketide synthases. There is precedent for a mechanism that links two polyketide chains, namely in the two orthologous biosynthetic pathways that produce lovastatin and compactin in which diketide side chains made by the PKSs, LDKS, or MLCB, respectively, are attached to nonaketides, fashioned iteratively by the PKSs, LNKS, or MLCA, respectively (Abe *et al.*, 2002; Kennedy *et al.*, 1999). In lovastatin biosynthesis, the monomodular PKS [LNKS (LOVB), encoded by lovB] produces the nonaketide in an iterative, highly controlled way, differentially utilizing reducing domains at each step. A diketide produced by LDKS [(LOVF), encoded by lovF] is then attached to the modified LNKS product by means of an ester bond.

It has been suggested (Baker *et al.*, 2006) that the variously reduced β-keto groups of T-toxin result from highly controlled and coordinated differential use, in each iterative cycle, of the various reducing domains in PKS1 (and/or PKS2), as with LNKS in lovastatin biosynthesis. Furthermore, the lovastatin model for building T-toxin requires 2 and only 2 PKSs, in line with the observation that only 2 (*PKS1* and *PKS2*) of the 25 PKS-encoding genes found in the *C. heterostrophus* genome are unique to race T. However, the proposed mechanism for joining the two T-toxin polyketides must be different from that required for the lovastatin pathway. With lovastatin, a dehydration, catalyzed by the transesterase lovD (Kennedy *et al.*, 1999), results in a cyclized, branched polyketide. The T-toxin family members remain as linear polyketides; the major component is a C_{41} chain with four groups of three unreduced to partially reduced β-ketones, separated by one or two completely reduced β-ketones. The structure of T-toxin suggests that the two polyketides produced by PKS1 and PKS2 are joined via a carbon–carbon bond, not by an ester bond as in lovastatin.

The following hypothesis for T-toxin synthesis (Fig. 6.6) is proposed. T-toxin synthesis is similar to lovastatin synthesis in that two PKSs are involved, both of which are monomodular, and given the differential processing of the β-keto groups in the final product, likely exhibit a high degree of control, per iterative cycle, over the use of the various reducing domains. The relative contribution of each PKS to T-toxin synthesis is, however, unknown. As noted above, T-toxin lacks the ester group formed in lovastatin through the concatenation of the two polyketide chains, and the putative PKS1 and PKS2 gene products appear to be joined by a carbon–carbon bond, reminiscent of the mechanism of aflatoxin synthesis (Hitchman *et al.*, 2001; Yabe and Nakajima, 2004). Synthesis of the aflatoxin backbone requires the interaction of a fatty acid synthetase (FAS) and a PKS, two proteins with the same domain structure, encoded by members of the same gene superfamily (Hopwood, 1997). The FAS synthesizes a C_6 starter unit for the PKS (Hitchman *et al.*, 2001). PKS1 and PKS2 could interact in a similar way. One of the PKSs could synthesize a precursor, used subsequently by the other PKS as starter unit, to add additional carbons. This mechanism is also reminiscent of the action of modular bacterial PKSs in which the ACP of the upstream protein subunit passes the nascent chain to the KS domain of the downstream subunit (Staunton and Weissman, 2001). Variability in the length of the final carbon backbone C_{35-49} could be achieved by varying the number of condensation cycles, and/or use of different starter and/or extender units, or might be due to decarboxylation of the penultimate product, catalyzed by DEC1, as proposed earlier (Rose *et al.*, 2002).

The dual PKS requirement for T-toxin biosynthesis is exciting because, to our knowledge, a mechanism in which two polyketides are stitched together via a carbon–carbon bond and in which the polyketide produced by one of the PKSs is used as a starter for the second PKS to produce the final metabolite has not been described for filamentous fungi. Most characterized Type I PKSs from filamentous fungi act iteratively, that is, a single monomodular PKS is used repeatedly to build the full length final molecule or set of molecules. Two exceptions, described above, are the pairs of PKSs involved in lovastatin and compactin biosynthesis. In these cases, however, the two polyketides are joined by an ester, not a carbon–carbon bond. Bacterial polyketides, on the other hand, are usually synthesized by noniterative Type I polyketide synthase complexes, encoded contiguously as modules in the bacterial genome. The linear, modular arrangement of bacterial *PKS* genes makes it possible to deduce the encoded polyketide synthases, their enzymatic domain organizations, the nature of the growing polyketide that is passed along to the adjacent module, and, ultimately, the biochemical structure of the metabolite produced. How a growing polyketide chain is processed is not obvious with iterative PKSs; reducing domains if present, may or may not be employed in any given cycle. In addition, although we now know that biosynthesis of T-toxin requires two PKSs, the corresponding *PKS* genes are not adjacent to each other, although they do appear to be linked at *Tox1A*.

IX. THE EVOLUTION OF POLYKETIDE-MEDIATED FUNGAL SPECIFICITY FOR T-CYTOPLASM CORN

Initial studies on the epidemiology of SCLB disease led to the hypothesis that race T arose from race O (Alcorn, 1975; Leonard, 1973, 1977), and that the origin was recent. Evidence for the recent origin of race T included the following: (1) When race T was first isolated, all isolates were *MAT1–1* while *MAT* distribution in race O was 50% *MAT1–1*/50% *MAT1–2*. A few years later, presumably after mating with race O, race T was also 50% *MAT1–1*/50% *MAT1–2* (Leonard, 1973). (2) A spore killer gene was identified in race O (∼50% of field isolates) and not in race T, suggesting long-term interbreeding among race O, but not race T, isolates. (3) The extant collection of race T field isolates shows fewer RFLPs than a corresponding collection of race O isolates. (4) All field isolates of race T show the same reciprocal translocation and "extra" DNA linked to the *Tox1* locus (Chang and Bronson, 1996).

Did race T originate in the Philippines? Increased susceptibility of corn to *C. heterostrophus* was reported there prior to the US epidemic (Mercado and Lantican, 1961); however, this report did not infer a new race. It has been suggested that race T may have been in the field in the Philippines prior to the introduction of T-cms corn in 1957, since race T symptoms appeared so soon after the host arrived (Mercado and Lantican, 1961; Wise *et al.*, 1999). In contrast, it is considered unlikely that race T was in the US field population prior to the widespread planting of T-cms corn, unless it occupied a specific, as yet unknown, ecological niche, since it is less fit than race O, as evidenced by its rapid decline after the withdrawal of its highly susceptible host (Klittich and Bronson, 1986; Wise *et al.*, 1999). Did race T originate just once, in the Philippines, and was it introduced from there to the United States? Alternatively, were there multiple origins of the race T genotype? The ideal approach to answering this question would come from direct examination of isolates of race

T collected in the Philippines prior to the US epidemic, and comparison of these isolates with those from the United States. Unfortunately, all Philippine isolates of race T have been lost, apparently.

Earlier, a model for race T evolution was proposed, based on MAT gene distribution over time, the structural features of the *Tox1* genes, the unusual chromosome rearrangement at *Tox1*, and the lack of these genes in any other species (Turgeon and Berbee, 1998; Yang *et al.*, 1996). It was proposed that *Tox1* DNA inserted, via horizontal gene transfer from an unknown source, into chromosome 6 or 12 of a *C. heterostrophus MAT1–1* race O isolate. The large insertion (about two-thirds the size of the race O chromosome) then caused instability, promoting a reciprocal translocation between chromosomes 6 and 12, which placed *Tox1*A on one chromosome (12;6) and *Tox1*B on another (6;12), and created the extant form we call race T. (Note that none of the *Tox1* genes found thus far is physically located at the translocation break points.)

Since *D. zeae-maydis* and *C. heterostrophus* race T appeared simultaneously, it was not unreasonable, before the genes responsible for polyketide production were available, to expect that these two fungi shared genes for toxin biosynthesis. The evidence, however, does not support this idea. First, although the general structures of *ChPKS1* and *MzPKS1* are similar, the corresponding proteins have only 60% overall amino acid identity. Second, the neighborhoods of these genes are different. *MzPKS1* is part of a cluster that includes at least two genes (*MzRED1* and *MzRED2*), encoding two different keto-reductases, both of which are required for PM-toxin production; *ChRED1*, *RED2*, and *RED3* have no nucleotide or amino acid similarity with *MzRED1* or *MzRED2*. When *D. zeae-maydis* genomic DNA is probed with *ChDEC1*, no signal can be detected even at low stringency. Attempts to amplify a *ChPKS2* equivalent from *D. zeae-maydis* by PCR have been unsuccessful (Inderbitzin, unpublished data). The discovery that ChPKS1 and MzmPKS1 are only 60% similar allows us to rule out the possibility that the two fungi acquired genes for toxin production, horizontally, from the same organism recently (e.g., at the time of the epidemic in 1970), and also to rule out the possibility that one of these fungi transferred genes to the other, recently. If the transfer were recent, there were two different sources of genes. If the transfer were ancient, or if the genes were present in the ancestors of both *C. heterostrophus* and *D. zeae-maydis*, the genes have diverged substantially, subsequently.

On this latter point, the genome-wide analyses of *PKS* genes in fungal genomes (Baker *et al.*, 2006; Kroken *et al.*, 2003) reveal that this type of gene is among the fastest evolving in the genome; there is enormous diversity of sequence identity within a single genome and between genomes. The 25 *C. heterostrophus* PKSs group in all 8 clades described by Kroken *et al.* (2003) and, as noted above, *PKS1* and *PKS2* are squarely nested in fungal branches, not in bacterial. There are few cases where the gene relationships mirror species relationships. Furthermore, these genes are often found near the telomeric ends of chromosomes, an environment that may favor genetic rearrangement. In light of genome-wide data, there appears to be nothing special about the *PKSs* for T- or PM-toxin production that would suggest a different evolutionary history from the other *PKSs* in the genome.

How did race T and *Tox1* evolve? One of our speculations (Kroken, Turgeon, and Baker, unpublished data) to account for the unusual sequence composition and complexity and the dual nature of the *Tox1* locus is that a small, dispensable chromosome (~1.2 Mb the size of *Tox1*) "hopped" into one of the race O chromosomes involved in the reciprocal translocation (6 or 12) destabilizing it. This idea runs into trouble when we remember that all race O strains examined to date lack this 1.2 Mb, including strains collected before the SCLB epidemic. It is in line, however, with what we know about C. *carbonum* race 1 isolates and HC-toxin production (Ahn *et al.*, 2002), and with the fact that genes for HSTs produced by *Alternaria alternata* appear to be on conditionally dispensable chromosomes (Johnson *et al.*, 2001).

X. CONCLUSIONS

Although much progress has been made toward understanding the genetic nature of the *Tox1* locus of C. *heterostrophus*, the evolutionary origin remains unknown 35 years after the emergence of race T. With respect to structure of the locus, the facts are as follows: (1) The *Tox1* locus is about 1.2 Mb of DNA distributed on two chromosomes reciprocally translocated with respect to their counterparts in race O. (2) T-toxin production requires at least six genes (*PKS1, PKS2, DEC1, RED2, RED3,* and *LAM1*) and possibly *RED1* and others (Asvarak, 2003; P. Inderbitzin and Turgeon, unpublished data), all of which map to the defined *Tox1* locus. (3) Known genes are scattered and embedded in highly A + T rich DNA. (4) Known genes have characteristics typical of other genes involved in secondary metabolism in fungi, they are discontinuously distributed and fast evolving. Phylogenetic analyses, in general, suggest a history of vertical transmission. There is no compelling evidence for horizontal transfer of the *Tox1* genes. (5) Genes mapping to the *Tox1* locus are not found in race O. Of the 25 *PKS* genes in C. *heterostrophus*, only *PKS1* and *PKS2* are unique to race T. (6) Genes affecting T-toxin production unlinked to *Tox1* have been identified (Lu, 1998). These genes are found in both races T and O. (7) The current hypothesis for assembly of the family of T-toxin molecules is that the backbone is built by a process in which the carbon chain product of one PKS acts as a starter for the second PKS (Baker *et al.*, 2006). (8) In the genome assembly of strain C4 race T DNA, known *Tox1* genes map to the smallest scaffolds, highlighting the fact that the region is difficult to sequence, due to the A + T rich, repeated nature of the regions surrounding known genes.

How will we know when we have all the genes for T-toxin production? Most may be in hand; however, it is likely that there are one or more additional genes required for reduction of the oxygenated groups on the carbon backbone. Definitive proof would be conversion of race O into race T.

Acknowledgments

The authors thank all current and former lab members for their thoughtful and lively input into the T-toxin project.

References

Abe, Y., Suzuki, J., Mizuno, T., Ono, C., Iwamoto, K., Hosobuchi, M., and Yoshikawa, H. (2002). Effect of increased dosage of the ML-236B (compactin) biosynthetic gene cluster on ML-236B production in *Penicillium citrinum*. *Mol. Genet. Genomics* **268**, 130–137.

Ahn, J. H., Cheng, Y. Q., and Walton, J. D. (2002). An extended physical map of the *TOX2* locus of *Cochliobolus carbonum* required for biosynthesis of HC-toxin. *Fungal Genet. Biol.* **35**, 31–38.

Akamatsu, H., Itoh, Y., Kodama, M., Otani, H., and Kohmoto, K. (1997). AAL-toxin-deficient mutants of *Alternaria alternata* tomato pathotype by restriction enzyme-mediated integration. *Phytopathology* **87**, 967–972.

Alcorn, J. L. (1975). Race-mating type associations in Australian populations of *Cochliobolus heterostrophus*. *Plant Dis. Rep.* **59**, 708–711.

Amberg, D. C., Botstein, D., and Beasley, E. M. (1995). Precise gene disruption in *Saccharomyces cerevisiae* by double fusion polymerase chain reaction. *Yeast* **11**, 1275–1280.

Arny, D. C., and Nelson, R. R. (1971). *Phyllosticta maydis* species nova, the incitant of Yellow Leaf Blight of maize. *Phytopathology* **61**, 1170–1172.

Arny, D. C., Worf, G. L., Ahrens, R. W., and Lindsey, M. F. (1970). Yellow Leaf Blight of maize in Wisconsin: Its history and the reactions of inbreds and crosses to the inciting fungus (*Phyllosticta* sp.). *Plant Dis. Rep.* **54**, 281–285.

Asvarak, T. (2003). Functional analysis of genes at the *Cochliobolus heterostrophus Tox1* locus and evaluation of a REMI mutant altered in conidium development. Ph.D. Thesis, Cornell University, Ithaca NY.

Baker, S., Kroken, S., Inderbitzin, P., Asvarak, T., Li, B.-Y., Shi, L., Yoder, O. C., and Turgeon, B. G. (2006). Two polyketide synthase-encoding genes are required for biosynthesis of the polyketide virulence factor, T-toxin, by *Cochliobolus heterostrophus*. *Mol. Plant Microbe Interact.* **19**, 139–149.

Berbee, M. L. (2001). The phylogeny of plant and animal pathogens in the Ascomycota. *Physiol. Mol. Plant Pathol.* **59**, 165–187.

Berbee, M. L., Sandhu, B. S., Bhatia, R. S., Singh, A., and Singh, G. (1996). Loculoascomycete origins and evolution of filamentous ascomycete morphology based on 18S rRNA gene sequence data. *Mol. Biol. Evol.* **13**, 462–470.

Berbee, M. L., Pirseyedi, M., and Hubbard, S. (1999). *Cochliobolus* phylogenetics and the origin of known, highly virulent pathogens, inferred from ITS and glyceraldehyde-3-phosphate dehydrogenase gene sequences. *Mycologia* **91**, 964–977.

Bhullar, B., Daly, J., and Rehfeld, D. (1975). Inhibition of dark carbon dioxide fixation and photosynthesis in leaf discs of corn susceptible to the host-specific toxin produced by *Helminthosporium maydis* race T. *Plant Physiol.* **56**, 1–7.

Boelker, M., Boehnert, H. U., Braun, K. H., Goerl, J., and Kahmann, R. (1995). Tagging pathogenicity genes in *Ustilago maydis* by restriction enzyme-mediated integration (REMI). *Mol. Gen. Genet.* **248**, 547–552.

Bronson, C. R. (1988). Ascospore abortion in crosses of *Cochliobolus heterostrophus* heterozygous for the virulence locus *Tox1*. *Genome* **30**, 12–18.

Bronson, C. R., Taga, M., and Yoder, O. C. (1990). Genetic control and distorted segregation of T-toxin production in field isolates of *Cochliobolus heterostrophus*. *Phytopathology* **80**, 819–823.

Catlett, N., Lee, B.-N., Yoder, O., and Turgeon, B. (2003). Split-marker recombination for efficient targeted deletion of fungal genes. *Fungal Genet. Newsl.* **50**, 9–11.

Chang, H. R., and Bronson, C. R. (1996). A reciprocal translocation and possible insertion(s) tightly associated with host-specific virulence in *Cochliobolus heterostrophus*. *Genome* **39**, 549–557.

Ciuffetti, L., Yoder, O., and Turgeon, B. G. (1992). A microbiological assay for host-specific fungal polyketide toxins. *Fungal Genet. Newsl.* **39**, 18–19.

Comstock, J., Martinson, C., and Gengenbach, B. (1972). Characteristics of a host-specific toxin produced by *Phyllosticta maydis*. *Phytopathology* **62**, 1107.

Comstock, J. C., Martinson, C. A., and Gengenbach, B. G. (1973). Host specificity of a toxin from *Phyllosticta maydis* for Texas cytoplasmically male sterile maize. *Phytopathology* **63**, 1357–1360.

Danko, S. J., Kono, Y., Daly, J. M., Suzuki, Y., Takeuchi, S., and McCrery, D. A. (1984). Structural and biological activity of a host-specific toxin produced by the fungal corn pathogen *Phyllosticta maydis*. *Biochemistry* **23**, 759–766.

Dasgupta, M. K. (1984). The Bengal famine, 1943 and the brown spot of rice—an inquiry into their relations. *Hist. Agric.* **2**, 1–18.

Debuchy, R., and Turgeon, B. G. (2006). Mating-type structure, evolution, and function in Euascomycetes. *In* "The Mycota" (U. Kues and R. Fischer, eds.), Vol. 1, pp. 293–324. Springer-Verlag.

Dewey, R. E., Siedow, J. N., Timothy, D. H., and Levings, C. S., III (1988). A 13-kilodalton maize mitochondrial protein in *E. coli* confers sensitivity to *Bipolaris maydis* toxin. *Science* **239**, 293–295.

Doyle, J. (1985). "Altered Harvest: Agriculture, Genetics, and the Fate of the World's Food Supply." Viking Penguin, NY, 502 p.

Drechsler, C. (1925). Leafspot of maize caused by *Ophiobolus heterostrophus* n. sp., the ascigerous stage of a *Helminthosporium* exhibiting bipolar germination. *J. Agric. Res.* **31**, 701–726.

Drechsler, C. (1927). An emendation of the description of *Ophiobolus heterostrophus*. *Phytopathology* **17**, 414.

Drechsler, C. (1934). Phytopathological and taxonomic aspects of *Ophiobolus*, *Pyrenophora*, *Helminthosporium*, and a new genus, *Cochliobolus*. *Phytopathology* **24**, 953–983.

Fairhead, C., Thierry, A., Denis, F., Eck, M., and Dujon, B. (1998). "Mass-murder" of ORFs from three regions of chromosome XI from *Saccharomyces cerevisiae*. *Gene* **223**, 33–46.

Forde, B. G., Oliver, R. J. C., and Leaver, C. J. (1978). Variation in mitochondrial translation products associated with male-sterile cytoplasms in maize. *Proc. Natl. Acad. Sci.* **75**, 3841–3845.

Gengenbach, B., Koeppe, D., and Miller, R. (1972). Mitochondrial reactions indicating the influence of kaempferol on corn blight toxin effects. *Plant Physiol.* **49**, 10.

Gracen, V., Forster, M., and Grogan, C. (1971). Reactions of corn (*Zea mays*) genotypes and cytoplasms to *Helminthosporium maydis* toxin. *Plant Dis. Rep.* **55**, 938–941.

Hitchman, T. S., Schmidt, E. W., Trail, F., Rarick, M. D., Linz, J. E., and Townsend, C. A. (2001). Hexanoate synthase, a specialized type I fatty acid synthase in aflatoxin B-1 biosynthesis. *Bioorg. Chem.* **29**, 293–307.

Hooker, A. L. (1974). Cytoplasmic susceptibility in plant disease. *Annu. Rev. Phytopathol.* **12**, 167–179.

Hooker, A. L., Smith, D., Lim, S., and Beckett, J. (1970a). Reaction of corn seedlings with male-sterile cytoplasm to *Helminthosporium maydis*. *Plant Dis. Rep.* **54**, 708–712.

Hooker, A. L., Smith, D. R., Lim, S. M., and Musson, M. D. (1970b). Physiological races of *Helminthosporium maydis* and disease resistance. *Plant Dis. Rep.* **54**, 1109–1110.

Hopwood, D. A. (1997). Genetic contributions to understanding polyketide synthases. *Chem. Rev.* **97**, 2465–2498.

Hopwood, D. A., and Sherman, D. H. (1990). Molecular genetics of polyketides and its comparison to fatty acid biosynthesis. *Annu. Rev. Genet.* **24**, 37–66.

Horwitz, B. A., Sharon, A., Lu, S. W., Ritter, V., Sandrock, T. M., Yoder, O. C., and Turgeon, B. G. (1999). A G protein alpha subunit from *Cochliobolus heterostrophus* involved in mating and appressorium formation. *Fungal Genet. Biol.* **26**, 19–32.

Huang, J., Lee, S. H., Lin, C., Medici, R., Hack, E., and Myers, A. M. (1990). Expression in yeast of the T-URF13 protein from Texas male-sterile maize mitochondria confers sensitivity to methomyl and to Texas-cytoplasm-specific fungal toxins. *EMBO J.* **9**, 339–347.

Johnson, L. J., Johnson, R. D., Akamatsu, H., Salamiah, A., Otani, H., Kohmoto, K., and Kodama, M. (2001). Spontaneous loss of a conditionally dispensable chromosome from the *Alternaria alternata* apple pathotype leads to loss of toxin production and pathogenicity. *Curr. Genet.* **40**, 65–72.

Karr, A., and Hsu, L. (1975). Host-specif toxins: *H. maydis*. *Plant Physiol.* **56**, 53.

Kennedy, J., Auclair, K., Kendrew, S. G., Park, C., Vederas, J. C., and Hutchinson, C. R. (1999). Modulation of polyketide synthase activity by accessory proteins during lovastatin biosynthesis. *Science* **284**, 1368–1372.

Kimura, M., Kamakura, T., Tao, Q. Z., Kaneko, I., and Yamaguchi, I. (1994). Cloning of the blasticidin S deaminase gene (BSD) from *Aspergillus terreus* and its use as a selectable marker for *Schizosaccharomyces pombe* and *Pyricularia oryzae*. *Mol. Gen. Genet.* **242**, 121–129.

Klittich, C. R. J., and Bronson, C. R. (1986). Reduced fitness associated with *Tox1* of *Cochliobolus heterostrophus*. *Phytopathology* **76**, 1294–1298.

Kodama, M., Rose, M. S., Yang, G., Yun, S. H., Yoder, O. C., and Turgeon, B. G. (1999). The translocation-associated *Tox1* locus of *Cochliobolus heterostrophus* is two genetic elements on two different chromosomes. *Genetics* **151**, 585–596.

Kono, Y., and Daly, J. M. (1979). Characterization of the host-specific pathotoxin produced by *Helminthosporium maydis* race T affecting corn with Texas male sterile cytoplasm. *Bioorg. Chem.* **8**, 391–397.

Kono, Y., Takeuchi, S., Kawarada, A., Daly, J. M., and Knoche, H. W. (1981). Studies on the host-specific pathotoxins produced in minor amounts by *Helminthosporium maydis* race T. *Bioorg. Chem.* **10**, 206–218.

Kono, Y., Danko, S. J., Suzuki, Y., Takeuchi, S., and Daly, J. M. (1983). Structure of the host-specific pathotoxins produced by *Phyllosticta maydis*. *Tetrahedron Lett.* **24**, 3803–3806.

Korth, K. L., and Levings, C. S. (1993). Baculovirus expression of the maize mitochondrial protein URF13 confers insecticidal activity in cell cultures and larvae. *Proc. Natl. Acad. Sci. USA* **90**, 3388–3392.

Kroken, S., Glass, N. L., Taylor, J. W., Yoder, O. C., and Turgeon, B. G. (2003). Phylogenomic analysis of type I polyketide synthase genes in pathogenic and saprobic ascomycetes. *Proc. Natl. Acad. Sci. USA* **100**, 15670–15675.

Leach, J., Lang, B. R., and Yoder, O. C. (1982). Methods for selection of mutants and *in vitro* culture of *Cochliobolus heterostrophus*. *J. Gen. Microbiol.* **128**, 1719–1729.

Leonard, K. J. (1973). Association of mating type and virulence in *Helminthosporium maydis*, and observations on the origin of the race T population in the United States. *Phytopathology* **63**, 112–115.

Leonard, K. J. (1977). Races of *Bipolaris maydis* in the Southeastern US from 1974–1976. *Plant Dis. Rep.* **61**, 914–915.

Levings, C. S. (1990). The Texas cytoplasm of maize: Cytoplasmic male sterility and disease susceptibility. *Science* **250**, 942–947.

Levings, C. S., and Siedow, J. N. (1992). Molecular basis of disease susceptibility in the Texas cytoplasm of maize. *Plant Mol. Biol.* **19**, 135–147.

Levings, C. S. I., Rhoads, D. M., and Siedow, J. N. (1995). Molecular interactions of *Bipolaris maydis* T-toxin and maize. *Can. J. Bot.* **73**, S483–S489.

Li, B.-Y., Kwan, W. K., Turgeon, G. B., Wu, J., Wang, X., Li, E., Zhu, T., and Shi, L. (2002). Analysis of differential gene expression by ligation specificity-based transcript profiling. *OMICS* **6**, 175–185.

Lim, S., and Hooker, A. (1971). Southern Corn Leaf Blight: Genetic control of pathogenicity and toxin production in race T and race O of *Cochliobolus heterostrophus*. *Genetics* **69**, 115–117.

Lim, S., and Hooker, A. (1972a). A preliminary characterization of *Helminthosporium maydis* Toxins. *Plant Dis. Rep.* **56**, 805–807.

Lim, S., and Hooker, A. (1972b). Disease determinant of *Helminthosporium maydis* Race T. *Phytopathology* **62**, 968–971.

Liu, Y. G., and Whittier, F. (1995). Thermal asymmetric interlaced PCR: Automatable amplification and sequencing of insert end fragment from P1 and YAC clones for chromosome walking. *Genomics* **25**, 674–681.

Lu, S. (1998). Molecular genetic analysis of general and specific pathogenesis factors in *Cochliobolus heterostrophus*. Ph.D. Thesis, Cornell University.

Lu, S. W., Lyngholm, L., Yang, G., Bronson, C., Yoder, O. C., and Turgeon, B. G. (1994). Tagged mutations at the *Tox1* locus of *Cochliobolus heterostrophus* using restriction enzyme-mediated integration. *Proc. Natl. Acad. Sci. USA* **91**, 12649–12653.

Mercado, A. C., and Lantican, R. M. (1961). The susceptibility of cytoplasmic male-sterile lines of corn to *Helminthosporium maydis*. *Philipp. Agric.* **45**, 235–243.

Miller, P., Wallin, J., and Hyre, R. (1970). Plans for forecasting corn blight epidemics. *Plant Dis. Rep.* **54**, 1134–1136.

Miller, R. J., and Koeppe, D. E. (1971). Southern Corn Leaf Blight: Susceptible and resistant mitochondria. *Science* **173**, 67–69.

Moore, W. (1970). Origin and spread of Southern Corn Leaf Blight in 1970. *Plant Dis. Rep.* **54**, 1104–1108.

Mukunya, D. M., and Boothroyd, C. W. (1973). *Mycosphaerella zeae-maydis* sp. n., the sexual stage of *Phyllosticta maydis*. *Phytopathology* **63**, 529–532.

Nelson, R. (1957). A major gene locus for compatibility in *Cochliobolus heterostrophus*. *Phytopathology* **47**, 742–743.

Orillo, F. (1952). Leaf spot of maize caused by *Helminthosporium maydis*. *Philipp. Agric.* **36**, 327–392.

Punithalingam, E. (1990). CMI descriptions of fungi and bacteria No. 1015: *Mycosphaerella zeae-maydis*. *Mycopathologia* **112**, 49–50.

Rose, M. R. (1996). Molecular genetics of polyketide toxin production in *Cochliobolus heterostrophus*. Ph.D. Thesis, Cornell University.

Rose, M. S., Yoder, O. C., and Turgeon, B. G. (1996). A decarboxylase required for polyketide toxin production and high virulence by *Cochliobolus heterostrophus*. *In* "Eighth International Symposium on Molecular Plant–Microbe Interactions," p. J-49. Knoxville, TN.

Rose, M. S., Yun, S. H., Asvarak, T., Lu, S. W., Yoder, O. C., and Turgeon, B. G. (2002). A decarboxylase encoded at the *Cochliobolus heterostrophus* translocation-associated *Tox1B* locus is required for polyketide biosynthesis and high virulence on maize. *Mol. Plant Microbe Interact.* **15,** 883–893.

Rutherford, K., Parkhill, J., Crook, J., Horsnell, T., Rice, P., Rajandream, M.-A., and Barrell, B. (2000). Artemis: Sequence visualisation and annotation. *Bioinformatics* **16,** 944–945.

Scheifele, G., Whitehead, W., and Rowe, C. (1970). Increased susceptibility to Southern leaf spot (*Helminthosporium maydis*) in inbred lines and hybrids of maize with Texas male-sterile cytoplasm. *Plant Dis. Rep.* **54,** 501–503.

Scheifele, G. L., and Nelson, R. R. (1969). The occurrence of *Phyllosticta* leaf spot of corn in Pennsylvania. *Plant Dis. Rep.* **53,** 186–189.

Sivanesan, A. (1984). "The Bitunicate Ascomycetes and Their Anamorphs." Strauss & Cramer, Hirschberg.

Smith, D., Hooker, A., and Lim, S. (1970). Physiologic races of *Helminthosporium maydis*. *Plant Dis. Rep.* **54,** 819–822.

Staunton, J., and Weissman, K. J. (2001). Polyketide biosynthesis: A millennium review. *Nat. Prod. Rep.* **18,** 380–416.

Straubinger, B., Straubinger, E., Wirsel, S., Turgeon, B. G., and Yoder, O. C. (1992). Versatile fungal transformation vectors carrying the selectable marker *bar* gene of *Streptomyces hygroscopicus*. *Fungal Gent. Newsl.* **39,** 82–83.

Suzuki, Y., Knoche, H. W., and Daly, J. M. (1982). Analogs of host-specific phytotoxin produced by *Helminthosporium maydis* race T. I. Synthesis. *Bioorg. Chem.* **11,** 300–312.

Suzuki, Y., Danko, S. J., Daly, J. M., Kono, Y., Knoche, H. W., and Takeuchi, S. (1983). Comparison of activities of the host-specific toxin of *Helminthosporium maydis* race T and a synthetic C-41 analog. *Plant Physiol.* **73,** 440–444.

Sweigard, J. (1996). A REMI primer for filamentous fungi. International Society for Molecular Plant-Microbe Interactions. *IS-MPMI Reporter Spring* 3–5.

Taga, M., Bronson, C. R., and Yoder, O. C. (1985). Nonrandom abortion of ascospores containing alternate alleles at the *Tox1* locus of the fungal plant pathogen *Cochliobolus heterostrophus*. *Can. J. Genet. Cytol.* **27,** 450–456.

Tasma, I. M., and Bronson, C. R. (1998). Genetic mapping of telomeric DNA sequences in the maize pathogen *Cochliobolus heterostrophus*. *Curr. Genet.* **34,** 227–233.

Tegtmeier, K. J., Daly, J. M., and Yoder, O. C. (1982). T-toxin production by near-isogenic isolates of *Cochliobolus heterostrophus* races T and O. *Phytopathology* **72,** 1492–1495.

Turgeon, B. G. (1998). Application of mating type gene technology to problems in fungal biology. *Annu. Rev. Phytopathol.* **36,** 115–137.

Turgeon, B. G., and Berbee, M. L. (1998). Evolution of pathogenic and reproductive strategies in *Cochliobolus* and related genera. *In* "Molecular Genetics of Host-Specific Toxins in Plant Disease" (K. Kohmoto and O. C. Yoder, eds.), Vol. 13, pp. 153–163. Kluwer, Dordrecht.

Turgeon, B. G., and Lu, S.-W. (2000). Evolution of host specific virulence in *Cochliobolus heterostrophus*. *In* "Fungal Pathology" (J. W. Kronstad, ed.), pp. 93–126. Kluwer, Dordrecht, The Netherlands.

Turgeon, B. G., Garber, R. C., and Yoder, O. C. (1985). Transformation of the fungal maize pathogen *Cochliobolus heterostrophus* using the *Aspergillus nidulans amdS* gene. *Mol. Gen. Genet.* **201,** 450–453.

Turgeon, B. G., Garber, R. C., and Yoder, O. C. (1987). Development of a fungal transformation system based on selection of sequences with promoter activity. *Mol. Cell. Biol.* **7,** 3297–3305.

Turgeon, B. G., Bohlmann, H., Ciuffetti, L. M., Christiansen, S. K., Yang, G., Schafer, W., and Yoder, O. C. (1993). Cloning and analysis of the mating type genes from *Cochliobolus heterostrophus*. *Mol. Gen. Genet.* **238,** 270–284.

Turgeon, B. G., Kodama, M., Yang, G., Rose, M. S., Lu, S. W., and Yoder, O. C. (1995). Function and chromosomal location of the *Cochliobolus heterostrophus Tox1* locus. *Can. J. Bot.* **73,** S1071–S1076.

Tzeng, T. H., Lyngholm, L. K., Ford, C. F., and Bronson, C. R. (1992). A restriction fragment length polymorphism map and electrophoretic karyotype of the fungal maize pathogen *Cochliobolus heterostrophus*. *Genetics* **130,** 81–96.

Ullstrup, A. J. (1970). History of Southern Corn Leaf Blight. *Plant Dis. Reptr.* **54,** 1100–1102.

Ullstrup, A. J. (1972). The impacts of the Southern Corn Leaf Blight epidemics of 1970–1971. *Annu. Rev. Phytopathol.* **10,** 37–50.

von Arx, J. A. (1987). *Didymella Sacc.,* Plant Pathogenic Fungi. *Beihefte zur Nova Hedwigia* J. Cramer, Berlin, 52.

Vonallmen, J. M., Rottmann, W. H., Gengenbach, B. G., Harvey, A. J., and Lonsdale, D. M. (1991). Transfer of methomyl and HmT-toxin sensitivity from T-cytoplasm maize to tobacco. *Mol. Gen. Genet.* **229,** 405–412.

Walton, J. D. (1996). Host-selective toxins: Agents of compatibility. *Plant Cell* **8,** 1723–1733.

Walton, J. D. (2000). Horizontal gene transfer and the evolution of secondary metabolite gene clusters in fungi: An hypothesis. *Fungal Genet. Biol.* **30,** 167–171.

Wheeler, H. (1977). Ultrastructure of penetration by *Helminthosporium maydis. Physiol. Plant Pathol.* **11,** 171–178.

Wirsel, S., Turgeon, B. G., and Yoder, O. C. (1996). Deletion of the *Cochliobolus heterostrophus* mating type (*MAT*) locus promotes function of *MAT* transgenes. *Curr. Genet.* **29,** 241–249.

Wise, R. P., Bronson, C. R., Schnable, P. S., and Horner, H. R. (1999). The genetics, pathology, and molecular biology of T-cytoplasm male sterility in maize. *Adv. Agron.* **65,** 79–130.

Wolpert, T. J., Dunkle, L. D., and Ciuffetti, L. M. (2002). Host-selective toxins and avirulence determinants: What's in a name? *Annu. Rev. Phytopathol.* **40,** 251–285.

Yabe, K., and Nakajima, H. (2004). Enzyme reactions and genes in aflatoxin biosynthesis. *Appl. Microbiol. Biotechnol.* **64,** 745–755.

Yang, G., Turgeon, B. G., and Yoder, O. C. (1994). Toxin-deficient mutants from a toxin-sensitive transformant of *Cochliobolus heterostrophus. Genetics* **137,** 751–757.

Yang, G., Rose, M. S., Turgeon, B. G., and Yoder, O. C. (1996). A polyketide synthase is required for fungal virulence and production of the polyketide T-toxin. *Plant Cell* **11,** 2139–2150.

Yoder, O., and Mukunya, D. (1972). A host-specific toxic metabolite produced by *Phyllosticta maydis. Phytopathology* **62,** 799.

Yoder, O., Payne, G., Gregory, P., and Earle, E. (1976). Relative sensitivities of bioassays for *Helminthosporium maydis* race T toxin. *Proc. Amer. Phytopath. Soc.* **3,** 281.

Yoder, O. C. (1973). A selective toxin produced by *Phyllosticta maydis. Phytopathology* **63,** 1361–1365.

Yoder, O. C. (1976). Evaluation of the role of *Helminthosporium maydis* race T toxin in Southern Corn Leaf Blight. *In* "Biochemistry and Cytology of Plant Parasite Interaction" (K. Tomiyama, J. M. Daly, I. Uritani, H. Oku, and S. Ouchi, eds.), pp. 16–24. Elsevier, New York.

Yoder, O. C. (1980). Toxins in pathogenesis. *Annu. Rev. Phytopathol.* **18,** 103–129.

Yoder, O. C., and Gracen, V. E. (1975). Segregation of pathogenicity types and host-specific toxin production in progenies of crosses between races T and O of *Helminthosporium maydis* (*Cochliobolus heterostrophus*). *Phytopathology* **65,** 273–276.

Yoder, O. C., and Gracen, V. E. (1977). Evaluation of a chemical method for assay of *Helminthosporium maydis* race T toxin. *Plant Physiol.* **59,** 792–794.

Yoder, O. C., Payne, G. A., Gregory, P., and Gracen, V. E. (1977). Bioassays for detection and quantification of *Helminthosporium maydis* race T toxin: A comparison. *Physiol. Plant Pathol.* **10,** 237–245.

Yoder, O. C., Valent, B., and Chumley, F. (1986). Genetic nomenclature and practice for plant pathogenic fungi. *Phytopathology* **76**, 383–385.

Yoder, O. C., Yang, G., Rose, M. S., Lu, S. W., and Turgeon, B. G. (1994). Complex genetic control of polyketide toxin production by *Cochliobolus heterostrophus*. *In* "Advances in the Molecular Genetics of Plant–Microbe Interaction" (M. J. Daniels, J. A. Downie, and A. E. Osbourn, eds.), Vol. 3, pp. 223–230. Kluwer, Dordrecht.

Yoder, O. C., Macko, V., Wolpert, T. J., and Turgeon, B. G. (1997). *Cochliobolus* spp. and their host-specific toxins. *In* "The Mycota, Vol. 5: Plant Relationships, Part A" (G. Carroll and P. Tudzynski, eds.), Vol. 5, pp. 145–166. Springer-Verlag, Berlin.

Yu, T. (1933). Studies on *Helminthosporium* leaf spot of maize. *Sinensia* **3**, 273–318.

Yun, S. H. (1998). Molecular genetics and manipulation of pathogenicity and mating determinants in *Cochliobolus heterostrophus* and *Mycosphaerella zeae-maydis*. Ph.D. Thesis, Cornell University, Ithaca, NY.

Yun, S. H., Turgeon, B. G., and Yoder, O. C. (1998). REMI-induced mutants of *Mycosphaerella zeae-maydis* lacking the polyketide PM-toxin are deficient in pathogenesis to corn. *Physiol. Mol. Plant Pathol.* **52**, 53–66.

Zhu, X. (1999). Molecular analysis of loci controlling T-toxin biosynthesis in *Cochliobolus heterostrophus*. Ph.D. Thesis, Cornell University, Ithaca, NY.

Fungal Genomics: A Tool to Explore Central Metabolism of *Aspergillus fumigatus* and Its Role in Virulence

Taylor Schoberle and Gregory S. May
Division of Pathology and Laboratory Medicine
The University of Texas M. D. Anderson Cancer Center
Houston, Texas 77030

ABSTRACT

Aspergillus fumigatus is an opportunistic pathogenic fungus that primarily infects neutropenic animal hosts. This fungus is found throughout the world, can utilize a wide range of substrates for carbon and nitrogen sources, and is capable of growing at elevated temperatures. The ability to grow at high temperatures and utilize a range of nutrient substrates for growth potentially contributes to this being the number one human pathogenic mold worldwide. The recently completed genome sequence for this fungus creates an opportunity to examine how central metabolic pathways and their regulation contribute to pathogenesis. A review of the existing literature illustrates that genes involved in the biosynthesis of key nutrients are essential for pathogenesis in *A. fumigatus*. In addition, nutrient sensing and regulation of biosynthetic pathways also contribute to

Advances in Genetics, Vol. 57
0065-2660/07 $35.00
DOI: 10.1016/S0065-2660(06)57007-5

fungal pathogenesis. The advent of improved methods for manipulating the genome of A. *fumigatus*, along with the completed genome sequence, now make it feasible to investigate the role of all metabolic pathways and control of these pathways in fungal virulence.

I. INTRODUCTION

Maintenance of a constant intracellular environment is essential for viability of any organism and its competitiveness with other organisms in its ecological niche. Thus, a wide range of regulatory mechanisms has evolved to control cellular homeostasis in response to nutrient availability and environmental change. Among these systems are those that contribute to carbon or nitrogen source utilization, ambient pH response, and amino acid biosynthesis. Additional pathways also regulate cell growth in response to the availability of specific nutrients such as iron, a nutrient essential for microbial growth that is frequently limiting.

Aspergillus fumigatus is a saprophytic fungus that is involved primarily in the degradation of plant material. A. *fumigatus* is also able to grow at elevated temperatures, making it a dominant organism during high-temperature composting. A. *fumigatus* is also the most common cause of invasive mold infections in humans. While this fungus is the cause of significant morbidity and mortality, it is primarily the cause of infections in immunocompromised hosts. Invasive disease caused by A. *fumigatus* has a high mortality rate in the absence of antifungal drug therapy and even with treatment mortality still remains at least 50% (Brakhage and Langfelder, 2002; Krappmann et al., 2004; Latge, 1999; Liebmann et al., 2004a; Liebmann et al., 2004b; Sheppard et al., 2005; Wasylnka and Moore, 2003). In addition to being thermotolerant, A. *fumigatus* is able to use a wide array of carbon and nitrogen sources. The combination of nutritional versatility and the ability to grow at elevated temperatures have been cited as possible virulence traits (Bhabhra et al., 2004; Brown et al., 2000; D'Enfert et al., 1996; Krappmann et al., 2004; Liebmann et al., 2004a; Liebmann et al., 2004b; Panepinto et al., 2003; Sandhu et al., 1976). The recent completion of the genome sequence for A. *fumigatus* and other species provides new avenues by which to investigate the genetic contributions to the virulence traits of this important fungus (Galagan et al., 2005; Machida et al., 2005; Nierman et al., 2005). It is the goal of this chapter to assess how metabolic pathways and their regulation contribute to virulence in A. *fumigatus*. Some of these pathways may also represent targets for the development of novel antifungal therapies.

II. NUTRITIONAL AUXOTROPHY AND FUNGAL GENETICS

The ability to respond to changes in the availability of nutrients is an essential attribute for a number of successful pathogens. Some host environments may lack essential nutrients required for the survival of these pathogens. Other host systems may produce these nutrients, but have areas throughout the body where the nutrients are not readily available. Adapting to this kind of nutritionally unfavorable environment is crucial for the *in vivo* metabolism and proliferation of any disease causing organism. Some studies have been able to link nutrient-regulated signaling pathways to pathogenesis in many species of fungi, yet the relevance of these pathways in the virulence of A. *fumigatus* is largely unexplored (Panepinto *et al.*, 2003). Nutrient biosynthesis in bacterial pathogens has been studied extensively. Loss of biosynthetic ability in many of these organisms has been shown to have an attenuating effect on their virulence (Sandhu *et al.*, 1976). So why have not more biosynthesis studies involving pathogenic fungi been done?

The *pabaA* gene, which is involved in the folate synthesis pathway (Fig. 7.1), encodes *p*-aminobenzoic acid synthase (Brown *et al.*, 2000; Liebmann *et al.*, 2004). Disruption of this gene causes mutants to be auxotrophic for *p*-aminobenzoic acid (Brown *et al.*, 2000; D'Enfert *et al.*, 1996; Krappmann *et al.*, 2004; Liebmann *et al.*, 2004a; Sandhu *et al.*, 1976; Tang *et al.*, 1994). PABA-requiring mutants show a complete loss of pathogenicity in A. *fumigatus*, as well as A. *nidulans*, in murine infection models of invasive pulmonary aspergillosis (Sandhu *et al.*, 1976; Tang *et al.*, 1994). The virulence of these *pabaA* deletion strains can be restored by adding *p*-aminobenzoic acid to the drinking water of animals (D'Enfert *et al.*, 1996; Sandhu *et al.*, 1976; Tang *et al.*, 1994). The importance of the folate synthesis pathway for *in vivo* survival of A. *fumigatus* can be confirmed by the inability of *pabaA* deletion strains to cause lethal infections in murine models.

Does this avirulence prove that a fungal strain, such as A. *nidulans*, truly requires PABA to colonize and germinate or could it be that PABA is only required for fungal germination in airways and for initiation of infection? If this is the case and PABA biosynthesis is not actually crucial for fungal growth beyond this initial stage, inhibitors of this pathway might prove to be ineffective in the treatment of an established infection. Tang *et al.* (1994) demonstrated that PABA is not only required for initial germination, but is also needed once the infection has been established. These researchers inoculated a group of animals with a *pabaA* deletion strain, and provided them with PABA supplementation for 3 days postinoculation. Histological examination of lung tissue at this stage showed germination of mutant conidia, along with extension of hyphae into the lung parenchyma. Once the PABA supplementation that had been provided for the first 3 days was withdrawn, all signs of infection in the

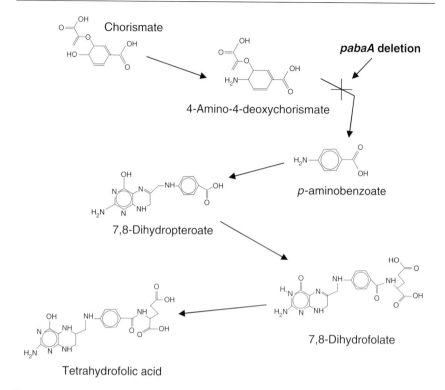

Figure 7.1. Folate biosynthesis pathway of *Saccharomyces cerevisiae* (Hong et al., 1994). There are two branches to this pathway, one beginning with chorismate (pictured) and the other beginning with GTP (not pictured). The step disrupted by deletion of *pabaA* in *A. fumigatus* is indicated.

animals disappeared. This experimental design demonstrates that the requirement of PABA for growth and survival does indeed extend beyond the initial stages of germination.

This same trend is also seen for the uridine monophosphate (UMP) biosynthesis pathway (Fig. 7.2). The *pyrG* deletion mutant of *A. fumigatus* is defective in orotidine-5′-monophosphate decarboxylase and lacks the ability to germinate in the lung of infected mice (D'Enfert *et al.*, 1996; Krappmann *et al.*, 2004; Weidner *et al.*, 1998). Histological examinations from studies by D'Enfert *et al.* (1996) have revealed that wild-type *A. fumigatus* conidiospores are able to achieve a significant rate of germination within the first 24 h following inhalational challenge, while mutant conidia tend to remain almost entirely ungerminated even after 42 h of infection. This UMP-requiring mutant is consequently nonpathogenic, indicating that uridine/uracil is limited in the lung of a mouse. Supplementation of uridine/uracil in the drinking water restores virulence, as seen for *pabaA* mutants.

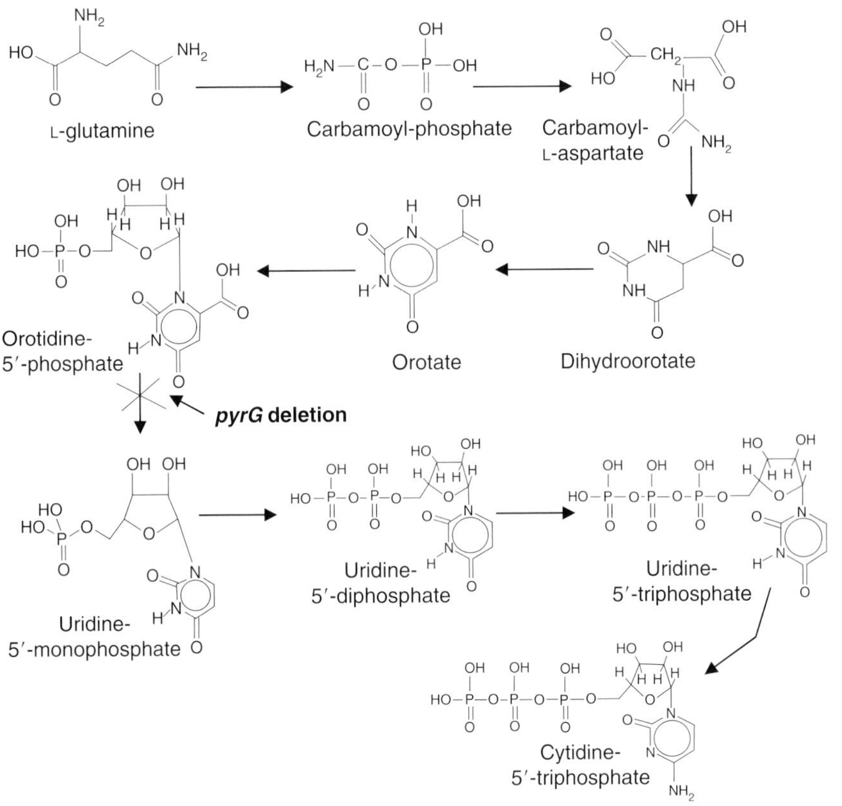

Figure 7.2. UMP biosynthesis pathway of *Saccharomyces cerevisiae* (Hong *et al.*, 1994). This pathway is one branch of a network of pathways leading to histidine, purine, and pyrimidine biosynthesis. The step disrupted by the *pyrG* deletion mutant of *A. fumigatus* is marked.

Being able to vary the level of available uridine in mouse models, thereby controlling virulence of *A. fumigatus pyrG* mutants, could serve as a powerful tool to understand the interactions of the conidia with host cells either *in vivo* or in cellular assays that mimic some steps of invasive aspergillosis. These types of mutants could also be used for *in vitro* systems as selectable markers for transformation experiments. This type of nutritional marker, which is quickly becoming a common selectable trait in transformations, overcomes some of the problems associated with use of dominant antibiotic selective markers (Weidner *et al.*, 1998). Some strains can easily become contaminated with antibiotic resistant strains, while some other species are just naturally resistant to antifungal antibiotics.

    Aspergillus oryzae has been reported to be resistant to antifungal antibiotics (Jin *et al.*, 2004). This poses a problem for *A. oryzae* transformation systems, since antifungal antibiotics are commonly used as selectable markers (Weidner *et al.*, 1998). Over the years, many auxotrophic and dominant selectable markers have been developed to overcome this obstacle (Jin *et al.*, 2004; Weidner *et al.*, 1998). To use these markers as selective tools, scientists must use a positive selection method using reagents that allow the growth of only those strains deficient in specific metabolic processes or a filtration method to remove wild-type mycelia grown in liquid media (Jin *et al.*, 2004). For some researchers, these methods can be laborious. By creating a mutant of A. *oryzae* that is auxotrophic for adenine, Jin and colleagues (2004) have developed a more convenient way of screening for transformants. Two mutants were created lacking either the *adeA* or *adeB* gene of the purine biosynthetic pathway. The colonies that grow from these mutants have a distinct red pigment due to the polymerization and oxidation of an accumulated intermediate, 4-amino-imidazole ribotide. This intermediate accumulates because of a blockade in the adenine biosynthetic pathway. Virulence studies have not been done on A. *oryzae* strains auxotrophic for adenine, but this nutritional biosynthetic mutation does still serve an important purpose as a model system to further study the effects of mutants auxotrophic for other essential nutrients. Along with A. *oryzae*, researchers have found it very difficult to screen for transformants in other fungal species. Using this adenine-requiring mutant with these difficult strains of fungi can provide researchers with an easier method of screening.

Being nutritionally prototrophic is advantageous to invading pathogens. This has been shown to greatly affect virulence because if the organism cannot grow, it is unable to overcome a host's immune defenses (D'Enfert *et al.*, 1996; Liebmann *et al.*, 2004a; Sandhu *et al.*, 1976; Tang *et al.*, 1994). The importance of these nutritional biosynthetic pathways could lead to potential targets for new antifungal drugs. These nutritional genes can also serve as selectable markers in transformation experiments for many fungal species. Mutants lacking *pyrG* are commonly used in transformation systems involving A. *fumigatus* and other species (Weidner *et al.*, 1998). This concept of nutritional markers can be especially useful for certain species, such as A. *oryzae*, that are resistant to common dominant selectable antibiotic markers used in other species (Jin *et al.*, 2004). Development of molecular techniques that will make fungi amenable to genetic analyses could ultimately contribute to the search for novel antifungal medications.

III. REGULATION OF AMINO ACID BIOSYNTHESIS

All free-living microbes need to respond to changing nutrient availability in the environment in order to maintain proper cellular physiology. As a result, regulatory networks have evolved that control changes in gene expression that regulate metabolic pathways and maintain a relatively constant cellular physiology. In order for a pathogenic fungal species to successfully infect a human host, it must have the ability to produce certain nutrients that are not readily available. There are eight amino acids that cannot be produced by the human body. These essential amino acids are tryptophan, lysine, methionine, phenylalalanine, threonine, valine, leucine, and isoleucine.

Fungal mutants auxotrophic for essential nutrients, such as uridine/uracil, p-aminobenzoic acid, and adenine, have been shown to have reduced virulence or in some cases a complete loss of virulence (Krappmann et al., 2004; Liebmann et al., 2004a; Tang et al., 1994). Since nutritional auxotrophy affects virulence, one might assume that amino acid auxotrophs would show the same trend. So far, studies have only shown that the lysine biosynthetic pathway plays a role in fungal virulence (Liebmann et al., 2004a). There are seven more essential amino acids humans cannot produce, yet this topic of amino acid auxotrophy does not get the recognition it deserves.

Lysine is one of the essential amino acids that fungi synthesize via the α-aminoadipate pathway (Fig. 7.3) (Garrad and Bhattacharjee, 1992; Liebmann et al., 2004a). Since humans lack the ability to produce this amino acid, the enzymes required for lysine biosynthesis are nonexistent in humans (Liebmann et al., 2004a). There have been at least seven biosynthesis enzymes identified, five of which have been found to be active in wild-type cells of A. fumigatus. These five enzymes, homocitrate synthase, homoisocitrate dehydrogenase, α-aminoadipate reductase, saccharopine reductase, and saccharopine dehydrogenase, are all part of the α-aminoadipate pathway (Garrad and Bhattacharjee, 1992; Liebmann et al., 2004a). In A. fumigatus, lysF encodes homoaconitase, which activates the conversion of homoaconitate to homoisocitrate. Deletion of lysF disrupts the α-aminoadipate pathway, leading to auxotrophy of lysine (Liebmann et al., 2004a). Liebmann et al. (2004a) reported that lysF mutants of A. fumigatus have reduced virulence in low-dose murine infection models of invasive aspergillosis.

Interestingly, another group, Tang et al. (1994), published data showing that mutants lacking lysA2 of A. nidulans, encoding saccharopine dehydrogenase, have a slight reduction in virulence, but this reduction is not statistically significant. What would cause this kind of inconsistency? They both used the same technique when inoculating the mice with Aspergillus conidia, which was described by Smith et al. (1994). The genes used by each study are located at different points in the lysine biosynthetic pathway. LysF acts upstream of the α-aminoadipate branch point, while LysA serves as a catalyst of the final step of the lysine-specific biosynthesis branch. Alternatively, these two studies could reflect differences between A. fumigatus and A. nidulans. It is true that A. fumigatus causes more cases of infectious pulmonary aspergillosis than A. nidulans. Furthermore, the number of conidia from A. nidulans required to

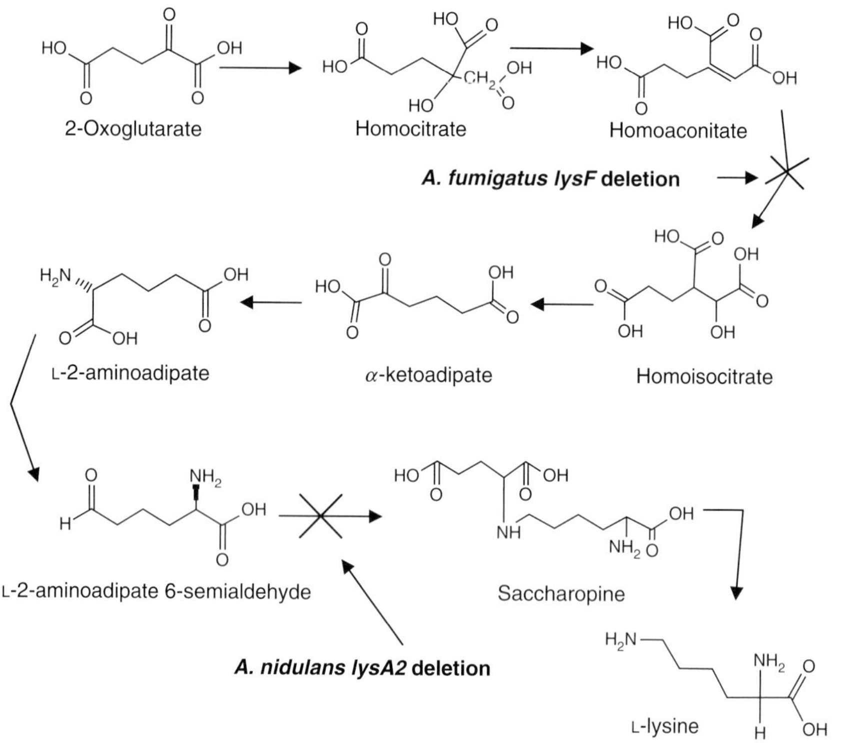

Figure 7.3. Lysine biosynthesis α-amino adipate pathway of *Saccharomyces cerevisiae* (Hong et al., 1994). The steps disrupted by the *lysF* deletion mutant of A. *fumigatus* and the *lysA2* deletion mutant of A. *nidulans* are noted.

cause infection in neutropenic mice is at least 1000-times higher than for A. *fumigatus* (Liebmann et al., 2004a). This is reflected in both papers, since 2×10^6 conidia from A. *nidulans* were needed to properly infect neutropenic mice, while only 5×10^3 conidia from A. *fumigatus* were used. Also, both groups used different strains of mice, which lead to different concentrations of antibiotics and other supplements to render the mice neutropenic (Liebmann et al., 2004a; Tang et al., 1994). CD1 mice were used by Tang et al. (1994), while Liebmann et al. (2004a) used BALB/c mice. Both groups also did mixed-inoculum studies, which supported the results each group published. Thus, there are a number of reasons why both groups came up with conflicting results as to the effect of lysine auxtrophy on fungal virulence.

Tryptophan is another essential amino acid that fungi are able to synthesize. There are four genes that have been characterized in A. *nidulans* encoding enzymes within this biosynthetic pathway, that is, *trpA*, *trpB*, *trpC*, and *trpD* (Eckert *et al.*, 1999). Eckert *et al.* (1999) demonstrated that deletion of any of these four genes leads to a dependence of the mutant strain on exogenously supplied tryptophan for growth and differentiation. These deletion strains were not tested in relation to virulence in murine models, but this auxotrophic strain does exhibit the same characteristics of the lysine-dependent mutants, and therefore would be ideal candidates for virulence studies.

The cross-pathway control system is a global regulatory system modulating fungal amino acid biosynthesis. Although this cross-pathway control system has been intensively studied in the baker's yeast *Saccharomyces cerevisiae*, the regulation of this pathway in species of *Aspergillus* has been studied to a lesser extent (Hoffmann *et al.*, 2001; Krappmann *et al.*, 2004). A study by Hoffmann *et al.* (2001) on the *cpcA* gene in *A. nidulans* demonstrated that on amino acid starvation, CpcA protein levels are dramatically increased to activate transcription of a large number of target genes. In *Saccharomyces cerevisiae*, these target genes encode proteins involved in purine biosynthesis, aminoacyl-tRNA synthetases, and most importantly, amino acid biosynthetic enzymes. Indeed, *cpcA* mutants of *A. nidulans* showed severely retarded growth rates when exposed to histidine or tryptophan starvation conditions compared to wild-type strains. This sensitivity could be reversed by the addition of the respective amino acid to the growth medium. Further studies indicated that mRNA and protein levels of the transcriptional regulator remained at or near basal levels in amino acid starvation conditions.

Since *A. fumigatus* mutants auxotrophic for lysine have been reported to show attenuated virulence in murine models (Liebmann *et al.*, 2004a), this cross-pathway control system seems like a very promising candidate for further *A. fumigatus* virulence studies. Krappmann *et al.* (2004) tested this hypothesis by disrupting the *cpcA* gene encoding the transcriptional activator of the cross-pathway control system in *A. fumigatus*. The *cpcA* mutant showed the same phenotype as the mutant from *A. nidulans* in the presence of amino acid starvation. When the *A. fumigatus cpcA* mutant was compared to a wild-type and a restored strain in a neutropenic murine model of invasive pulmonary aspergillosis, reduced virulence could be seen with the deletion strain. The authors noted that on histological inspection of lung tissue, there were no differences with respect to the extent of growth or invasiveness of fungal foci, suggesting that an auxotrophic factor is not responsible for this reduced virulence.

If the *cpcA* deletion mutants grow *in vivo*, then why do they have reduced virulence? The authors suggest the pattern seen in these murine models could be attributed to a requirement of CpcA for basal transcription of a virulence-determining gene in *A. fumigatus* (Hoffmann *et al.*, 2001; Krappmann *et al.*, 2004). Perhaps a failure to balance amino acid concentrations and respond to stressful environments has an effect on the ability of the fungal pathogen to actually penetrate the lung tissue and create disease. Even though some studies have reported *cpcA* to be responsible for biosynthetic pathways other than amino acid biosynthesis, there has not been any evidence that the cross-pathway control response is induced by any stress conditions other than amino acid starvation (Krappmann *et al.*, 2004). Deletion of *cpcA* from *A. fumigatus* does not leave the fungus auxotrophic for amino acids, but rather the major role of the cross-pathway control system seems to be in sensing and responding to environmental stress conditions. Because this cross-pathway control system has been shown to be highly conserved among fungi, further studies into the mechanism by which CpcA controls virulence could prove to be quite valuable (Hoffmann *et al.*, 2001; Krappmann *et al.*, 2004).

A. *fumigatus* as well as other species have demonstrated a strong versatility in regards to adapting to the nutritional needs of the surrounding environment. Fungal strains defective in the lysine biosynthetic pathway and the cross-pathway control system have proven the need of amino acid biosynthesis for survival (Hoffmann *et al.*, 2001; Krappmann *et al.*, 2004; Liebmann *et al.*, 2004a). There are seven more essential amino acids that cannot be produced by humans. It seems that the ability to synthesize these amino acids would be crucial for fungal pathogens to adapt to such a harsh environment, since humans lack the essential amino acid biosynthetic machinery. Further research into what kind of effect these other essential amino acids have on fungal pathogenesis would be relevant to the search for new and improved ways of protecting ourselves against these pathogens, especially for immunocompromised patients.

IV. REGULATION OF AMBIENT pH RESPONSE

In addition to balancing vital nutrients to maintain cell viability, fungal pathogens must also be able to adapt to pH levels present in a host's environment. On colonization of lung tissue, or other organs, cells of the immune system are alerted of the invasion and travel to the site of infection to fight off the intruders. Foreign pathogens entering the human body first come into contact with phagocytic cells such as neutrophils, macrophages, and dendritic cells. On arrival, the phagocytic cells engulf the foreign pathogens. Inside these cells of the immune system, vacuoles containing highly acidic components are used to denature and destroy the invading organisms. Fungal pathogens wishing to survive these attacks must respond to the acidic conditions to which they are exposed. Studies have been done that demonstrate the ability of *A. fumigatus* to remain viable in an acidic pH as low as 3. What, though, would happen if fungal conidia were exposed to extreme alkaline environments? Human mucosal surfaces have a more alkaline pH, when compared to other parts of the body. These surfaces, therefore, do not promote growth of foreign organisms, since these conditions are less favorable to most pathogens.

There is a six-component pH signal transduction system, which is largely conserved in fungi (Fig. 7.4). These pH regulators include PalA, PalB, PalC, PalF, PalH, and PalI. PalH is a seven-transmembrane protein that researchers believe serves as a pH sensor, possibly with the aid of PalI, a four-transmembrane protein. Once pH conditions are recognized as stressful, PalA will bind to PacC, causing the first proteolytic cleavage. This process is believed to be catalyzed by PalB, a calpain-like cysteine protease. PacC is a transcription factor that mediates gene regulation by ambient pH. Alkaline conditions result in a two-step proteolytic processing activation of PacC. Alkaline-induced activation of PacC leads to the generation of the active form of the protein, $PacC^{27}$, which acts positively on expression of alkaline-responsive genes, and negatively on acidic-responsive genes. The functions of *palC* and *palF* still remain unclear, although studies have shown that PalF interacts with PalH (Bignell *et al.*, 2005; Herranz *et al.*, 2005; Tilburn *et al.*, 2005).

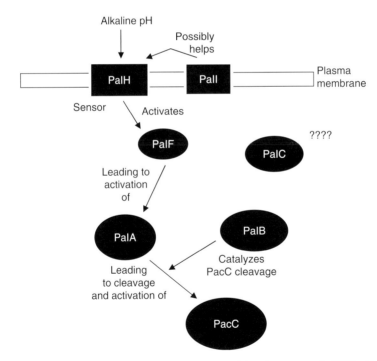

Figure 7.4. Model for the pH response pathway of A. *nidulans* (Herranz *et al.*, 2005). Although most of the components of this pathway have a known function, the purpose of PalC remains elusive.

Bignell *et al.* (2005) performed a study on the influence of *pacC* and *palB* on fungal virulence in A. *nidulans*. Different mutated strains were made to test the requirement of PacC, proteolysis of PacC, and pH signaling through PalB. It is important to note that it is quite difficult to select for transformants defective in pH response. Three loss-of-function strains were created: a *pacC* null mutant, a mutant without the ability to perform proteolysis on PacC, and a *palB* null mutant. The *pacC* null mutant showed decreased radial growth within a pH range of 6-8, while the other two mutants were not able to grow at all above pH 7.2 *in vitro*. All three mutants showed significant attenuation of virulence in murine models of invasive pulmonary aspergillosis. Although growth of the mutant strains within lung cavities seemed comparable to wild type, there was a significant decrease in capability of the mutant conidia to penetrate the lung tissue to cause an invasive infection. As a result, recruitment of immune system cells to the lungs was minimal compared to wild-type strains. Given these results, it seems unlikely that the reduced pathogenicity associated with the *pacC* and *palB* mutants is solely attributable to the general growth defect demonstrated on a pH range of 6–8. Could PacC be a true virulence factor in *Aspergillus*?

Bignell *et al.* (2005) also did murine IPA studies using a mutational truncation of the PacC protein rendering it permanently accessible to the pH-independent processing protease. This truncated version of PacC was constructed in a wild-type background and also a pH signaling deficient background. Both strains showed considerable increase in virulence with respect to the reconstituted strain. Histological examination illustrated enhanced ability to penetrate surrounding lung tissue in both mutant strains. These experiments suggest that PacC does indeed have an important role in fungal pathogenicity.

Fungal pathogens, whether living in soil or a human host, must face a wide pH range as well as rapid changes in pH following phagocytosis by macrophages or exposure to neutrophil vacuole contents. PacC has certainly been shown to contribute to the virulence of a fungal pathogen (Bignell *et al.*, 2005). Most studies dealing with the ambient pH response system, though, have focused solely on alkaline conditions. The environment inside a human host may be neutral or slightly alkaline, but what about acidic conditions? When a fungal pathogen is engulfed by phagocytic cells, they are thrown into a highly acidic environment. Fungal species that have been exposed to acidic solutions *in vitro* have been able to remain viable with pH reaching as low as 3, giving evidence of an acidic pH signal transduction pathway. If a fungal pathogen is forced to overcome these harsh conditions to cause infection, one could hypothesize that these acidic signal transduction pathway components play an important role in pathogenicity. Additional studies are needed to elucidate how fungi respond to acid stress conditions and the role this plays in pathogenesis.

V. REGULATION OF NITROGEN RESPONSE PATHWAYS

Nitrogen metabolism is regulated by two main pathways in fungi. By using nitrogen catabolite repression (NCR), an organism has the ability to discriminate between rich and poor nitrogen sources. Transcription of genes necessary for utilization of poor nitrogen sources is suppressed by NCR in the presence of a rich nitrogen source, such as ammonium. When a fungus is in the presence of a poor nitrogen source, NCR is released (Panepinto et al., 2003). A second pathway, the cross-pathway control system, or general control pathway, regulates the ability of a fungal pathogen to adapt to amino acid starvation, among other stresses (Hoffmann et al., 2001; Krappmann et al., 2004; Panepinto et al., 2003). Nitrogen response pathways seen in yeast have been compared to pathways in filamentous fungi, in an attempt to figure out the exact components of the nitrogen response pathways in these fugal species (Panepinto et al., 2002). This may prove to be an unfavorable method, however, since there are significant differences between yeast and filamentous fungi. Despite these difficulties, researchers have been able to identify some genes in A. fumigatus that play a role in nitrogen sensing and utilization. Being able to recognize and adapt to a wide range of nitrogen sources could be important in relation to fungal virulence.

Researchers have characterized the areA gene in A. nidulans as a positive-acting transcription factor. This gene has been shown to be required for the utilization of a broad range of nitrogen sources other than ammonium and glutamine. Deletion strains of A. nidulans lacking a functional areA allele lose the ability to consume a broad range of nitrogen sources (Hensel et al., 1998). Hensel et al. (1998) used the corresponding gene (AfareA) of A. fumigatus to investigate the role of areA in virulence. The growth rate of the areA null mutant was only significant in the presence of ammonium or glutamine on minimal medium containing glucose as a carbon source. Furthermore, the growth rate of the mutant became identical to that of the wild-type strain when an excess of ammonium or glutamine was added to the media. When protein was supplied as skimmed milk or collagen, the mutant strain lacked the ability to degrade and use these protein sources for growth. Assays for proteolytic activity in growth culture supernatants illustrated the lack of secreted alkaline protease and secreted acidic protease activities.

Studies of these mutants in neutropenic murine models suggest that the nitrogen source(s) available to the pathogen in the lung tissue is unlikely to be ammonium and/or glutamine. These results indicate that a functional areA gene gives a fungal pathogen selective advantage for growth in lung tissue. Mice infected with A. fumigatus areA deletion strains did not show a significant survival rate compared to wild-type strains, but the onset of mortality was delayed for the mutant strains. Indeed, given these different results, areA of

A. fumigatus does seem to contribute to, but is not essential for, fungal growth and virulence in neutropenic mice (Hensel *et al.*, 1998).

Similar to the effects of *areA* on fungal growth and virulence, *sakA* of *A. fumigatus* (Fig. 7.5) also seems to have an effect on fungal growth and hyphal formation in the presence of different nitrogen sources and high osmolarity. When grown in liquid media containing poor nitrogen sources such as sodium nitrate or sodium nitrite, conidia of *sakA* deletion mutants germinated faster than those of the wild-type strain. In spite of the conidial germination rate, mutants lacking *sakA* are unable to form hyphae as quickly as the wild-type strain. This was seen in liquid media with or without 1-M NaCl (Xue *et al.*, 2004). Since hyphal elongation is required for a fungal pathogen to penetrate

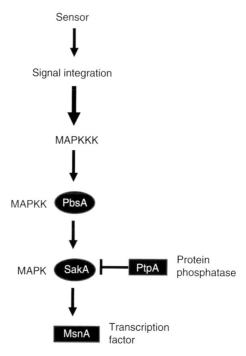

Figure 7.5. Model for the mitogen-activated *sakA* pathway of *A. fumigatus*. Sensing of an outside signal leads to the activation of a MAP kinase kinase kinase, followed by a MAP kinase kinase, followed by a MAP kinase. The MAP kinase in this case is SakA, which can also be called HogA. The activation of SakA subsequently motivates downstream transcription factors which turn on or turn off necessary genes in response to the environmental cue.

the surrounding tissue and cause disease, *sakA* seems like a good candidate for virulence studies. Only limited studies have been done in relation to *sakA* deletion strains and murine mortality, though a *hog1* deletion strain was shown to have reduced virulence in *Candida albicans* (Alonso-Monge et al., 1999). Future studies involving a *sakA* deletion strain, a wild-type strain, and a reconstituted strain of *A. fumigatus* in murine models may prove quite useful in helping researchers understand the role of *sakA* in fungal virulence.

Another gene, *rhbA*, is expressed in response to nitrogen starvation in *A. fumigatus*. This is the first rheb gene reported in filamentous fungi. Rheb proteins consist of a group of Ras-related proteins, all of which share sequence changes in the conserved ras domains. This feature separates the rheb family of proteins from other ras proteins. These rheb proteins have been shown to be conserved from lower eukaryotes to mammals. In *Saccharomyces cerevisiae*, deletion of *rhb1* leads to hypersensitivity to canavanine, a toxic arginine analogue. In *Schizosaccharomyces pombe*, the rheb homologue (*rhb1*) is required for growth in response to extremely limited nitrogen sources (Panepinto et al., 2002, 2003). Similar to *rhb1* of *Schizosaccharomyces pombe*, *rhbA* message levels increase dramatically on nitrogen starvation in *A. fumigatus* (Panepinto et al., 2002, 2003). When *rhbA* mRNA levels were tested after carbon starvation, no change could be detected, suggesting that *rhbA* responds solely to nitrogen stress (Panepinto et al., 2002). Contrary to *areA* mutants, mice inoculated with *rhbA* knockout conidia showed a significant increase in survival when compared to the wild-type or the reconstituted strains. When grown on different nitrogen sources, the *rhbA* mutant showed a significantly reduced growth rate on minimal medium containing nonpreferred nitrogen sources (proline, histidine, or nitrate) compared to the wild-type and rescued strain. When the *rhbA* mutant of *A. fumigatus* was exposed to ammonium, a preferred nitrogen source, the growth rate was similar to that of the wild-type and rescued strains. These comparable growth rates could also be seen on Sabouraud dextrose agar, which contains a pancreatic digest of casein and neopeptone as sources of nitrogen (Panepinto et al., 2003). These results suggest that *rhbA* may be an essential component to the nitrogen response pathway and therefore an essential component in fungal pathogenesis.

VI. REGULATION OF CARBON RESPONSE PATHWAYS

Carbon response pathways have not been studied nearly as extensively as nitrogen response pathways. Fungal pathogens need carbon sources to produce energy for conidiation and growth, yet these response pathways remain largely unexplored in *A. fumigatus*. Early studies have shown that *Saccharomyces cerevisiae* ascospore germination is most efficient in the presence of a carbon source that is readily fermentable such as glucose (Lafon *et al.*, 2005). This suggests that the ability to germinate is linked to carbon nutrient availability. Despite the lack of investigation in carbon source sensing, a few proteins in *A. fumigatus* have been identified that contribute to carbon source sensing and signaling.

Many organisms use the cAMP/protein kinase A (PKA) signaling pathway (Fig. 7.6) to activate necessary biological processes in response to environmental changes. Some studies have linked cAMP/PKA signaling to virulence in certain species of fungus (D'Souza *et al.*, 2001; Hicks *et al.*, 2004). On stimulation, adenyl cyclase is activated by a G-protein–coupled receptor. ATP is converted to cAMP, a process which is catalyzed by adenyl cyclase. The accumulation of cAMP in the cell leads to the activation of PKA. Once the PKA regulatory subunits bind to cAMP, autophosphorylation of the PKA catalytic subunits leads to the phosphorylation of downstream targets. Environmental stress, particularly nutritional stress, has been shown to contribute to activation of the cAMP/PKA signaling pathway in *Saccharomyces cerevisiae*. The presence of glucose in the medium leads to the repression of genes involved in metabolizing nonglucose carbon sources by the PKA signaling pathway in *Saccharomyces cerevisiae*. Cells lacking the PKA regulatory subunit, which contains the cAMP binding domain, exhibit an inability to grow on glycerol and ethanol, carbon sources that are nonfermentable and nonrepressing. Galactose, a weakly fermentable carbon source, also provoked this phenotype. *A. fumigatus* was shown to have elevated transcript levels of the PKA regulatory subunit in the presence of endothelial cells, suggesting a role for the PKA signaling pathway in this environment (Oliver *et al.*, 2002). Despite these studies, little is known about the PKA response pathway in *A. fumigatus*.

To test the response to different carbon sources in the presence of cAMP, Oliver *et al.* (2002) grew a wild-type strain of *A. fumigatus* on medium containing fructose, galactose, glycerol, and acetate, all of which vary in their ability to be fermented and to participate in carbon catabolite repression. There was no inhibition of growth when *A. fumigatus* was grown in the presence of cAMP on fructose, a weakly repressing carbon source. The growth rate of *A. fumigatus* was severely inhibited by cAMP on galactose, glycerol, and acetate, all nonrepressing carbon sources, suggesting that carbon catabolite repression may be involved in the response to exogenous cAMP.

Liebmann *et al.* (2004b) wanted to test the virulence of PKA mutant strains of *A. fumigatus* in murine models. Two genes were selected for deletion, *gpaB*, the G-protein α-subunit, and *pkaC1*, encoding a PKA catalytic subunit. Interestingly, an inhalation method, where the mice are exposed to a certain amount of airborne spores, was used in this particular study. This method of murine inhalation is more apt to mimic the natural mode of infection compared to the intranasal method where liquid containing spores is placed in the nose of

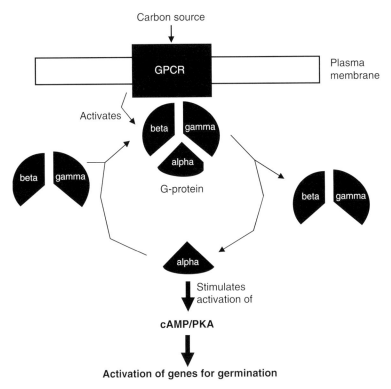

Figure 7.6. Model for the PKA pathway in *A. nidulans* (Lafon *et al.*, 2005). Carbon source sensing through a possible G-protein–coupled receptor leads to the activation and conformational change of the heterotrimeric G-protein GanB(Gα)-SfaD(Gβ)-GphA(γ). Once GDP is converted to GTP on the GanB(Gα) subunit, the other two subunits are expelled. Activation of the GanB(Gα) subunit leads to the activation of the cAMP/PKA pathway, making this subunit the primary signaling element. Once the cAMP/PKA pathway has served its purpose, the downregulation of GanB activity recycles the Gα-subunit, allowing its reattachment to the Gβ- and Gγ-subunits for future activation.

anesthetized mice. Both the *gpaB* and *pkaC1* deletion mutants of A. *fumigatus* showed a reduction in virulence compared to the wild-type strain. Histological examination of murine lung tissue illustrated the ability of the *gpaB* deletion strain to germinate in a similar fashion to the wild-type strain. Mutants lacking *gpaB*, therefore, were able to grow normally in lung tissue, but seemed to persist without invasive growth. This suggests that the G-protein α-subunit upstream of adenyl cyclase contributes to virulence. This was not the case for *pkaC1* mutants, though. These strains were almost avirulent because they grew very poorly in the mouse lung tissue.

Although the PKA catalytic subunit does contribute to virulence, is this a true virulence factor? Of course a fungal pathogen will lose its pathogenicity if it cannot grow *in vivo*, but this is not the only issue to take into account when trying to identify virulence factors. One should determine if a particular species still has the capability to grow on complete medium. Complete medium contains a wide variety of nutrients that the fungal pathogen can use to germinate. If a fungal pathogen lacks the ability for growth even on complete medium, this fungus will obviously not be able to sustain itself in an animal host. This trend was seen with the *pkaC1* deletion strain. The *pkaC1* mutant had significantly reduced growth even on maltose extract agar, used as complete medium (Liebmann *et al.*, 2004b). Of course, the *pkaC1* deletion strain did have a significant reduction in virulence, but can this gene really be called a virulence factor?

The mitogen activated protein kinases (MAPK) are another set of protein kinases that respond to environmental changes. Environmental stresses, such as increased osmolarity, heat shock, high concentrations of heavy metals, reactive oxygen species, and nutrient limitations, as well as other environmental cues activate MAPK pathways within fungi. MAPKs have been shown to contribute to virulence in plant pathogenic fungi, making this MAPK signaling pathway the target for agricultural antifungal agents (Hou *et al.*, 2002; Jenczmionka *et al.*, 2003; May *et al.*, 2005). This evidence makes MAPK signaling pathways a subject of interest for future antifungal drugs for animal fungal pathogens, though these pathways remain less studied. MAP kinases serve as terminal proteins in a kinase cascade involving other protein kinases acting in series. MAP kinase kinase kinase (MAPKKK), activates MAP kinase kinase (MAPKK), which in turn activates MAP kinase. Transcription factors controlling changes in gene expression serve as downstream targets of these MAP kinase cascades most of the time (Gustin *et al.*, 1998; May *et al.*, 2005; Millar, 1999).

There are four MAP kinase proteins in *A. fumigatus*: SakA/HogA, MpkA, MpkB, and MpkC. MpkA and MpkB are suspected of regulating cell wall integrity and pheromone signaling, respectively, based on homology with other fungal MAPK genes. SakA/HogA, discussed earlier, functions within a nitrogen response pathway, as well as regulating growth in high osmolarity. MpkC regulates the utilization of different carbon sources, though the other components within this pathway are currently unknown (Reyes *et al.*, 2006). In the presence of preferred carbon sources, such as glucose, Δ*mpkC* strains of *A. fumigatus* have a growth rate similar to wild type. However, deletion of *mpkC* causes a significant reduction in growth on sorbitol and mannitol, suggesting MpkC plays a role in the regulation of a carbon response pathway (Reyes *et al.*, 2006). Virulence studies have not been done on *mpkC*, since our knowledge of this MAP kinase is still in its infancy. Further research on *mpkC* and other MAP kinases may lead to a new group of antifungal drugs, once we understand the different signaling pathways used.

VII. CONCLUDING REMARKS

A. fumigatus is the main fungal pathogen involved in invasive pulmonary aspergillosis. The pathogenicity of A. fumigatus can be partially explained by its exceptional physiological versatility. Thermotolerance and the ability to consume a broad range of different carbon and nitrogen sources as nutrients are just a few examples of this specie's adaptability. Researchers have studied a number of genes involved in signal transduction in several fungal systems. These signal transduction genes have been shown to play a role in the regulation of virulence in many of the fungal systems studied, yet signaling pathways and their contributions to virulence have not been fully explored in A. fumigatus.

Responding to a changing environment to maintain a relatively constant cellular physiology is vital to an organism's survival. Wild-type strains of A. fumigatus, as well as A. nidulans, have no specific nutritional requirements. They have numerous biosynthetic pathways that are turned on and off in response to stressful conditions or absence of an essential nutrient, and can therefore produce any nutritional supplement needed, including amino acids. When one of these pathways becomes disrupted due to mutation, the survival capability of that particular strain can sometimes be compromised. Finding conserved genes that are essential for one of these important pathways should be a top priority in the fight against invasive pulmonary aspergillosis. With mortality rates being greater than 50% with current antifungal therapy, the search for new and aggressive antifungal treatments is very important. Targeting these conserved biosynthetic pathway enzymes may prove to be useful in developing new antifungal drugs.

References

Alonso-Monge, R., Navarro-Garcia, F., Molero, G., Diez-Orejas, R., Gustin, M., Pla, J., Sanchez, M., and Nombela, C. (1999). Role of the mitogen-activated protein kinase Hog1p in morphogenesis and virulence of Candida albicans. J. Bacteriol. 181(10), 3058–3068.

Bhabhra, R., Miley, M. D., Mylonakis, E., Boettner, D., Fortwendel, J., Panepinto, J. C., Postow, M., Rhodes, J. C., and Askew, D. S. (2004). Disruption of the Aspergillus fumigatus gene encoding nucleolar protein CgrA impairs thermotolerant growth and reduces virulence. Infect. Immun. 72 (8), 4731–4740.

Bignell, E., Negrete-Urtasun, S., Calcagno, A. M., Haynes, K., Arst, H. N., Jr., and Rogers, T. (2005). The Aspergillus pH-responsive transcription factor PacC regulates virulence. Mol. Microbiol. 55(4), 1072–1084.

Brakhage, A. A., and Langfelder, K. (2002). Menacing mold: The molecular biology of Aspergillus fumigatus. Annu. Rev. Microbiol. 56, 433–455.

Brown, J. S., Aufauvre-Brown, A., Brown, J., Jennings, J. M., Arst, H., Jr., and Holden, D. W. (2000). Signature-tagged and directed mutagenesis identify PABA synthetase as essential for Aspergillus fumigatus pathogenicity. Mol. Microbiol. 36(6), 1371–1380.

D'Enfert, C., Diaquin, M., Delit, A., Wuscher, N., Debeaupuis, J. P., Huerre, M., and Latge, J. P. (1996). Attenuated virulence of uridine-uracil auxotrophs of Aspergillus fumigatus. Infect. Immun. 64(10), 4401–4405.

D'Souza, C. A., Alspaugh, J. A., Yue, C., Harashima, T., Cox, G. M., Perfect, J. R., and Heitman, J. (2001). Cyclic AMP-dependent protein kinase controls virulence of the fungal pathogen *Cryptococcus neoformans*. *Mol. Cell. Biol.* **21**(9), 3179–3191.

Eckert, S. E., Hoffmann, B., Wanke, C., and Braus, G. H. (1999). Sexual development of *Aspergillus nidulans* in tryptophan auxotrophic strains. *Arch. Microbiol.* **172**(3), 157–166.

Galagan, J. E., Calvo, S. E., Cuomo, C., Ma, L. J., Wortman, J. R., Batzoglou, S., Lee, S. I., Basturkmen, M., Spevak, C. C., Clutterbuck, J., Kapitonov, V., Jurka, J., *et al.* (2005). Sequencing of *Aspergillus nidulans* and comparative analysis with *A. fumigatus* and *A. oryzae*. *Nature* **438** (7071), 1105–1115.

Garrad, R. C., and Bhattacharjee, J. K. (1992). Lysine biosynthesis in selected pathogenic fungi: Characterization of lysine auxotrophs and the cloned LYS1 gene of *Candida albicans*. *J. Bacteriol.* **174**(22), 7379–7384.

Gustin, M. C., Albertyn, J., Alexander, M., and Davenport, K. (1998). MAP kinase pathways in the yeast *Saccharomyces cerevisiae*. *Microbiol. Mol. Biol. Rev.* **62**(4), 1264–1300.

Hensel, M., Arst, H. N., Jr., Aufauvre-Brown, A., and Holden, D. W. (1998). The role of the *Aspergillus fumigatus areA* gene in invasive pulmonary aspergillosis. *Mol. Gen. Genet.* **258**(5), 553–557.

Herranz, S., Rodriguez, J. M., Bussink, H. J., Sanchez-Ferrero, J. C., Arst, H. N., Jr., Penalva, M. A., and Vincent, O. (2005). Arrestin-related proteins mediate pH signaling in fungi. *Proc. Natl. Acad. Sci. USA* **102**(34), 12141–12146.

Hicks, J. K., D'Souza, C. A., Cox, G. M., and Heitman, J. (2004). Cyclic AMP-dependent protein kinase catalytic subunits have divergent roles in virulence factor production in two varieties of the fungal pathogen *Cryptococcus neoformans*. *Eukaryot. Cell* **3**(1), 14–26.

Hoffmann, B., Valerius, O., Andermann, M., and Braus, G. H. (2001). Transcriptional autoregulation and inhibition of mRNA translation of amino acid regulator gene *cpcA* of filamentous fungus *Aspergillus nidulans*. *Mol. Biol. Cell* **12**(9), 2846–2857.

Hong, E., Balakrishnan, R., Christie, K., Costanzo, M., Dwight, S., Engel, S., Fisk, D., Hirschman, J., Livstone, M., Nash, R., Park, J., Oughtred, R., *et al.* (1994). *Saccharomyces* Genome Database. http://www.yeastgenome.org/

Hou, Z., Xue, C., Peng, Y., Katan, T., Kistler, H. C., and Xu, J. R. (2002). A mitogen-activated protein kinase gene (*MGV1*) in *Fusarium graminearum* is required for female fertility, heterokaryon formation, and plant infection. *Mol. Plant Microbe Interact.* **15**(11), 1119–1127.

Jenczmionka, N. J., Maier, F. J., Losch, A. P., and Schafer, W. (2003). Mating, conidiation and pathogenicity of *Fusarium graminearum*, the main causal agent of the head-blight disease of wheat, are regulated by the MAP kinase *gpmk1*. *Curr. Genet.* **43**(2), 87–95.

Jin, F. J., Maruyama, J., Juvvadi, P. R., Arioka, M., and Kitamoto, K. (2004). Adenine auxotrophic mutants of *Aspergillus oryzae*: Development of a novel transformation system with triple auxotrophic hosts. *Biosci. Biotechnol. Biochem.* **68**(3), 656–662.

Krappmann, S., Bignell, E. M., Reichard, U., Rogers, T., Haynes, K., and Braus, G. H. (2004). The *Aspergillus fumigatus* transcriptional activator CpcA contributes significantly to the virulence of this fungal pathogen. *Mol. Microbiol.* **52**(3), 785–799.

Lafon, A., Seo, J. A., Han, K. H., Yu, J. H., and d'Enfert, C. (2005). The heterotrimeric G-protein GanB(alpha)-SfaD(beta)-GpgA(gamma) is a carbon source sensor involved in early cAMP-dependent germination in *Aspergillus nidulans*. *Genetics* **171**(1), 71–80.

Latge, J. P. (1999). *Aspergillus fumigatus* and aspergillosis. *Clin. Microbiol. Rev.* **12**(2), 310–350.

Liebmann, B., Muhleisen, T. W., Muller, M., Hecht, M., Weidner, G., Braun, A., Brock, M., and Brakhage, A. A. (2004a). Deletion of the *Aspergillus fumigatus* lysine biosynthesis gene *lysF* encoding homoaconitase leads to attenuated virulence in a low-dose mouse infection model of invasive aspergillosis. *Arch. Microbiol.* **181**(5), 378–383.

Liebmann, B., Muller, M., Braun, A., and Brakhage, A. A. (2004b). The cyclic AMP-dependent protein kinase a network regulates development and virulence in *Aspergillus fumigatus*. *Infect. Immun.* **72**(9), 5193–5203.

Machida, M., Asai, K., Sano, M., Tanaka, T., Kumagai, T., Terai, G., Kusumoto, K., Arima, T., Akita, O., Kashiwagi, Y., Abe, K., Gomi, K., *et al.* (2005). Genome sequencing and analysis of *Aspergillus oryzae*. *Nature* **438**(7071), 1157–1161.

May, G. S., Xue, T., Kontoyiannis, D. P., and Gustin, M. C. (2005). Mitogen activated protein kinases of *Aspergillus fumigatus*. *Med. Mycol.* **43**(Suppl. 1), S83–S86.

Millar, J. B. (1999). Stress-activated MAP kinase (mitogen-activated protein kinase) pathways of budding and fission yeasts. *Biochem. Soc. Symp.* **64,** 49–62.

Nierman, W. C., Pain, A., Anderson, M. J., Wortman, J. R., Kim, H. S., Arroyo, J., Berriman, M., Abe, K., Archer, D. B., Bermejo, C., Bennett, J., Bowyer, P., *et al.* (2005). Genomic sequence of the pathogenic and allergenic filamentous fungus *Aspergillus fumigatus*. *Nature* **438**(7071), 1151–1156.

Oliver, B. G., Panepinto, J. C., Askew, D. S., and Rhodes, J. C. (2002). cAMP alteration of growth rate of *Aspergillus fumigatus* and *Aspergillus niger* is carbon-source dependent. *Microbiology* **148**(Pt. 8), 2627–2633.

Panepinto, J. C., Oliver, B. G., Amlung, T. W., Askew, D. S., and Rhodes, J. C. (2002). Expression of the *Aspergillus fumigatus* rheb homologue, *rhbA*, is induced by nitrogen starvation. *Fungal Genet. Biol.* **36**(3), 207–214.

Panepinto, J. C., Oliver, B. G., Fortwendel, J. R., Smith, D. L., Askew, D. S., and Rhodes, J. C. (2003). Deletion of the *Aspergillus fumigatus* gene encoding the Ras-related protein RhbA reduces virulence in a model of invasive pulmonary aspergillosis. *Infect. Immun.* **71**(5), 2819–2826.

Reyes, G., Romans, A., Nguyen, C. K., and May, G. S. (2006). Novel mitogen-activated protein kinase MpkC of *Aspergillus fumigatus* is required for utilization of polyalcohol sugars. *Eukaryot. Cell* **5**(11), 1934–1940.

Sandhu, D. K., Sandhu, R. S., Khan, Z. U., and Damodaran, V. N. (1976). Conditional virulence of a p-aminobenzoic acid-requiring mutant of *Aspergillus fumigatus*. *Infect. Immun.* **13**(2), 527–532.

Sheppard, D. C., Doedt, T., Chiang, L. Y., Kim, H. S., Chen, D., Nierman, W. C., and Filler, S. G. (2005). The *Aspergillus fumigatus* StuA protein governs the up-regulation of a discrete transcriptional program during the acquisition of developmental competence. *Mol. Biol. Cell* **16**(12), 5866–5879.

Smith, J. M., Tang, C. M, Van Noorden, S., and Holden, D. W. (1994). Virulence of *Aspergillus fumigatus* double mutants lacking restriction and an alkaline protease in a low-dose model of invasive pulmonary aspergillosis. *Infect. Immun.* **62**(12), 5247–5254.

Tang, C. M., Smith, J. M., Arst, H. N., Jr., and Holden, D. W. (1994). Virulence studies of *Aspergillus nidulans* mutants requiring lysine or p-aminobenzoic acid in invasive pulmonary aspergillosis. *Infect. Immun.* **62**(12), 5255–5260.

Tilburn, J., Sanchez-Ferrero, J. C., Reoyo, E., Arst, H. N., Jr., and Penalva, M. A. (2005). Mutational analysis of the pH signal transduction component PalC of *Aspergillus nidulans* supports distant similarity to BRO1 domain family members. *Genetics* **171**(1), 393–401.

Wasylnka, J. A., and Moore, M. M. (2003). *Aspergillus fumigatus* conidia survive and germinate in acidic organelles of A549 epithelial cells. *J. Cell Sci.* **116**(Pt. 8), 1579–1587.

Weidner, G., d'Enfert, C., Koch, A., Mol, P. C., and Brakhage, A. A. (1998). Development of a homologous transformation system for the human pathogenic fungus *Aspergillus fumigatus* based on the *pyrG* gene encoding orotidine 5′-monophosphate decarboxylase. *Curr. Genet.* **33**(5), 378–385.

Xue, T., Nguyen, C. K., Romans, A., and May, G. S. (2004). A Mitogen-activated protein kinase that senses nitrogen regulates conidial germination and growth in *Aspergillus fumigatus*. *Eukaryot. Cell* **3**(2), 557–560.

Index

A

ABC transporters, 28, 108, 110
AFUT1, 157
Agrobacterium tumefaciens-mediated
    transformation (ATMT), 205–206
Agroinfection, 117–118
Agroinfiltration, 117–118
ALB1, 182, 201, 229
Allele-specific phenotypes, annotation of, in
    Neurospora crassa, 75–78
  allele database, 76
  allele phenotypic characterization and, 76–77
  allele summaries, 76–77
  phenotypic effects, 76
a locus mating, in *Ustilago maydis*
  alleles, *a1* and *a2*, 4–6
  idiomorphs, 4–5
  lga2, rba1, rga2, mrb1, 5
  pheromone precursors, 5
α-aminoadipate pathway
  in *Aspergillus nidulans*, 269–271
  lysF and, 269
  lysine biosynthesis, in *Aspergillus fumigatus*
    and, 269–270
Amino acid biosynthesis regulation, in fungus
    pathogens, 268–272
Aminoacyl-tRNA synthetases, 271
Anaphase-promoting complex (APC), 20–21
Annotation and Genomics Group, 71, 76
Annotation version 3, in *Neurospora crassa*, 67
APC. *See* Anaphase-promoting complex
API. *See* Application programming interface
Application programming interface (API), 71
Appressorium, of *Magnaporthe oryzae*
  morphogenesis of
    cAMP signaling and surface
      recognition, 178–182
    CaM signaling and, 184
    endocytosis, conidium hydration and, 177
    formation and maturation, 182–184

fungal attachment and germination, 177–178
genes expressed during, 187
integrin, 177
PMK1, 182, 184
PMK1 MAP kinase pathway, 184–187
penetration of, 189–190
BCK1, 189
MMK2, 189
MPS1, 189
nutrient sensing and, 189
PLS1, 189–190
turgor generation, 188–189
Appressorium defect, in *P. infestans*, 126
Appressorium-like structures, 22
ArrayOligoSelector, 78
ARS. *See* Autonomously replicating sequences
Asexual sporulation, in *Phytophthora*, 124
Aspergillus flavus, 144
  genetic variation in, 161
  HMG box transcription factor, 158
  mating type locus in, 158
Aspergillus fumigatus, 144
  AFUT1, 157
  bipolar mating systems in, 149
  fungal genomics and
    ambient pH response, regulation of,
      272–274
    amino acid biosynthesis, regulation
      of, 268–272
    cAMP/PKA signaling pathway
      in, 278–280
    carbon response pathways,
      regulation of, 277–280
    crosspathway control system in, 271
    nitrogen response pathways, regulation
      of, 275–277
    nutritional auxotrophy and, 265–268
  genetic variation in, 161
  mating-type loci in, 157
  polymorphisms in, 161

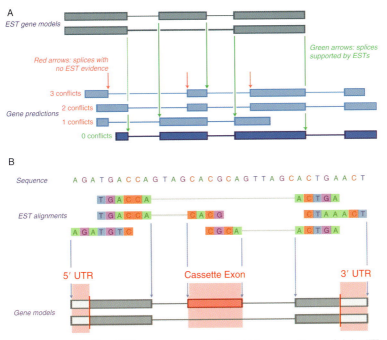

Chapter 2, Figure 2.6. How EST sequences are used for (A) gene calling and (B) UTR and alternative splice prediction.

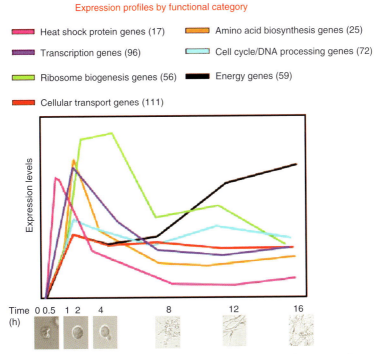

Chapter 2, Figure 2.10. Changes in expression of different categories of genes over the course of the first 16-h growth in Neurospora as elucidated by microarray analysis.

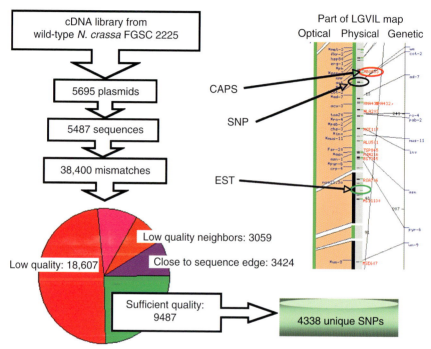

Chapter 2, Figure 2.12. Flow chart for the preliminary SNP generation and mapping project. On the left is shown the progressive stages of data generation and screening, and on the right an example of the final product. This shows a fragment of chromosome VI L including parts of two contigs (green and black). Shown are the optical, physical, and genetic maps. Green dashes in the EST band indicate regions with EST coverage (at 3-kb resolution), black dashes are unconfirmed SNPs, and validated CAPS are identified by a three-letter enzyme designator and a number in red (see http://www.broad.mit.edu/annotation/genome/neurospora/maps/ViewMap.html?sp=5).

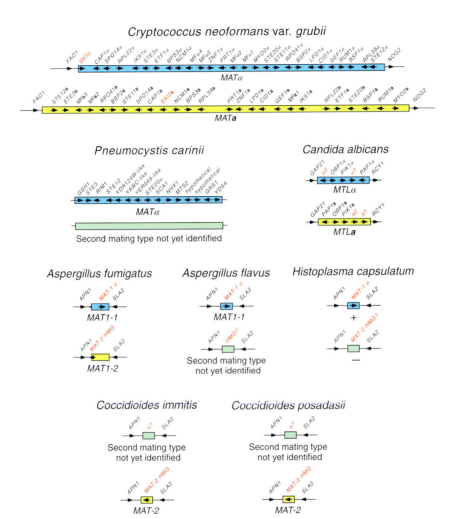

Chapter 4, Figure 4.3. Mating-type loci in the human pathogenic fungi. Genome sequencing has identified mating-type loci in all of the predominant human pathogenic fungi. Both alleles of the mating-type locus have been identified in *Cryptococcus neoformans*, *Candida albicans*, and *Aspergillus fumigatus*, and are designated by yellow and blue boxes. Only one allele has been identified in several human pathogenic fungi based on homology and synteny to known mating-type loci in other organisms. A hypothetical second allele for these mating-type loci has been proposed (green boxes) based on homology and synteny with loci from closely related organisms. In cases where only one mating-type allele has been identified, the boundaries of the locus are speculative. Homeodomain, α-box, or HMG box-encoding genes are designated in red. Other genes present in the mating-type locus are designated in black. Not to scale.

Chapter 6, Figure 6.4. (A) Race T, inoculated on T-cms corn (right), has a much more devastating effect than when inoculated on N-cytoplasm corn (left). (B) Culture filtrate of Race T, injected into T-cms corn (left), causes massive yellowing of the leaves, in contrast to the same filtrate injected into N-cytoplasm corn (right).